Some Complex Phenomena in Fluid and Solid Mechanics

Mathematical and Mechanical Engineering Set

coordinated by
Abdelkhalak El Hami

Volume 13

Some Complex Phenomena in Fluid and Solid Mechanics

Edited by

Abdelghani Saouab
Stéphane Champmartin
Jaafar Khalid Naciri

iSTE

WILEY

First published 2024 in Great Britain and the United States by ISTE Ltd and John Wiley & Sons, Inc.

ISTE Ltd
27-37 St George's Road
London SW19 4EU
UK

www.iste.co.uk

John Wiley & Sons, Inc.
111 River Street
Hoboken, NJ 07030
USA

www.wiley.com

Library of Congress Control Number: 2024943365

British Library Cataloguing-in-Publication Data
A CIP record for this book is available from the British Library
ISBN 978-1-78630-990-7

Contents

Preface

This preface aims to briefly outline the career path and portrait of the professor, researcher and exceptional human being Abdelhak Ambari.

Abdelhak Ambari (1950–2022) passed away at the age of 71. His life was shared between his native country, Morocco, and France. Five main periods of his life can be identified: (i) a Moroccan period of his youth, when he completed his secondary education and the preparatory classes for admission to *grandes écoles*; (ii) a second period of nine years in France, where he completed his engineering studies and his PhD; (iii) a third period (of five years) essential to his career, during which he brilliantly prepared his doctoral thesis; (iv) a fourth period of 12 years, during which Abdelhak Ambari chose to serve his native country Morocco as a research professor at EHTP (Hassania School of Public Works in Casablanca); (v) finally, the last period of 20 years, during which he was a Professor at ENSAM of Angers (*École nationale supérieure des Arts et Métiers*).

Abdelhak Ambari was born in Casablanca, Morocco, on September 10, 1950. From a very young age, he proved to be an outstanding student. Between 1970 and 1972, he followed the "Preparatory classes for admission to *grandes écoles*" in Advanced and Special Mathematics at one of the most prestigious high schools in Morocco at Casablanca, *lycée Lyautey*. He then joined ESTP (*École Supérieure des Travaux Publics*) in Paris, where he earned his engineering degree in 1975. At that time, given his passion for sciences and research, he decided to embrace a research professor career. He then enrolled in doctoral studies with one of the best universities in France: Pierre and Marie Curie University – Paris VI. In 1976, he earned a Diploma of Advanced Studies in Fluid Mechanics. He then continued by pursuing a PhD in Fluid Mechanics specializing in non-Newtonian fluids, and he defended his thesis on June 27, 1979. Given his passion for research, he continued

Preface written by Abdelghani SAOUAB, Stéphane CHAMPMARTIN and Jaafar Khalid NACIRI.

his research activities and he earned the title of State Doctor in Physics on January 17, 1986. Throughout his studies, in order to finance his doctoral studies, he performed freelance activities and worked as a teaching assistant at ESPCI (*École supérieure de physique et de chimie industrielles*) in Paris and at EHTP in Casablanca.

Figure P.1. *Ambari at l'ESPCI (early 1980s)*

After earning his doctoral thesis, he decided to return to Morocco in order to promote and develop research in his native country. He held a Senior Lecturer position at EHTP and became a Professor in 1990. During this period, he initiated several actions that contributed to the boom of research in Morocco: the development of the hydraulics department, the implementation of the Diploma of Advanced Studies in Mechanics and Automation and the supervision of several doctoral thesis works. Abdelhak Ambari dedicated a lot of time and energy to the development and organization of research in Mechanics in Morocco: (i) he set up a research laboratory at EHTP with very limited resources; (ii) he initiated the creation of the school of Fluid Mechanics in Casablanca, which gave rise to the International Congress of Mechanics that is now at its 15th edition; (iii) he contributed to training a new generation of research professors and built up several bridges between Moroccan and French researchers.

In 1998, he decided to return to France for the position of Professor at ENSAM in Angers, and started once again from scratch, striving to build a new fluid mechanics research team. He was appointed the head of the Energy Department of ENSAM nationwide for several years, and he trained several PhD students and post-doctoral researchers. For 20 years, he played an essential role in emulating and leading the research in fluid mechanics at ENSAM by building collaborations, particularly with laboratories in Paris and Bordeaux. Being equally invested in research and education, he developed courses that contributed to the training of generations of student engineers in Angers. Combining the demanding nature of science and the pragmatic approach to industrial problems, he instilled an interest for fluid mechanics in many students, some of whom pursued academic careers in the field. He retired in 2019, leaving a lasting legacy and the memory of a profoundly human and altruistic person.

Throughout his entire career, Professor Abdelhak Ambari shared (with enormous generosity) his knowledge with his students, PhD students and postdoctoral researchers or with his colleagues. According to Professor Etienne Guyon (former Director of the Laboratory of Hydrodynamics and Physical Mechanics of ESPCI, former Director of *Palais de la Découverte* and the *École normale supérieure* in Paris), "Generosity is the word that comes to my mind when I think of Abdelhak, both with respect to human relations and to his spontaneous knowledge sharing". Ambari was highly appreciated by his engineering students at EHTP in Casablanca and at ENSAM in Angers. He represented a father figure to them. A lot of his every day time was dedicated to exchanges and discussions with his research collaborators, and he often said that "he learned from his students as much as they learnt from them". On the news of his passing, many of his students posted testimonials on the Internet, expressing gratitude and admiration. According to these testimonials, "(i) He was a highly competent, very generous and very modest professor; (ii) he was an outstanding professor; a great scientist, a methodical person who was listening to his students. (iii) An outstanding and kind professor who left no one indifferent".

Professor Abdelhak Ambari was also an outstanding scientist. He had a very broad scientific culture and he worked on several research themes in Fluid Mechanics and in the Physics of Polymers. His PhD thesis, conducted under the supervision of Professor André Fortier, focused on the determination of rheological parameters of polymer solutions. His State Doctoral thesis, which focused on the study of the interaction between hydrodynamics and particle dynamics, was supervised by the distinguished Professor Pierre-Gilles de Gennes (Nobel Prize in Physics in 1991). For his outstanding work for the doctoral thesis, Abdelhak Ambari was awarded the Alain Brelot Prize by the French Society of Physics in 1987.

Figure P.2. *A. Ambari receiving the Alain Brelot Prize of the French Society of Physics in July 1987. For a color version of this figure, see www.iste.co.uk/saouab/mechanics.zip*

Abdelhak Ambari had a lot of respect for his mentors. P.G. de Gennes had a remarkable influence on his work: he knew how to avoid the pitfalls of highly theoretical approaches and enjoyed formulating a physical problem in terms of simple concepts and images. Given his broad scientific culture, he perceived the analogies between various branches of physics. The most recent research themes of Abdelhak Ambari concerned several subjects: (i) low Reynolds number-flows in Newtonian and non-Newtonian fluids; (ii) transport phenomena in porous media; (iii) characterization of porous media; (iv) rheology of complex fluids; (v) capillary phenomena; and (vi) hydrodynamic interactions. Abdelhak Ambari was considered as one of the best specialists in the field of "Hydrodynamic Interactions".

Many works of Abdelhak Ambari led to various national and international collaborations. Due to his empathic nature and open-mindedness, Abdelhak Ambari established contacts with bright scientific personalities, such as Professors Abdu Salam (Nobel Prize in Physics in 1979) whom he met in 1989 at the International Centre for Theoretical Physics at Trieste in Italy, Daniel D. Joseph of the University of Minnesota in the United States, J. Hinch whom he met at Cambridge, in the United Kingdom, and Professor Chhabra of the Indian Institute of Technology, whom he invited twice to ENSAM in Angers.

Figure P.3. *During the defense of his doctoral thesis in 1986. For a color version of this figure, see www.iste.co.uk/saouab/mechanics.zip*

Abdelhak Ambari dedicated his doctoral thesis "to the memory of those who fought for the abolition of ignorance in developing countries". He is now part of this family of noble people. This book is a tribute to him, and also to his family, his wife Annick and his son Mehdi, who contributed directly or indirectly to his work.

We would like to conclude with this beautiful testimonial:

> The loss of Professor Abdelhak Ambari caused great sadness shared by the academic mechanics community in France, in Morocco and elsewhere. Abdelhak Ambari was the embodiment of professional generosity, quiet knowledge with immeasurable availability and great modesty. There is no greater loss for humanity than that of a human being with a well-rounded brain. People pass away, and loyal friends have a duty to continue their work.

Rest in peace, dear friend. You will be missed, and will forever have a place in our hearts.

This book has been written as a tribute to the memory of Professor Abdelhak Ambari. It is divided into 10 chapters authored by his French and Moroccan colleagues who had the chance to meet or collaborate with him.

Figure P.4. *A. Ambari at Angers in 2020. For a color version of this figure, see www.iste.co.uk/saouab/mechanics.zip*

The contributions presented in this book focus on research themes that were of interest to Abdelhak Ambari during his career, and to which he brought his valuable experimental and theoretical contributions:

– rheology of complex fluids and polymers;

– hydrodynamic interactions;

– low Reynolds number-flows;

– characterization of porous media;

– flows in porous media;

– instabilities in hydrodynamics and in solid mechanics;

– metrology by electrochemical methods.

We wish to thank all the authors for their contributions, and ISTE-Wiley for having kindly agreed to publish this book dedicated to the memory of our colleague and friend, Professor Abdelhak Ambari.

Abdelghani SAOUAB,
Stéphane CHAMPMARTIN,
Jaafar Khalid NACIRI.
August 2024

1

Kinetics of Microphase Separation in Interpenetrating Polymer Gels

The physical system considered here is an interpenetrating gel composed of two chemically different polymers, cross-linked in the high-temperature single-phase region. When the bath temperature is sufficiently lowered, below the usual critical temperature, the mixture tends to separate into two distinct phases; however, this does not occur, due to the permanent presence of cross-links. Below the spinodal temperature, a microphase separation takes place, accompanied by the appearance of microphase domains, contrary to the usual macrophase separation (in the absence of cross-links), where the phase domains are macroscopic.

This chapter summarizes the accomplishments in studying the kinetics of microphase separation in interpenetrating gels by a progressive reduction of the bath temperature. We particularly discuss the appearance of instabilities at the critical point. The chapter ends with a discussion on the effects of the presence of a good or a theta solvent, the initial composition fluctuations or charges in these kinetics.

1.1. Introduction

When the temperature is lowered, the polymer blends separately into two distinct phases, due to the chemical difference between the present species. This phase separation occurs even in the presence of a solvent. When the polymer chains are of high molar mass, the translational entropy is very small, and then, the chemical segregation between chemically different monomers is significantly amplified. In the temperature-composition phase diagram, a low or high critical separation temperature is then observed (de Gennes 1979a; Pincus 1981; Binder 1983) in

Chapter written by Mabrouk BENHAMOU.

certain cases. In the vicinity of this temperature, there are significant composition fluctuations, and there is a divergence in the correlation length that characterizes the spatial extent of these fluctuations.

Phase separation can be avoided by a chemical crosslinking of blend in the single-phase region to obtaining a gel termed *interpenetrating gel*. This gel can also be obtained by mixing, from the beginning, chemically different monomers and then initiating the polymerization process. In this case, the chemical cross-links are replaced by topological entanglements.

Naturally, when the temperature is reduced, the cross-linked polymer blend tends to separate in two phases, but the cross-links prevent the appearance of a macrophase separation, due to the elastic forces resulting from the presence of cross-links. As result of this competition, a microphase separation takes place (de Gennes 1979b; Binder and Frisch 1984; Battachy et al. 1991); this was theoretically studied for the first time by de Gennes (1979b) using a mean-field approach. The author has shown that the size of the microphase domains depends on the crosslinking rates and approximately corresponds to the average distance between two consecutive cross-links. A second important effect of cross-links is that it renders the polymer blend more compatible, and there is a deviation of the critical temperature that also depends on the cross-link density, defined as the number of cross-links per unit volume. Microphase separation in interpenetrating gels has been experimentally observed by Briber and Bauer (1988), who had used small-angle neutron scattering for the study of the structure of microphase domains in a cross-linked blend, which is composed of polystyrene and poly(vinyl methyl ether).

Let us consider an interpenetrating gel with high cross-link density. The typical distance scale of microphase separation is very small in comparison with the gyration-radius of uncross-linked polymer chains. Therefore, different kinetics are expected for non-cross-linked and cross-linked cases. The non-cross-linked polymer blends have been studied by de Gennes (1979a), Pincus (1981) and Binder (1983). They have shown that *reptation* of polymer chains has a significant contribution to the kinetics of phase separation. However, for cross-linked polymer blends, a small typical time scale is expected, since large-scale motions of cross-linked polymer chains are frozen by the presence of permanent cross-links. It can therefore be anticipated that, for the considered system, only Rouse motion is important. In this context, it had been shown that the dynamic structure factor of a polymer blend susceptible to phase separate can be decomposed in two modes (de Gennes 1981; Pincus 1981), namely a rapid mode, of low amplitude, and a low mode, of high amplitude. The first one corresponds to a local motion, while the second one is associated with a large-scale reptation of polymer chains. In a cross-linked system, above the freezing point, the reptation phenomenon is absent, because of constraints. This is why only a local motion of Rouse type is expected, which is essentially responsible for the microphase separation kinetics. Rouse time corresponds to a motion occurring very early in the case of a non-cross-linked blend, which was discussed by the above-mentioned authors.

Section 1.2 reviews the main results concerning the static study (or at thermodynamic equilibrium) of the microphase separation. Section 1.3 is dedicated to the kinetic study of this microphase separation. Finally, in section 1.4, we discuss the results dealt with the kinetic study and compare them to those related to non-cross-linked blends.

1.2. Static study of microphase separation

Let us consider a blend made of two chemically different polymers, A and B. N denotes the common polymerization-degree of polymer-chains initially in the single-phase region of the phase diagram (high-temperature phase). The monomer composition is denoted by Φ. Due to the symmetry, $\Phi_c = 1/2$ is chosen as the critical monomer composition. Let us assume that the polymer chains are initially randomly distributed and there is no composition fluctuation. This is a high temperature approximation, which involves that the system is very far from its critical point. The blend is now strongly cross-linked, and by lowering the temperature, the interpenetrating gel is brought into the single-phase region. According to de Gennes (1979b), the free energy of such a system can be written, in a Landau approximation, as follows:

$$\frac{F[\phi]}{k_B T} = a^{-3} \int \left\{ (\chi_c - \chi)\,\phi^2(\vec{r}) + a^2 \left(\vec{\nabla}\phi(\vec{r}) \right)^2 + C\vec{P}^2(\vec{r}) \right\} d\vec{r} \qquad [1.1]$$

Here, χ denotes the Flory monomer–monomer interaction parameter, which is inversely proportional to temperature, $\chi_c = 2/N$ is the critical value of the interaction parameter of the non-cross-linked mixture, $\phi(\vec{r})$ is the local fluctuation of composition (or *order parameter*), defined by: $\phi(\vec{r}) = \Phi(\vec{r}) - \Phi_c$, where $\Phi(\vec{r})$ is the composition of a species of the non-cross-linked blend, a is the common size of monomers, k_B is the Boltzmann's constant and T is the absolute temperature of the cross-linked blend. There, the gradient term takes into account the local fluctuations of composition (Joanny 1978) and the last term, which is discussed below, is the elastic contribution due to cross-links. Critical parameters, $\chi_c = 2/N$ and $\Phi_c = 1/2$, are determined in the Flory–Huggins (FH) approximation (Flory 1953). It is important to note that relation [1.1] can be considered as the sum of the expansion of FH free energy in the vicinity of the critical point and the elastic term.

On the other hand, the resolution of equation [1.1] is equivalent to determining the effect of perturbations in the vicinity of the critical point. Previous studies on critical phenomena have shown that this effect can be either weak, causing a simple perturbation, or dramatic, leading to a change in the universality class of the problem. This will be the case in what follows.

The elastic contribution in equation [1.1] was introduced by de Gennes (1979b) by analogy with the polarization of a dielectric medium. The quantity C is the *rigidity constant* that was linked to the number, n, of monomers between two consecutive cross-links. According to de Gennes (1979b), its expression is given as:

$$C \cong n^{-2} \qquad [1.2]$$

In expression [1.1] of FH free energy, $\vec{P}$ is the *displacement vector* between the centers of mass of two strands A and B (a strand is the segment of a polymer chain between two neighboring cross-links). The polarization in the problem of the dielectric medium is equivalent to the distance between the centers of mass of various strands in the problem of polymer, and the charge fluctuations corresponding to the composition fluctuations. This involves that the quantities $\vec{P}$ and ϕ are not independent to each other, but linked by: $div\,\vec{P} = -\phi$, or in the reciprocal space, by:

$$i\vec{q}.\vec{P}(q) = -\phi(q) \qquad [1.3]$$

We are therefore led to consider only the longitudinal modes, and the free energy can be written as follows:

$$\frac{F[\phi]}{k_B T} = \sum_{\vec{q}} \left\{ (\chi_c - \chi) + q^2 + \frac{C}{q^2} \right\} \left| \phi_q \right|^2 \qquad [1.4]$$

Here, ϕ_q is the Fourier transform of the composition fluctuation:

$$\phi_q = \int \left[\Phi(\vec{r}) - \Phi_c \right] e^{i\vec{q}.\vec{r}}\, d\vec{r} \qquad [1.5]$$

The scattered intensity (or structure factor), $S(q)$, in a light or neutron scattering, is given as:

$$S(q) = \left\langle \left| \phi_q \right|^2 \right\rangle = \frac{1}{(\chi_c - \chi) + q^2 + \frac{C}{q^2}} \qquad [1.6]$$

It has a maximum for:

$$q^* = C^{1/4} \cong n^{-1/2} \qquad [1.7]$$

The value of the scattering intensity at this maximum diverges at spinodal for:

$$\chi_c - \chi_s \cong -C^{1/2} \cong -n^{-1} \qquad [1.8]$$

This deviation varies as the inverse of cross-links dose, n. Therefore, a divergence occurs for a non-zero value of the modulus of the transfer wave-vector, q, and this can be interpreted as a microphase separation. Then, there is a deviation of the critical temperature; the cross-linked system is more miscible than the non-cross-linked one.

The following section focuses on the study of kinetics of microphase in interpenetrating gels.

1.3. Kinetic study of microphase separation

As mentioned above, the main difference between demixing of a non-cross-linked polymer blend and a highly cross-linked polymer blend is that the distance scales being examined are highly different. Indeed, in the first case, the motions of polymer chains, at long distance, are controlled by reptation, while in the second case, there are local arrangements, over distances of the order of the size of microphase domains.

For a non-cross-linked polymer blend, de Gennes (1981) and Pincus (1981) have shown that the time dependence of the scattered intensity is the sum of *two* contributions. The first one is a rapid mode and it corresponds to a local reptation. The second having a high amplitude and being slower corresponds to a long distance reptation of the polymer chains and causes a disentanglement of species. In the present case, due to the permanent presence of cross-links, the latter should not be expected to bring their contribution, and longtime motions are then frozen. Local dynamics was not considered in the previous works, as it corresponds to kinetics during the first instants of the phase separation of blend. It is the only motion that is present here. It is the local Rouse dynamics that occur inside a tube formed by the other chains (de Gennes 1979a; Doi and Edwards 1986). This will appear in the expressions of Onsager coefficient that will be introduced below. Formally, we write equations similar to those related to a non-cross-linked blend (de Gennes 1980). More precisely, what we are looking for is the time evolution of the fluctuation at fixed wave-vector modulus, q. For this, the standard procedure that was thoroughly discussed by Binder (1983) is applied. This time evolution can be found by introducing a current, $\vec{j}$, for species A. The continuity equation is given as:

$$\frac{\partial \Phi}{\partial t} + div\,\vec{j} = 0 \qquad [1.9]$$

Here, t represents the time parameter. The current $\vec{j}$ is linked to a chemical potential, $\mu(\vec{r}) = \partial F / \partial \Phi(\vec{r})$, by the following relation:

$$\vec{j}(\vec{r}) = \int \Lambda(\vec{r} - \vec{r}\,')\vec{\nabla}\mu(\vec{r}\,')d\vec{r}\,' \qquad [1.10]$$

$\Lambda(\vec{r})$ being the *Onsager coefficient* that will be discussed below. Let us assume an exponential decrease of the fluctuation, as a function of the wave-vector, $\vec{q}$:

$$\delta\Phi(\vec{q},t) = \delta\Phi(\vec{q},0)e^{-t/\tau(q)} \qquad [1.11]$$

Here, $\tau(q)$ is the relaxation rate. Using relations [1.9] to [1.11] and keeping only the linear terms in the fluctuation of composition, the following result is found (Cahn 1968):

$$\tau^{-1}(q) = -\frac{1}{\delta\Phi_q}\frac{\partial(\delta\Phi_q)}{\partial t} = q^2\Lambda(q)\left[(\chi_c - \chi) + q^2 + \frac{C}{q^2}\right] \qquad [1.12]$$

Equation [1.12] then provides the relaxation rates for the first instants, when the linearization procedure is valid. Positive relaxation rates correspond to stable cases when fluctuation decreases in time, while negative rates correspond to unstable modes. Such fluctuations increase in time (Binder and Heermann 1985). In order to determine the relaxation rate, we need to know Onsager transport coefficients, $\Lambda(q)$, and therefore the nature of the strand motion. The shape of equation [1.12] indicates that the product of the last two terms is a diffusion-coefficient. As usual, this coefficient is the ratio of mobility, μ, to the correlation function, that is:

$$D(q) = \mu(q) / S(q) \qquad [1.13]$$

Comparing it with relation [1.6], we find that the Onsager coefficient can be interpreted as the mobility of strands. In a highly cross-linked system, such as considered here, the motion of polymer-chains is local, over a distance $d \cong n^{1/2}a$, where a is the common size of monomers and n is the number of monomers between consecutive cross-links (polymerization-degree of strands). Notice that, for highly cross-linked polymer blends, with $n < N_e$, where N_e is the distance between consecutive entanglements, only Rouse motion is present. Onsager transport coefficient, $\Lambda(q)$, is then proportional to Rouse mobility. It is emphasized that the case with $n >> N_e$ is discussed in section 1.4 (see below). For high cross-linked blends, the number $n(q)$ of monomers in a strand at distance q^{-1} is proportional to q^{-2}. Within the framework of Rouse approximation, the mobility is as follows:

$$\Lambda(q) \cong n^{-1}(q) \cong (qa)^2 \qquad [1.14]$$

Then, we have:

$$\tau^{-1}(q) = q^4\left[(\chi_c - \chi) + q^2 + \frac{C}{q^2}\right] \qquad [1.15]$$

Figure 1.1 shows the characteristic frequency, for various cases, $\Omega_1(q) = \tau^{-1}(q)/q^4$. Figure 1.1(a) describes the case in the absence of cross-links and for the first instants of dynamics. For $\chi < \chi_c$ (or $T > T_c$), all frequencies are positive and therefore there is no instability. In the opposite case, $T < T_c$, however, we find that all modes with:

$$q > q_c \cong (\chi - \chi_c)^{1/2} \qquad [1.16]$$

are stable, while the long wavelength modes with $q < q_c$ are unstable. Any fluctuation with such a wavelength increases in time. For a cross-linked polymer blend, for which the rigidity constant, C, is non-zero, the situation is quite different (Figure 1.1(b)). For $\chi < \chi_c$, the curves have a minimum that can be either positive or negative. For $\chi < \chi_s$, all frequencies are positive, and then, there is no instability. Let us recall that:

$$\chi_s - \chi_c \cong C^{1/2} \qquad [1.17]$$

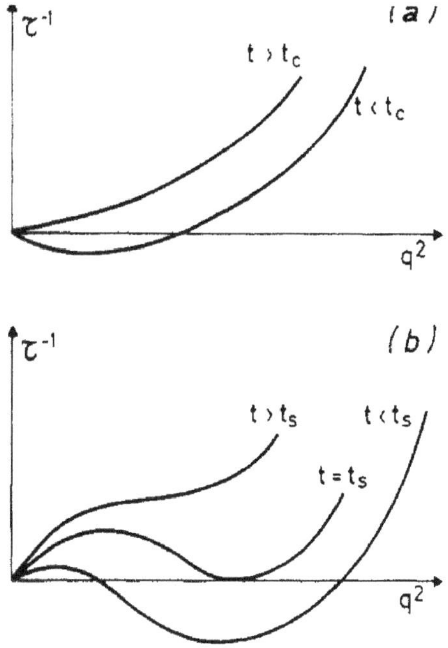

Figure 1.1. *(a) Relaxation rate, as a function of the wave number, for a non-cross-linked blend. For $\chi > \chi_c$, the small q modes are unstable. (b) Relaxation rate for a cross-linked blend. For $\chi_c > \chi > \chi_s$, the minimum is positive and then there is no instability. Above χ_s, the minimum becomes negative and the modes are unstable for a large q*

For $\chi = \chi_s$, the minimum of the characteristic frequency becomes *zero*, and the latter corresponds to a microphase separation. For low temperatures, $\Omega_1(q)$ has a negative part corresponding to the presence of instabilities in a finite interval of wave-vector, q. The most fast increasing instability corresponds to the minimum of the curve for:

$$q_{\min}^2 \cong \chi - \chi_c + \left[(\chi - \chi_c)^2 - 3C \right]^{1/2} \qquad [1.18]$$

The condition, $\chi \cong \chi_s$, corresponds to the size, q^{*-1}, of microdomains. As shown above, this can be written under the following scale form:

$$q^2 = (\chi_c - \chi) f\left(C / (\chi_c - \chi)^2\right) \qquad [1.19]$$

It is important to note that the fact that the short wavelength part, q, of the curves is canceled out is a consequence of the scattered intensity, $S(q)$, which vanishes at $q = 0$. This is in perfect agreement with the results of the neutron scattering experiment realized by Briber and Bauer. Therefore, the previous considerations should not be expected to be valid in this region. This question will be reexamined in the next section. Finally, near typical size, q^{*-1}, the characteristic frequency varies as follows:

$$\Omega\left(q^*\right) \cong q^{*4}\left(\chi_s - \chi\right) \qquad [1.20]$$

with:

$$\chi_s = \chi_c + q^{*2} \qquad [1.21]$$

This characteristic frequency vanishes when χ is close to its spinodal value, χ_s. Therefore, the modes are slower than those of Rouse. It is important to note that according to relation [1.15], for $\chi = \chi_c$, the characteristic frequencies vary as follows:

$$\Omega\left(q\right) \cong q^6 \qquad [1.22]$$

Similar results have been previously found for a common polymer blend and have been interpreted as a weighted average of two characteristic rates, namely, local motions of Rouse type and long chain reptation.

The remaining part corresponds to a slow decrease related to the presence of spinodal decomposition.

1.4. Discussion

The previous sections discussed kinetics over short periods of time of microphase separation occurring in blends made of two chemically incompatible polymers, cross-linked in the single-phase region and brought to the demixing region by a progressive temperature decrease. According to our predictions, only a local motion of Rouse type is possible, due to the presence of permanent cross-links. This same motion occurs inside microphase domains. As a result, we found that the characteristic rate varies with the wave-vector as q^6 at spinodal temperature.

Furthermore, in the single-phase region, instabilities increase with a characteristic time, $\Omega_q \cong q^4(\chi_s - \chi)$, where χ is the segregation parameter, which varies essentially as the inverse of the absolute temperature, and χ_s is its spinodal value.

For the present study, the approximation made is that there is only one temperature in the problem. More precisely, crosslinking of the random blend of species is produced at high temperature, far away from the critical point. It is important to note that the above-mentioned theory of microphase separation shows that the theoretical scattering intensity vanishes when the scattering angle is zero. However, Briber and Bauer have experimentally shown that this is not the case. This discrepancy between theory and experiment has been explained by a series of published works (Vilgis et al. 1993; Benmouna et al. 1994b; Bettachy et al. 1995). In fact, the origin of this discrepancy is the existence of initial composition fluctuations at the moment of blend crosslinking. These fluctuations exist even at very high temperature.

It is important to note that the above considerations can be extended to the case of weakly cross-linked polymer gels, and let us extend kinetics study to these gels. For long polymer chains, the number of monomers between consecutive cross-links, n, is larger than the number of entanglements, N_e, and in this case, kinetics results from a reptation inside tubes between crosslinking points, over long periods of time. In this case, Onsager coefficient, $\Lambda(q)$, no longer corresponds to the local motion of Rouse type, but to a local reptation motion of strands inside tubes. According to Pincus (1981), such a motion implies:

$$\Lambda(q) \cong (qn)^{-2} \qquad [1.23]$$

and therefore:

$$\tau^{-1}(q) \cong n^{-2}\left[(\chi_c - \chi) + q^2 + \frac{C}{q^2}\right] \qquad [1.24]$$

There is a change in behavior between relations [1.15] and [1.24] for $n \cong N_e$. According to relation [1.24], there is a fast instability for $q \cong q^*$, for which the characteristic frequency increases as follows:

$$\tau^{-1}\left(q^*\right) \cong n^{-2}(\chi_s - \chi) \qquad [1.25]$$

It is important to note that when the rigidity constant of gel, C, tends to zero, the previous relation becomes quantitatively similar to that of Pincus relative to a non-cross-linked polymer blend undergoing a macrophase separation:

$$\tau^{-1}\left(q^*\right) \cong N^{-2}\left(\chi_c - \chi\right) \cong N^{-2}t \qquad [1.26]$$

With the deviation from the critical temperature:

$$t = \left(\chi_c - \chi\right) / \chi_c \qquad [1.27]$$

This chapter, whose content is largely based on a previously published work (Battachy et al. 1992), focused on interpenetrating gels in a molten state. However, the extension of the present study has been achieved for interpenetrating gels in the presence of a solvent (Vilgis et al. 1993; Benmouna et al. 1994b; Bettachy et al. 1995; Benhamou et al. 1997; Derouiche et al. 2005; Benhamou et al. 2010). In this case, crosslinking was achieved at high temperature, therefore, under good solvent conditions. The essential change is that kinetics is this time governed by a local motion of Zimm type (Derouiche et al. 2005), where strands are subjected to hydrodynamic interactions.

Finally, it should be recalled that this study has been extended to the case of interpenetrating gels formed by a crosslinking of polyelectrolyte-chains (Benmouna et al. 1994a; Boussaid et al. 2009), and it has been shown that the sign of charges along chains and the ionic concentration of co-ions and counter-ions induce radical changes in kinetics of the microphase separation.

1.5. References

Battachy, A., Derouiche, A., Benhamou, M., Daoud, M. (1991). Phase separation of weakly cross-linked polymer blends. *Journal de Physique I (Paris)*, 1, 153–158.

Battachy, A., Derouiche, A., Benhamou, M., Daoud, M. (1992). Kinetics of microphase separation in cross-linked polymer blends. *Macromolecules*, 25, 7188–7191.

Benhamou, M., Derouiche, A., Bettachy, A. (1997). Microphase separation of crosslinked polymer blends in solution. *J. Chem. Phys.*, 106, 2513–2519.

Benhamou, M., El Fazni, A., Bettachy, A., Derouiche, A. (2010). Theory of microphase separation in crosslinked polymer blends immersed in theta-solvent. *The European Physical Journal E*, 32, 391–398.

Benmouna, M., Vilgis, T.A., Benhamou, M., Babaoui, M., Daoud, M. (1994a). Stability analysis and scattering properties of charged crosslinked blends in solution. *Macromolecular Theory and Simulations*, 3, 557–566.

Benmouna, M., Vilgis, T.A., Daoud, M., Benhamou, M. (1994b). Scattering and phase behavior of cross-linked blends. *Macromolecules*, 27, 1172–1176.

Bettachy, A., Derouiche, A., Benhamou, M., Benmouna, M., Vilgis, T.A., Daoud, M. (1995). Scattered intensity by a cross-linked polymer blend. *Macromolecular Theory and Simulations*, 4, 67–76.

Binder, K.J. (1983). Collective diffusion, nucleation and spinodal decomposition in polymer mixtures. *Chem. Phys.*, 79, 6387–6409.

Binder, K.J. and Frisch, H.L. (1984). Phase stability of weakly crosslinked interpenetrating polymer networks. *J. Chem. Phys.*, 81, 2126–2136.

Binder, K. and Heermann, D.W. (1985). Growth of domains and scaling in the late stages of phase separation and diffusion-controlled ordering phenomena. In *Scaling Phenomena in Disordered Systems*, Pynn, R., Skjeltorp, A. (eds). Springer, Boston. 207–230.

Boussaid, A., Benhamou, M., El Fazni, A. (2009). Theory of microphase separation in charged crosslinked polymer blends with arbitrary charge distribution. *The European Physical Journal E*, 30, 379–385.

Briber, R.M. and Bauer, B.J. (1988). Effect of crosslinks on the phase separation behavior of a miscible polymer blend. *Macromolecules*, 21, 3296–3303.

Cahn, J.W.T. (1968). Amorphous metallic alloys. *Trans. Metall. SOCA. IME*, 242, 166.

Derouiche, A., Benhamou, M., Bettachy, A. (2005). Kinetics of microphase separation in interpenetrated polymer networks in solution. *The European Physical Journal E*, 16, 365–371.

Doi, M. and Edwards, S.F. (1986). *The Theory of Polymer Dynamics*. Science Publications, Oxford.

Flory, P.J. (1953). *Principles of Polymer Chemistry*. Cornell University Press, Ithaca, New York.

de Gennes, P.-G. (1979a). Effect of cross-links on a mixture of polymers. *Journal de Physique Lettres*, 40, 69–72.

de Gennes, P.-G. (1979b). *Scaling Concept in Polymer Physics*. Cornell University Press, Ithaca, New York.

de Gennes, P.-G. (1980). Dynamics of fluctuations and spinodal decomposition in polymer blends. *J. Chem. Phys.*, 72, 4756–4763.

de Gennes, P.-G. (1981). Coherent scattering by one reptating chain. *Journal de Physique*, 42, 735–740.

Hilliard, J.E. (1970). Spinodal decomposition, In *Phase Transformations*, Aaronson, H.L. (ed.). American Society for Metals, Materials Park, Ohio, 497–560.

Joanny, J.-F. (1978). Thèse de 3ème cycle. Université Paris 6, Paris.

Pincus, P.J. (1981). Dynamics of fluctuations and spinodal decomposition in polymer blends. *Chem. Phys.*, 75, 1996–2000.

Vilgis, T.A., Benmouna, M., Daoud, M., Benhamou, M., Bettachy, A., Derouiche, A. (1993). The scattering behavior of crosslinked polymer blends. *Polym. Networks Blends*, 3, 59.

2

Physics of Hydrodynamic Interactions

A physicist might find it strange that a fluid mechanics scientist uses the term "interaction", as in the four classical fundamental interactions, to describe what happens in a number of complex flows. This term might be defined as the *mutual action exerted by two or several physical systems upon each other*. It certainly is what happens, for example, in a suspension of solid particles, where the global motion of the solution is governed by the coupling between the motion of each particle and that of its neighbors. For the sake of clarity, these are referred to as hydrodynamic interactions, meaning that they are propagated by a moving interstitial fluid. If inertia is negligible, these hydrodynamic interactions decrease very slowly, typically as the inverse of distance. It is therefore necessary to integrate them into the models developed over the last 150 years in order to understand the sometimes counterintuitive behaviors of some complex flows. This chapter presents an overview of the approaches used to solve these type of problems, ranging from a simple isolated particle to larger sets of particles. Mathematical tools, such as the resistance matrix and the method of reflections, will be illustrated for simple cases, to give an overview of the formalisms and ideas developed in this domain of fluid mechanics.

2.1. Introduction

In 1969, the physicist G.I. Taylor proposed a series of documentaries filmed at the initiative of the *National Committee for Fluid Mechanics Films*[1]. These films proposed a didactic illustration of the main branches of fluid mechanics. One of the

Chapter written by Stéphane CHAMPMARTIN.

1 Formed in 1961 by A. Shapiro, this committee of 11 personalities thought, and rightly so, that teaching fluid mechanics would be greatly improved if illustrated by didactic films contributing to an image-based understanding of the phenomena occurring in flows.

39 videos was entitled "Low-Reynolds-Number Flows" (Taylor 1969) and presented the main properties of flows with negligible inertia (another interesting source is the article "Life at Low Reynolds Number" by E.M Purcell (1977), which is also freely available online). This film and the accompanying text remain a valuable source of knowledge and an excellent starting point for an initiation in hydrodynamic interactions, even today. For example, it can be noted that a ball under sedimentation in a viscous fluid is moving more slowly in the vicinity of a wall than far away from it, as if the presence of the wall slows down the fall of the ball. This is a first effect due to the hydrodynamic interactions taking place, even at long distance. Similarly, while a ball that falls far away from any boundary will not rotate on itself, this is no longer the case when the ball sediments near a wall or in the vicinity of another particle. Other unexpected effects can also be encountered: the sedimentation velocity of a ball in a cylindrical tube is maximal for an offset position of the tube axis. On the other hand, when the ball sediments between two parallel planes, the symmetric position is the one that gives the maximal falling speed. These effects have been thoroughly explained in the works of Ambari et al. (1983, 1985). To take a step further, let us now imagine two identical balls that fall side-by-side: this time, the effect of hydrodynamic interactions of the balls will be an increase in the falling speed of the two balls: they will fall together at the same speed, without a change in the distance between them, but at a higher speed than if they sedimented separately. Let us consider the case of a larger number of particles. Will they continue to increase their collective speed? Up to a certain point only. When the number of particles becomes significant, everything happens as if the suspension became more viscous, leading to a lower sedimentation speed. In 1905, Einstein calculated the increase in the average viscosity μ of a diluted suspension (concentration $C \ll 1$) in an article dedicated to Brownian motion, but whose results are also applicable to non-Brownian motions:

$$\mu = \mu_0(1 + 2.5C) \tag{2.1}$$

where μ_0 is the viscosity of the solvent. For concentrations above a few percent, the distance between particles becomes sufficiently small (typically below 10 times their radius) for the hydrodynamic interactions to be noticeable. Einstein's relation must then be corrected by a term in C^2:

$$\mu = \mu_0(1 + 2.5C + \delta C^2) \tag{2.2}$$

whose coefficient δ depends, among others, on the type of flow (elongational, shear, etc.) to which the solution is subjected (typically $5.2 < \delta < 7.6$). Similarly, beyond a certain concentration, antagonistic effects will appear and will slow down the sedimentation velocity. These effects are due to fluid flows in the opposite direction to that of particles (referred to as return flows or backflows), which ensure flow rate conservation. Consequently, the sedimentation rate V_S will be reduced when concentration increases. Among the best known laws, it is important to cite the contribution of Batchelor (1972) valid for highly diluted suspensions of monodisperse particles, statistically homogeneous and far from any boundary (the influence of the walls of the container in which this sedimentation takes place is not taken into account):

$$V_S = V_\infty(1 - 6.55C) \quad [2.3]$$

where V_∞ is the sedimentation velocity of an isolated spherical particle:

$$V_\infty = \frac{2}{9}\frac{\Delta\rho g a^2}{\mu} \quad [2.4]$$

In this equation, $\Delta\rho$ is the difference between the densities of the particle and of its surrounding fluid, a is the radius of the sphere and g is the gravitational acceleration. While the case of a pair of particles can be solved analytically (for example, by the method of reflections), the behavior of a larger set of particles requires more complex statistical approaches. This type of situation is frequently encountered. For example, let us cite the test that is very commonly used in medicine, which consists of measuring the red cell sedimentation rate in a blood sample; red cells being slightly denser than plasma, their sedimentation rate depends on their concentration. Surprisingly, after a few minutes, a quite clear front can be observed between the lower part of the sample where red cells are concentrated and the higher part without blood cells. This makes it easier to follow up the evolution in time of this "boundary". Once again, this is a consequence of hydrodynamic interactions: if, for whatever reason, a red blood cell is "isolated" from the rest of cells, its velocity will tend to increase, which will help it reach the area with higher concentration. Once in this concentrated area, its velocity will once again be reduced. A well-defined sedimentation front can thus be observed. Things can get even more complicated, as the previous observations depend on the shape of the container in which they occur. During the study of red blood cells, the biologist A.E. Boycott (1920) observed a higher sedimentation rate when the test tube was tilted. In this case, red blood cells, with vertical sedimentation, are rapidly accumulating against the lower wall of the tube, thus leaving a lower concentration fluid zone against the higher wall of the tube. An ascending flow of "clear" fluid can then be observed in this zone, accompanied by a descending flow of "charged" fluid near the lower wall. Due to this separation, it is possible to accelerate the red cell sedimentation rate. If sedimentation takes place in a large radius spherical container, the same phenomenon occurs, and the sedimentation rate is once more increased, compared to the same phenomenon taking place in an ideal unlimited medium.

The law given by Batchelor underestimates the sedimentation velocity, which will rather be written as follows:

$$V_S = V_\infty(1 - 3.55C) \quad [2.5]$$

The previous examples illustrate the multitude of phenomena encountered in flows in which hydrodynamic interactions occur even at very low Reynolds numbers, a flow regime expected to present situations that are relatively easy to understand and solve. However, this is not the case in reality. The following section aims to provide details on some of these situations. From the mid-1980s, Professor A. Ambari dedicated his efforts to this specific branch of fluid mechanics. We will first present a summary of the fundamental equations of fluid mechanics required for solving this type of problem. The technique of the resistance (or mobility) matrix will then be introduced and will be used to solve several situations in which hydrodynamic interactions are important. Finally, the conclusion will provide the reader interested in more in-depth information with some ideas for consideration.

2.2. Equations of low Reynolds number fluid mechanics

Let us first recall the expression of the Reynolds number:

$$Re = \frac{\rho V L}{\mu} \quad [2.6]$$

where V and L are representative scales of velocity and the dimension of the system, and ρ and μ are the density and the dynamic viscosity of the fluid, respectively. This dimensionless number can be interpreted as the ratio of inertial force ($\rho V^2 L^2$) to viscosity force ($\mu V L$) in the flow. Low Reynolds number flows are therefore characterized by the predominance of viscous effects over inertia effects (it can also be interpreted as the ratio of a convection time L/V and a viscous diffusion time $\rho L^2/\mu$). A third interpretation was proposed by Purcell (1977): he noted that the term μ^2/ρ is homogeneous to a force and therefore any particle subjected to this force has a Reynolds number of the order of 1, irrespective of its shape. This occurs in the case of low displacement rate, or of very small size or very high viscosity, or for a combination of the previous conditions. Let us now consider how the equations of fluid mechanics are simplified when $Re \ll 1$. The flows of incompressible Newtonian fluids are governed by the continuity equation and Navier–Stokes equation:

$$\frac{\partial u_i}{\partial x_i} = 0 \quad [2.7]$$

$$\rho \frac{\partial u_i}{\partial t} + \rho u_j \frac{\partial u_i}{\partial x_j} = -\frac{\partial p}{\partial x_i} + \frac{\partial}{\partial x_j}\left(\mu \frac{\partial u_i}{\partial x_j}\right) + \rho f_i \quad [2.8]$$

When inertia terms are negligible compared to the viscous term, the Navier–Stokes equation can be written in the more simple form of unsteady Stokes equation:

$$\rho \frac{\partial u_i}{\partial t} = -\frac{\partial p}{\partial x_i} + \frac{\partial}{\partial x_j}\left(\mu \frac{\partial u_i}{\partial x_j}\right) + \rho f_i \qquad [2.9]$$

In this equation, in the absence of an intrinsic time scale T, the condition $Re \ll 1$ also implies that the local acceleration term can be neglected. Otherwise, another dimensionless number is used, namely the Stokes number St, which measures the ratio of the local acceleration and the viscosity term:

$$St = \frac{\rho L^2}{\mu T} \qquad [2.10]$$

For small size systems and/or for large viscosities, $St \ll 1$, which implies $T \gg \rho L^2/\mu$. In such conditions, the steady state is reached very rapidly (e.g. for a particle of size $L \sim 1mm$ in water ($\rho = 1{,}000\ \mathrm{kg/m^3}$ and $\mu = 1\ \mathrm{mPa.s}$), the typical time for momentum transport by viscous diffusion is of the order of 1 s). Hence, for very low Reynolds and Stokes numbers (we only consider these conditions), the Navier–Stokes equation takes the more simple form of the steady Stokes equation:

$$0 = -\frac{\partial p}{\partial x_i} + \frac{\partial}{\partial x_j}\left(\mu \frac{\partial u_i}{\partial x_j}\right) + \rho f_i \qquad [2.11]$$

The solution to this equation in incompressible fluid can then be used to calculate the stress tensor σ_{ij} and the force on the particle in the direction i:

$$\sigma_{ij} = -p\delta_{ij} + \mu\left(\frac{\partial u_i}{\partial x_j} + \frac{\partial u_j}{\partial x_i}\right) \qquad [2.12]$$

$$F_i = \int_S \sigma_{ij}\, n_j dS \qquad [2.13]$$

Among other things, the linearity of the Stokes equation implies the proportionality between causes and effects. For example, if the apparent weight of a particle is doubled, its sedimentation velocity will also double. Doubling the torque applied to a particle will lead to doubling its rotation speed. Doubling the pressure gradient in a pipe will double the flow rate in it. These equations have been used by Stokes (1851) to calculate the velocity and pressure fields around a fixed spherical particle of radius a in a flow of velocity $\vec{V}$ far away from it. If r is the distance to the center of the sphere and θ is the angle measured from the direction parallel to $\vec{V}$ (Figure 2.1), the components of the velocity field $\vec{u}$ write as follows:

$$u_r = V\left(1 - \frac{2a}{r} + \frac{a^3}{2r^3}\right)\cos\theta \qquad [2.14]$$

$$u_\theta = -V\left(1 + \frac{3a}{4r} + \frac{a^3}{4r^3}\right)\sin\theta \qquad [2.15]$$

One of the remarkable characteristics of this velocity field is its slow decrease with the distance as a function of $1/r$. As an illustration, even at a distance of $100\ a$, we only find V on the axis ($\theta = 0$) up to 1.5%. Therefore, the disturbances induced by a particle moving at $Re \ll 1$ are slowly dampened and propagate over very long distances. In the case of suspensions, the motion of a particle creates a long-range disturbance of the velocity field, which affects that of neighboring particles.

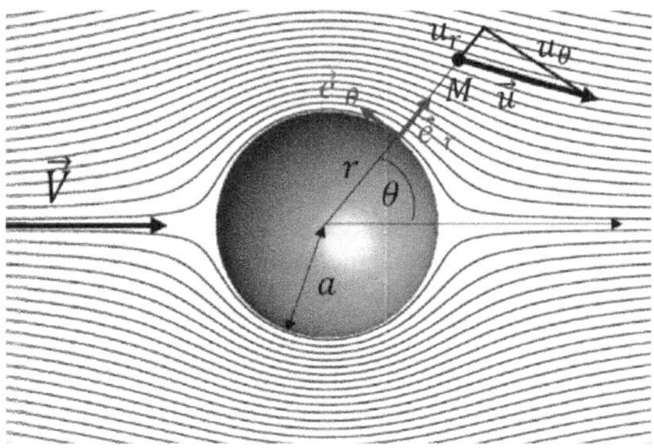

Figure 2.1. *Velocity field around a spherical particle at Re<<1. For a color version of this figure, see www.iste.co.uk/saouab/mechanics.zip*

These long-distance disturbances are commonly known as hydrodynamic interactions. All the solutions to the Stokes equation are of this type. More generally, due to the linearity of the Stokes equation, the flows that it describes have the following properties: uniqueness of solution, validity of the principle of superposition, flow reversibility, minimum energy dissipation rate and quasi-steadiness of flows (instantaneity principle). Due to these properties, many problems can be solved, and many mathematical techniques have been developed for this purpose (the books of Happel and Brenner (1983) or Clift et al. (1978) present a very significant number of such techniques and continue to be references).

The following section will illustrate how these properties can be used to solve several problems by introducing the technique of the resistance matrix.

2.3. Resistance matrix

This approach relies on the main properties stated in the previous section. Let us consider a particle of characteristic dimension L travelling in a fluid of viscosity μ and density ρ. Generally speaking, its motion is a combination of translational motion with velocity $\vec{V}$ and rotational motion with angular velocity $\vec{\Omega}$. Finally, let us consider $\vec{F}$ and $\vec{M}$ as the force and the torque that the fluid exerts on the particle. As long as $Re \ll 1$, there is a relation of proportionality between dynamic and kinematic variables. Therefore, the components of these variables verify the following relations:

$$F_i = -\mu\left(A_{ij}V_j + B_{ij}\Omega_j L\right) \quad [2.16]$$

$$\frac{M_i}{L} = -\mu(C_{ij}V_j + D_{ij}\Omega_j L) \quad [2.17]$$

Written in this form, the terms of the tensors between parentheses are homogeneous to a length. These relations can be obtained using the above mentioned principle of superposition. Tensors A_{ij} and D_{ij} connect, respectively, the force to the translational velocity and the torque to the rotational velocity. Tensors B_{ij} and C_{ij} connect, respectively, the force to the rotational velocity and the torque to the translational velocity. Furthermore, they have very general symmetry properties, independent from the shape of the particle, as follows:

$$A_{ij} = A_{ji}\ ,\ D_{ij} = D_{ji}\ ,\ B_{ij} = C_{ji} \quad [2.18]$$

For the particles with particular symmetry elements, there are other relations between these coefficients, as it will be seen below in the case of the spherical particle. These relations can be written as a matrix system in which the resistance matrix appears:

$$\begin{bmatrix} F_i \\ M_i/L \end{bmatrix} = -\mu \begin{bmatrix} A_{ij} & B_{ij} \\ C_{ij} & D_{ij} \end{bmatrix} \begin{bmatrix} V_j \\ \Omega_j L \end{bmatrix} \quad [2.19]$$

The terms of the resistance matrix are positive coefficients that depend only on the shape of the particle and on the boundary conditions. Let us first consider the simple case of a spherical particle of radius a moving in an unlimited medium.

When this is a translational motion with velocity $\vec{V}$, the force it is subjected to is given by the Stokes formula (Stokes 1851)[2].

$$\vec{F}_{\infty} = -6\pi\mu a\vec{V} \qquad [2.20]$$

A_{ij} is then a diagonal matrix with $A_{ii} = 6\pi a$. When the spherical particle is in rotational motion with velocity $\vec{\Omega}$, its torque is given by Stokes (1851):

$$\vec{M}_{\infty} = -8\pi\mu a^3\vec{\Omega} \qquad [2.21]$$

D_{ij} is then a diagonal matrix with $D_{ii} = 8\pi a$. Finally, all the other coefficients are zero ($B_{ij} = C_{ij} = 0$): there is total decoupling between the translational and rotational motions (this will be the case of all bodies having three perpendicular planes of symmetry). The previous results do not involve hydrodynamic interactions as they consider a single sphere and an unlimited medium, but they are a basis for going further. Let us now examine the manifestation of hydrodynamic interactions when the sphere moves along the axis of symmetry of a cylindrical container.

2.4. Sphere in hydrodynamic interaction in a cylindrical tube

Let us now imagine the same sphere of radius a moving on the axis of a cylindrical tube of radius b (Figure 2.2). There are now hydrodynamic interactions between the sphere and the lateral wall of the tube.

As long as the sphere remains on the axis of the tube, the problem remains axisymmetric and there is no torque or rotational velocity to be taken into account; the only hydrodynamic force $\vec{F}$ undergone by the sphere is a drag that is also axial. Two situations should then be studied: the sphere moving at constant velocity $\vec{V}$ and the fixed sphere in a Poiseuille flow of maximal velocity $\vec{V}_{\max}$ (Figure 2.2). In both cases, the presence of the sphere in the tube induces a disturbance ΔP of the pressure field that must be linearly connected to the characteristic velocities of the problem. Consequently, the resistance matrix can be written as follows (for the sake of clarity, we denote by R_{ij} the terms of the resistance matrix):

$$\begin{bmatrix} \vec{F} \\ \Delta P\pi b^2\vec{z} \end{bmatrix} = -\mu \begin{bmatrix} R_{11} & R_{12} \\ R_{21} & R_{22} \end{bmatrix} \begin{bmatrix} \vec{V} \\ -\vec{V}_{\max} \end{bmatrix} \qquad [2.22]$$

---

2 It is remarkable that this "simple" formula was used in the works rewarded by three Nobel Prizes, two in physics (R.A. Millikan in 1923 and J.B. Perrin in 1926) and one in chemistry (T. Svedberg in 1926).

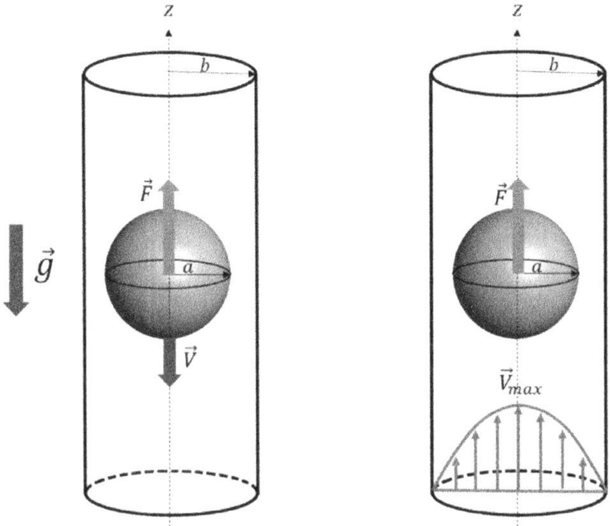

Figure 2.2. *Sphere in translational motion (left) and fixed sphere in Poiseuille flow (right). For a color version of this figure, see www.iste.co.uk/saouab/mechanics.zip*

Coefficients R_{11} and R_{12} result from calculations of the drag on a sphere in uniform translational motion and in Poiseuille flow. Coefficients R_{21} and R_{22} are calculated from the disturbance of pressure around the sphere (the matrix being symmetrical, $R_{12} = R_{21}$, one calculation is sufficient). Let us consider:

$$R_{11} = 6\pi a \times \alpha(k) \quad [2.23]$$

$$R_{12} = R_{21} = 6\pi a \times \beta(k) \quad [2.24]$$

$$R_{22} = 6\pi a \times \gamma(k) \quad [2.25]$$

where $k = a/b$ is the confinement factor ($0 \leq k \leq 1$). In the case of small Reynolds numbers, the effect of lateral confinement in the case of a sphere in axial motion in a tube has been asymptotically, numerically and experimentally studied by many authors, as discussed in the article by Ambari et al. (1985). A precise calculation of the correction factor $\alpha(k)$ of the Stokes force was numerically obtained by Ben Richou et al. (2003) and asymptotically confirmed for $k \approx 1$. These results are in perfect agreement with the asymptotic formula given by Bungay and Brenner (1973). The latter resulted from a combination of the asymptotic expansion obtained by Bungay and Brenner (1973) about $k = 1$ and of the asymptotic expansion about $k = 0$ given by Haberman and Sayre (1958):

$$\alpha(k) = \frac{3\pi\sqrt{2}}{8}(1-k)^{-\frac{5}{2}}\left[1 - \frac{73}{60}(1-k) + \frac{77.293}{50.4}(1-k)^2\right] - \frac{22.5083}{6\pi} - \frac{5.6117}{6\pi} - \frac{0.3363}{6\pi}k^2 - \frac{1.216}{6\pi}k^3 + \frac{1.647}{6\pi}k^4 \quad [2.26]$$

The coefficient $\beta(k)$ was numerically obtained by Ben Richou et al. (2003) for $0.05 \leq k \leq 0.95$. Bungay and Brenner (1973) also calculated it using the method of singular disturbances and gives an expression of this factor:

$$\beta(k) = \frac{3\pi\sqrt{2}}{16}(1-k)^{-\frac{5}{2}}[1 + \frac{7}{60}(1-k) - \frac{2.227}{50.4}(1-k)^2] + \frac{4.018}{12\pi} - \frac{3.9788}{12\pi}k - \frac{1.9215}{12\pi}k^2 + \frac{4.392}{12\pi}k^3 + \frac{5.006}{12\pi}k^4 \quad [2.27]$$

The same authors can be cited for the coefficient $\gamma(k)$, for which they gave the following expression:

$$\gamma(k) = \frac{3\pi\sqrt{2}}{16}(1-k)^{-\frac{5}{2}}[1 + \frac{29}{20}(1-k) - \frac{97.453}{50.4}(1-k)^2] - \frac{62.268}{12\pi} - \frac{18.509}{12\pi}k - \frac{12.156}{12\pi}k^2 + \frac{9.15}{12\pi}k^3 + \frac{43.17}{12\pi}k^4 \quad [2.28]$$

The resistance matrix is then complete and can be used to solve certain problems. Let us start by calculating the transport velocity of a neutrally buoyant sphere subjected to a Poiseuille flow. In this case, the hydrodynamic force $\vec{F}$ is zero (apparent weight is zero) and the resolution of the matrix system yields:

$$\frac{V}{V_{\max}} = \frac{R_{12}}{R_{11}} = \frac{\beta(k)}{\alpha(k)} \quad [2.29]$$

$$\frac{\Delta P\, \pi b^2}{\mu a V_{\max}} = \frac{1}{a}(R_{22} - \frac{R_{12}R_{21}}{R_{11}}) = 6\pi\left[\gamma(k) - \frac{\beta^2(k)}{\alpha(k)}\right] \quad [2.30]$$

Figure 2.3(a) shows the evolution of the transportation velocity normalized by the maximal velocity of Poiseuille flow. When $k \to 0$, the sphere is transported at local velocity $V_{\max}$. The stronger the confinement, the slower the transport velocity of the sphere. When $k \to 1$, the particle advances at 50% of $V_{\max}$, which is the average velocity of Poiseuille flow.

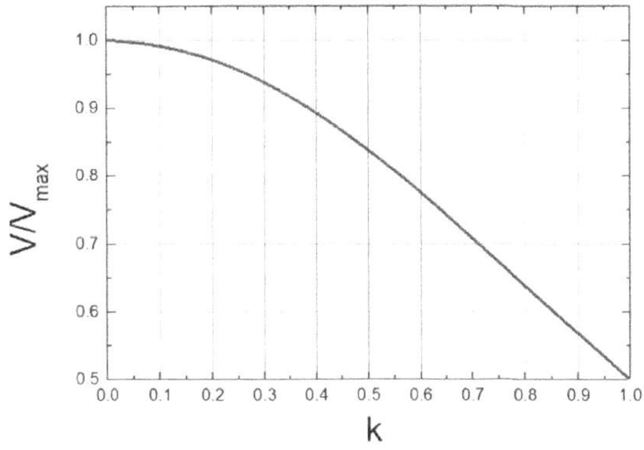

a)

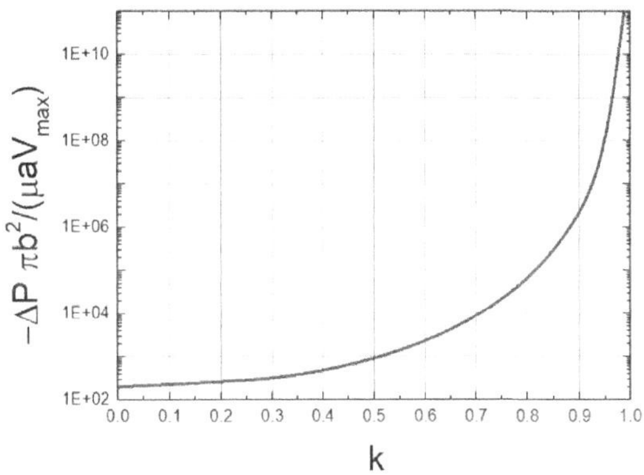

b)

Figure 2.3. *Transportation velocity (a) and pressure (b) as a function of confinement. For a color version of this figure, see www.iste.co.uk/saouab/mechanics.zip*

Figure 2.3(b) shows the evolution of the normalized pressure induced by the sphere, depending on confinement. The stronger the confinement of the sphere, the higher the head loss in the tube with a divergence when $k \to 1$. Let us now examine the sedimentation of a sphere of density ρ_s in a fluid of lower density ρ ($\Delta\rho = \rho_s - \rho > 0$). In this case, there is no Poiseuille flow ($\vec{V}_{max} = \vec{0}$) and the hydrodynamic force $\vec{F}$ is equal to the apparent weight of the particle:

$$\vec{F} = -\frac{4}{3}\pi a^3 \Delta\rho \vec{g} \qquad [2.31]$$

Confinement variation can be operated in two ways that are not equivalent: a sphere of constant radius in a tube of variable radius can be considered, or the opposite. In the first case, the resolution of the matrix system yields:

$$\frac{V}{V_\infty} = \frac{6\pi}{R_{11}} = \frac{1}{\alpha(k)} \qquad [2.32]$$

In this expression, the sedimentation velocity in infinite medium (equation [2.4]) was used to normalize the sedimentation velocity, as a is constant.

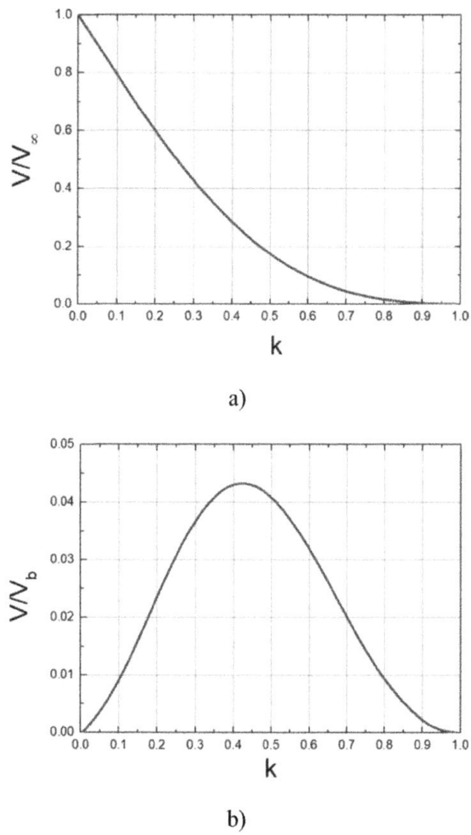

Figure 2.4. *Sedimentation for a sphere of constant radius (a) and of variable radius (b), depending on confinement. For a color version of this figure, see www.iste.co.uk/saouab/mechanics.zip*

It is logical that the more confined the sphere, the lower its sedimentation velocity (Figure 2.4(a)). When $k \to 0$, it sediments at the classical velocity V_∞, and when $k \to 1$, its velocity tends to 0. If we now consider the sedimentation of a sphere of variable radius in a tube of fixed radius, another normalization should be chosen. If the velocity scale is:

$$V_b = \frac{2}{9}\frac{\Delta\rho g b^2}{\mu}, \qquad [2.33]$$

then:

$$\frac{V}{V_b} = \frac{6\pi k^2}{R_{11}} = \frac{k^2}{\alpha(k)} \qquad [2.34]$$

An entirely different behavior is then observed in Figure 2.4(b). When $k \to 0$, the apparent weight of the sphere is zero, and no sedimentation is possible. The ascending part of the curve corresponds to a range where the increase in apparent weight of the sphere predominates over the increase in the effects of confinement. Beyond a certain confinement (for $k > 0.42$), the opposite behavior is manifest: hydrodynamic interactions increase at a higher rate than the apparent weight, which leads to a decreasing sedimentation rate, until its cancellation for $k \to 1$. The velocity of maximum fall is therefore reached when the radius of the sphere is 42% of that of the tube and is approximately 4.5 % of the velocity V_b. Let us compare the maximum velocity of this same sphere with the one it might have in an infinite medium:

$$\frac{V(k=0.42)}{V_b} = 0.045 \qquad [2.35]$$

$$\frac{V(k=0.42)}{V_\infty} = 0.045\frac{V_b}{V_\infty} = \frac{0.045}{k^2} = \frac{0.045}{0.42^2} = 0.255 \qquad [2.36]$$

In a confined medium, the velocity of a sphere would never exceed 25.5% of the sedimentation velocity it would have in the absence of confinement, at least as long as it is in a centered position.

Two aspects of confinement should be taken into consideration for sedimentation. The first one is related to the interactions with the wall, which tend to decrease the fall velocity: the stronger the confinement, the greater this effect. The second effect is due to the backflow of the fluid, whose consequence is also a decrease in the sedimentation velocity. Indeed, when the sphere travels, the volume of fluid it sets into motion must flow in the opposite direction to ensure the conservation of mass (Figure 2.5). As a result of this backflow, which is all the more significant as the confinement is larger, the constraints on the sphere increase and they slow it down even further.

What happens when the ball is not in axial position? Ambari et al. (1993) conducted an experimental study of this configuration using a rheometer with magnetic levitation and obtained an apparently surprising result: the sedimentation velocity is not maximal when the sphere is on the axis of the tube. Figure 2.6, extracted from this article, shows the evolution of the drag force normalized by the force in axial position. The parameter e featuring in this figure represents the relative eccentricity of the sphere (distance between the center of the sphere and the axis of the tube, normalized by the radius of the tube). The confinement in this experiment is $k = 0.44$. It can be noted that at relative eccentricity $e \sim 0.33$, the drag is reduced by 10% (or a 10% increase in the sedimentation velocity) compared to that at the center. This effect increases when the confinement is stronger, as shown by other experimental and numerical results. Replicating the same experiment for a sphere confined by two parallel planes, this effect disappears, as it can be seen in Figure 2.6(a) extracted from the same article.

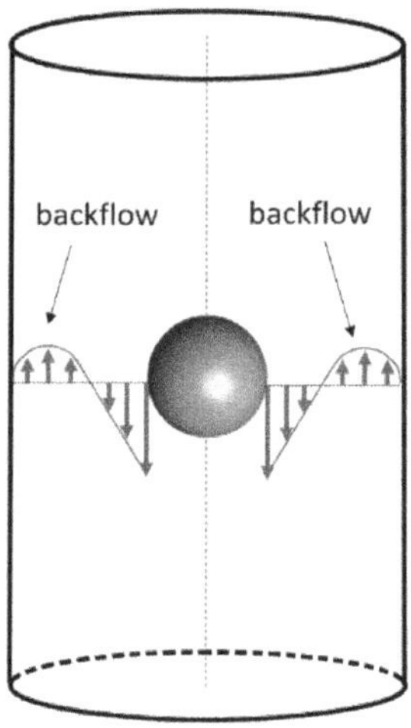

Figure 2.5. *Backflow induced by the displacement of the sphere. For a color version of this figure, see www.iste.co.uk/saouab/mechanics.zip*

Two questions arise: Why does an eccentric position give a maximal sedimentation velocity in a tube? Why does that not occur between two planes? The answers are related to backflow. This is no longer symmetric when the ball sediments off the axis of the tube. Most of the backflow then passes in the largest passage section (in the part left free by the sphere). On average, the detrimental effect of the backflow is less important in eccentric situation. On the other hand, when the eccentricity becomes too high and the sphere is very close to the wall, this positive effect is compensated by stronger interactions with the wall. The drag force increases once again. The answer to the second question is also due to the return flow. In the case of parallel planes, the "backflow" is not completely confined by the walls, and it can be indefinitely distributed in the directions parallel to these planes. In other terms, the backflow between two parallel planes is not influenced by the eccentric position of the sphere. It is not the case in a cylindrical tube, where the return flow is confined in all transversal directions. Therefore, the "backflow" in eccentric situation might be more favorable to inducing a weaker drag force in a tube but not between two parallel planes.

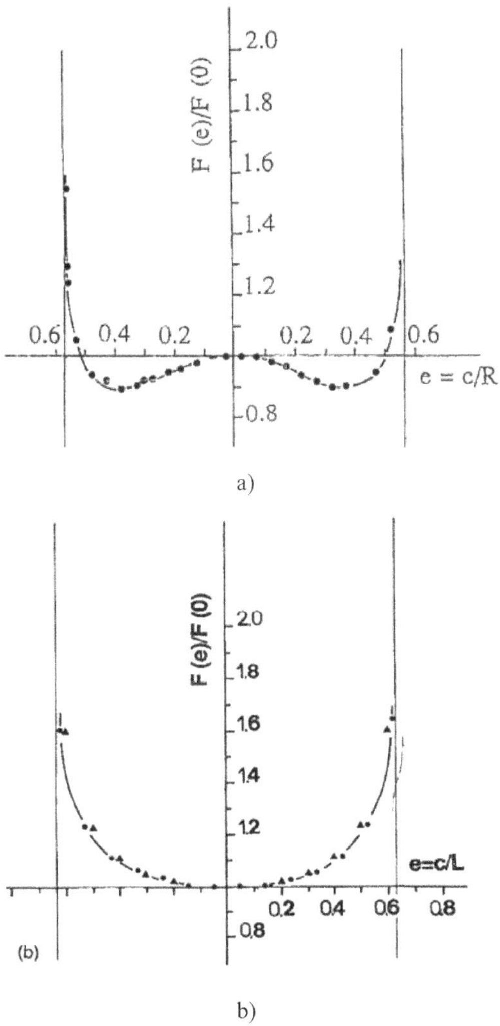

Figure 2.6. *Distribution of the drag force as a function of the lateral position of the sphere in a tube (a) and between two planes (b)*

2.5. Pair of spheres in hydrodynamic interaction

As a conclusion to this brief overview of hydrodynamic interactions, let us examine the case of those existing between two spheres of the same density that sediment (Figure 2.7) in unlimited medium.

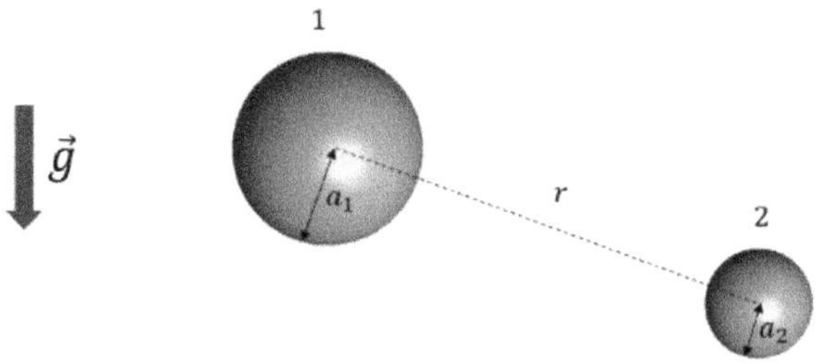

Figure 2.7. *Pair of spheres in sedimentation. For a color version of this figure, see www.iste.co.uk/saouab/mechanics.zip*

In the absence of inertia, each particle is subjected to its apparent weight $\vec{F}_i^e$ and to a drag force $\vec{F}_i^h$ that balance one another:

$$\vec{F}_i^e = \frac{4}{3}\Delta\rho\vec{g}\pi a_i^3 = -\vec{F}_i^h \qquad [2.37]$$

If the particles are very far apart, they do not interact, and the drag force on each particle is given by the Stokes formula $\vec{F}_i^h = -6\pi\mu a_i\vec{V}_i$. Then, the resistance matrix is simply written as follows:

$$\begin{bmatrix} \vec{F}_1^e \\ \vec{F}_2^e \end{bmatrix} = \mu \begin{bmatrix} R_{11} & 0 \\ 0 & R_{22} \end{bmatrix} \begin{bmatrix} \vec{V}_1 \\ \vec{V}_2 \end{bmatrix} \qquad [2.38]$$

where $R_{ii} = 6\pi a_i$. Using this approach, it is more convenient to introduce the mobility matrix, rather than the resistance matrix:

$$\begin{bmatrix} \vec{V}_1 \\ \vec{V}_2 \end{bmatrix} = \frac{1}{\mu} \begin{bmatrix} M_{11} & 0 \\ 0 & M_{22} \end{bmatrix} \begin{bmatrix} \vec{F}_1^e \\ \vec{F}_2^e \end{bmatrix} \qquad [2.39]$$

with $M_{ii} = 1/R_{ii} = 1/(6\pi a_i)$. Each sphere will sediment in the direction of gravity with the velocity:

$$\vec{V}_i = \frac{2}{9}\frac{\Delta\rho\vec{g}a_i^2}{\mu} \qquad [2.40]$$

This happens provided that the distance r between the two particles is very long due to the slow decrease of hydrodynamic interactions as a function of $1/r$. When this is no longer the case, there is hydrodynamic coupling between the motion of particles ($M_{12} = M_{21} \neq 0$) but the linearity between the dynamic and kinematic variables remains. Once again, the following can be written:

$$\begin{bmatrix} \vec{V}_1 \\ \vec{V}_2 \end{bmatrix} = \frac{1}{\mu} \begin{bmatrix} M_{11} & M_{12} \\ M_{21} & M_{22} \end{bmatrix} \begin{bmatrix} \vec{F}_1^e \\ \vec{F}_2^e \end{bmatrix} \quad [2.41]$$

Besides the shape and size of particles, coefficients M_{ij} are also a function of the interparticle distance r. The terms of the mobility matrix can be calculated using the method of reflections introduced by Smoluchowski (1911) and described in many works such as the books by Happel and Brenner (1983) and Kim and Karrila (1991). This method relies on the idea of considering a single particle, say particle 1, and the disturbance of the velocity field it induces is calculated to the position where particle 2 should be located. This disturbance is then added to the velocity field of particle 2. The operation is then repeated, by considering this time only particle 2, and calculating the disturbance it induces to the position where particle 1 should be located. This procedure can be repeated several times, each round trip being a new reflection. If isolated, particle 1 sediments at the already calculated velocity. Let us denote by $\vec{V}_1^0$ this velocity obtained without reflection:

$$\vec{V}_1^0 = \frac{2}{9} \frac{\Delta\rho \vec{g} a_1^2}{\mu} = \frac{\vec{F}_1^e}{6\pi\mu a_1} = \frac{1}{\mu} M_{11} \vec{F}_1^e \quad [2.42]$$

The disturbance $\vec{u}_1^0$ of the velocity field that it induces in its motion can be calculated as a function of the gravity force $\vec{F}_1^e$ (Stokes 1851):

$$\vec{u}_1^0 = \frac{1}{8\pi\mu} \left[\left(\frac{I}{r} + \frac{\vec{r}.\vec{r}}{r^3}\right). \vec{F}_1^e + \left(\frac{I}{3r^3} - \frac{\vec{r}.\vec{r}}{r^5}\right). a_1^2 \vec{F}_1^e \right] \quad [2.43]$$

Here, I is here the identity tensor and the position vector $\vec{r}$ has its origin at the center of particle 1. Using Faxén's (1922) law, it is possible to calculate the velocity of particle 2 undergoing this disturbance:

$$\vec{V}_2^1 = \vec{u}_1^0 + \frac{a_2^2}{6} \nabla^2 \vec{u}_1^0 \quad [2.44]$$

If the particles are not very close, the terms of the order $O(1/r^3)$ can be neglected:

$$\vec{V}_2^1 = \frac{1}{8\pi\mu} \left(\frac{I}{r} + \frac{\vec{r}.\vec{r}}{r^3}\right). \vec{F}_1^e + O(1/r^3) \quad [2.45]$$

Using the principle of superposition, the total velocity of particle 2 will be the sum of velocity $\vec{V}_2^0$ (particle 2 isolated) and $\vec{V}_2^1$ (disturbance due to particle 1):

$$\begin{aligned}\vec{V}_2 = \vec{V}_2^0 + \vec{V}_2^1 &= \frac{\vec{F}_2^e}{6\pi\mu a_2} + \frac{1}{8\pi\mu}\left(\frac{I}{r} + \frac{\vec{r}.\vec{r}}{r^3}\right).\vec{F}_1^e + O(1/r^3) \\ &= \frac{1}{\mu}\left[M_{21}\vec{F}_1^e + M_{22}\vec{F}_2^e\right] + O(1/r^3)\end{aligned} \quad [2.46]$$

For particle 1, a similar expression can be obtained:

$$\begin{aligned}\vec{V}_1 = \vec{V}_1^0 + \vec{V}_1^1 &= \frac{\vec{F}_1^e}{6\pi\mu a_1} + \frac{1}{8\pi\mu}\left(\frac{I}{r} + \frac{\vec{r}.\vec{r}}{r^3}\right).\vec{F}_2^e + O\left(\frac{1}{r^3}\right) \\ &= \frac{1}{\mu}\left[M_{11}\vec{F}_1^e + M_{12}\vec{F}_2^e\right] + O(1/r^3)\end{aligned} \quad [2.47]$$

In the mobility matrix, the diagonal terms are unchanged and the other two are at present non-zero:

$$M_{11} = \frac{1}{6\pi a_1} \quad M_{22} = \frac{1}{6\pi a_2} \quad [2.48]$$

$$M_{12} = M_{21} = \frac{1}{8\pi}\left(\frac{I}{r} + \frac{\vec{r}.\vec{r}}{r^3}\right) \quad [2.49]$$

This first stage is the first reflection, and it might be improved by other reflections according to the same procedure. The terms $M_{12} = M_{21}$ being positive, then a pair of spheres sediments more rapidly than an isolated particle. Moreover, if the particles have the same radius, they sediment at the same velocity and the distance between them does not change. Finally, unless the spheres are vertically or horizontally aligned, the sedimentation velocities have a component that is not aligned with gravity due to the second term of M_{12} or M_{21}. Hence, the two particles sediment by drifting from the vertical direction (this component tends toward 0 when $r \to \infty$). For two identical spheres of radius a vertically aligned, the calculation that stopped at the first reflection gives the following sedimentation velocity:

$$V = V_\infty\left(1 + \frac{3a}{2r}\right) \quad [2.50]$$

Horizontally aligned, this velocity is written as:

$$V = V_\infty\left(1 + \frac{3a}{4r}\right) \quad [2.51]$$

These velocities are represented in Figure 2.8.

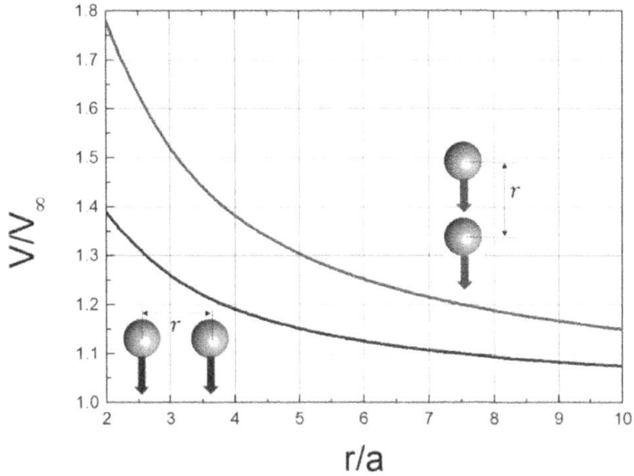

Figure 2.8. *Sedimentation velocity of a pair of identical spherical particles that are aligned horizontally (blue) or vertically (red). For a color version of this figure, see www.iste.co.uk/saouab/mechanics.zip*

The same calculation can be made for three spheres, but there is a radical change in behavior. The middle sphere sediments more rapidly than the surrounding spheres, as its velocity increases due to the contributions of its two neighbors, while for the two external spheres, these contributions are smaller. The problem then becomes nonstationary, and numerical approaches must be used for its complete resolution. For a very large number of particles, these methods are no longer applicable and statistical, and numerical approaches must be used to calculate the average sedimentation velocity of suspensions. The average velocities obtained have the following form:

$$V = V_\infty f(C) \qquad [2.52]$$

where C is the volume concentration in particles. The function $f(C)$, known as hindered settling function, is a decreasing function of concentration. The formulas presented in the introduction (equation [2.3] and [2.5]) are of this type. Other forms have been proposed, such as those of Richardson and Zaki (1954) written as $f(C) = (1 - C)^n$ with $n \approx 5$. All these expressions are no longer valid for too high concentrations (from just a few percent for that of Batchelor and of the order of 50% for that of Richardson).

2.6. Conclusion

This brief chapter uses a non-exhaustive approach to presenting the importance of hydrodynamic interactions that occur in very low Reynolds number flows. The slow decrease of velocity field in these regimes implies that these interactions are important even at long distance. They manifest not only among particles, but also between the particles and the walls of the containers they are stored in. Given the linearity of the Stokes equation and the resulting properties of solutions, a number of problems can be solved, a few examples of which are discussed in this chapter. An example of a case study is that of non-spherical particles, such as the ellipsoid, a model that approximates quite well the fibers encountered in some types of flows. Specific theories have been developed in the case of very "slender" particles (slender body theory) and they contributed to solving problems of great scientific and technological importance. These theories have been applied even in biology (particularly for cell motion; Lauga 2020). Another case worth mentioning is that of particles in translational motion parallel or perpendicular to a solid plane wall. When a particle gets very close to a wall, the domain is that of hydrodynamic lubrication, and solutions of the Stokes equation can be obtained by asymptotic expansions. Another interesting approach involves getting the particle size tend to 0 (or enormously distancing the particles). The velocity field is then referred to as a "Stokeslet", equivalent to that obtained by a punctually applied force. This approach contributed to many advances in the domain of the dynamics of suspensions. The interested reader is referred to the book of Guazzelli and Morris (2011), which is an excellent introduction to all these subjects. Due to his numerous and original contributions brought for more than 30 years to this domain, Professor Ambari elucidated some of the mysteries of flows, in which hydrodynamic interactions are involved. We express our deepest gratitude for this.

2.7. References

Ambari, A., Manuel, B.G., Guyon, E. (1983). Effect of a plane wall on a sphere moving parallel to it. *Journal de Physique Lettres*, 44(4), 143–146.

Ambari, A., Gauthier-Manuel, B., Guyon, E. (1985). Direct measurement of tube wall effect on the Stokes force. *Physics of Fluids*, 28, 1559–1561.

Ambari, A., Gauthier-Manuel, B., Guyon, E. (1993). Backflow effect in model sedimentation experiments. *Physics of Fluids A: Fluid Dynamics*, 5(8), 2061–2063.

Batchelor, G.K. (1972). Sedimentation in a dilute dispersion of spheres. *Journal of Fluid Mechanics*, 52(2), 245–268.

Ben Richou, A., Ambari, A., Naciri, J.K. (2003). Correction factor of the Stokes force undergone by a sphere in the axis of a cylinder in uniform and Poiseuille flows. *Eur. Phys. J. Phys.*, 24, 153–165.

Boycott, A.E. (1920). Sedimentation of blood corpuscles. *Nature*, 104(2621), 532.

Bungay, P.M. and Brenner, H. (1973). The motion of a closely-fitting sphere in a fluid-filled tube. *Int. J. Multiphase Flow*, 1, 25–56.

Clift, R., Grace, J.R., Weber, M.E. (1978). *Bubbles, Drops and Particles*. Academic Press, London.

Einstein, A. (1905). On the motion of small particles suspended in liquids at rest required by the molecular-kinetic theory of heat. *Annalen der Physik*, 17, 549–560.

Faxén, H. (1922). Der Widerstand gegen die Bewegung einer starren Kugel in einer zähen Flüssigkeit, die zwischen zwei parallelen ebenen Wänden eingeschlossen ist. *Annalen der Physik*, 373(10), 89–119.

Guazzelli, E. and Morris, J.F. (2011). *A Physical Introduction to Suspension Dynamics*. Cambridge University Press, Cambridge.

Haberman, W.L. and Sayre, R.M. (1958). Motion of rigid and fluid spheres in stationary and moving liquids inside cylindrical tubes. *David W. Taylor Model Basin Reports*, 1143.

Happel, J. and Brenner, H. (1983). *Low Reynolds Number Hydrodynamics: With Special Applications to Particulate Media*. Springer, Dordrecht.

Kim, S. and Karrila, S.J. (1991) *Microhydrodynamics: Principles and Selected Applications*. Buttertworth-Heineman, Boston.

Lauga, E. (2020). *The Fluid Dynamics of Cell Motility*. Cambridge University Press, Cambridge.

Purcell, E.M. (1977). Life at low Reynolds number. *American Journal of Physics*, 45(1), 3–11.

Richardson, J.F. and Zaki, W.N. (1954). Sedimentation and fluidization: Part I. *Trans. Inst. Chem. Eng.*, 32, 33–53.

Smoluchowski, M. (1911). Über die Wechselwirkung von Kugeln die sich in einer zähen Flüssigkeit bewegen. *Bull. Acad. Sci. Cracow*, 1a, 28.

Stokes, G.G. (1851). On the effect of the internal friction of fluids on the motion of pendulums. *Trans. Camb. Phi. Soc., Part II*, 9, 8–106.

Taylor, G.I. (1969). Notes for the film “Low Reynolds number flows” [Online]. Available at: http://web.mit.edu/hml/ncfmf/07LRNF.pdf.

3

Stokes Flows – Applications to Creeping Flows Around Obstacles and to Flows in Porous Media

This chapter proposes the resolution of the biharmonic equation for the streamfunction of a low Reynolds number permanent flow, known as Stokes flow. The equation is solved in various systems of coordinates to be adapted to the shape of the flow domain or to the shape of flow obstacles. The solution found in Cartesian coordinates serves for modeling the flow in a channel and in a network of channels. Applying certain boundary conditions can generate cell-structured flows. The latter can also be created between obstacles. Then, we shall present the case of flow between circles obtained by solving the biharmonic equation in polar coordinates. The latter is solved in a system of elliptical coordinates to determine the Stokes flow around an elliptical obstacle. Furthermore, the Stokes equation coupled with Darcy's equation can be used to calculate the permeability of model porous media. Applications to periodic fiber models are thus presented. The study then extends to a three-dimensional granular and fibrous porous medium. Finally, we shall analyze the influence of microstructure on permeability.

3.1. Introduction

There is a rich literature on the theoretical and experimental analysis of phenomena of separation and formation of cell-structured two-dimensional flows characterized by closed streamlines surrounding one or several points of zero

---

Chapter written by Mustapha HELLOU, Mohammed Khaled BOURBATACHE and Franck LOMINÉ.

velocity. In particular, when inertia forces are negligible (Stokes' regime), the analytical solutions to the equation of two-dimensional motion allow a theoretical analysis of the separation and development of the cellular motion (Moffatt 1964; Happel and Brenner 1973; Taneda 1979; O'Neill 1983; Bourot 1984; Hellou and Coutanceau 1992; Pozrikidis 1992; Moreau and Bourot 1993; Jeong 2001; Shankar 2007; Champmartin et al. 2007). Originally, the existence of these motions was proven in the case of the flow near a sharp corner, which is now known in the literature as Moffatt cells (Moffatt 1964). The latter has also predicted the possibility of the formation of these motions between parallel planes. Later on, O'Neill (1983) and Bourot (1984) proved the existence of an infinite series of cells in the case of flow between parallel planes. This result confirms Moffatt's (1964) predictions and is in agreement with the experimental visualization conducted in the case of a large elongation cavity (Taneda 1979; Hellou and Coutanceau 1992).

The study of the mass transfer by these cellular motions has attracted a lot of interest in the last few decades, thanks to the property of these motions of being controlled by the geometry of the domain and the boundary conditions. Many applications can thus be considered. These applications can be found in various domains, such as geosciences, microfluidics, medical field, chemistry and materials.

The particular case of flow between parallel planes, which will sometimes be called channels or flow branches, attracted interest in the field of geosciences, particularly for modeling flow and solute transport in fractured media. In order to comprehend this transport, many authors have used the model of networks of channels (Berkowitz et al. 1994; Adler and Thovert 1999; Bruderer and Bernabé 2001; Park et al. 2001; Mourzenko et al. 2002; Li 2002; Hellou and Bour 2014; Berre et al. 2019).

Flows in networks of branches are also of interest in highly specialized domains, such as microfluidics, in relation to medical engineering and process engineering (Brody et al. 1996; Stone et al. 2004; Mittal et al. 2013; Freund 2014; Guan et al. 2015; Corona et al. 2018; Huber et al. 2018; Toschi and Sega 2019). This model of networks allows for the simulation of small flows in respiratory systems, micro-pumps, units of electric batteries and particle separators (Stone et al. 2004; Mittal et al. 2013). One of the relatively recent models is that of the flow in the junction between four tubes produced by four rotating cylinders, a system known as FFoRM (Fluidic Four Roll Mill). This system can be used to produce a wide variety of Stokes flows depending on the applied boundary conditions. A summary of the possibilities offered by this system can be found in published articles (Toschi and Sega 2019; Corona et al. 2018).

Stokes flows are also of great interest for the analysis of fluid behavior in confined geometries, such as circles in relative motion. These configurations are found in the lubrication domain (Hajjam 1998). It is also important to mention the flows around assemblies of various forms of obstacles (Sangani and Acrivos 1982; Zick and Homsy 1982; Wang 1996; Higdon and Ford 1996; Wang 2002; Hellou et al. 2004; Tamayol and Bahrami 2011; Hellou et al. 2021), inside capillary tubes or even in the junctions of these tubes. This contributes to the comprehension of mechanisms that govern the flows in porous or fractured materials by determining the velocity and pressure field depending on microstructure (Hellou and Bach 2011; Cachile et al. 2012; Charpentier et al. 2018; Hellou and Bach 2018). Moreover, coupling the results of the pressure field to macroscopic laws can lead to an estimation of macroscopic quantities such as permeability depending on the characteristics of the porous medium such as porosity, specific area or tortuosity (Costa 2006; Wang 2010; Valdès-Parada et al. 2011; Yazdchi et al. 2011; Hommel et al. 2018; Bourbatache et al. 2019).

This chapter presents the solution of the biharmonic equation resulting from the Stokes equation for a permanent two-dimensional flow of an incompressible Newtonian fluid in various systems of coordinates. Then, it presents applications to the flow in a junction between four semi-infinite channels to the flow around circular and elliptical obstacles. These various configurations are used to simulate the flow in a periodic porous medium.

3.2. Several solutions to the Stokes equation in two-dimensional flow

When the effects of inertia are negligible, as in the case of low Reynolds number flow, the equation of motion is simplified and becomes:

$$\vec{\nabla} p = \mu \vec{\Delta} V \qquad [3.1]$$

where p is the pressure and $\vec{V}$ is the velocity vector.

In the case of a permanent incompressible two-dimensional flow, the introduction of the streamfunction ψ leads to the biharmonic equation:

$$\Delta\Delta\psi = 0 \qquad [3.2]$$

which can be solved by successively looking for solutions to equations: $\Delta\psi_h = 0$ and $\Delta\psi = \psi_h$.

3.2.1. *Solution to the Stokes equation in Cartesian coordinates*

Applying the variable separation method, $\psi_h = \phi(x)\omega(y)$ leads to the system of equations:

$$\begin{cases} \phi''(x) - \varepsilon^2\phi(x) = 0 \\ \omega''(y) + \varepsilon^2\omega(y) = 0 \end{cases} \qquad [3.3]$$

where ε is a real or complex constant. The solution to the system of equations [3.3] is readily obtained.

The following expressions of ψ_h are obtained for $\varepsilon = 0$ and $\varepsilon \neq 0$, respectively:

$$\psi_{h0} = a_0 + b_0x + c_0y + d_0xy \qquad [3.4]$$

$$\psi_{h\varepsilon} = \sum_{\varepsilon\neq 0}(a_\varepsilon e^{\varepsilon x} + b_\varepsilon e^{-\varepsilon x})(c_\varepsilon e^{i\varepsilon y} + de^{-i\varepsilon y}) \qquad [3.5]$$

where $a_\varepsilon, b_\varepsilon, c_\varepsilon, d_\varepsilon$ for $\varepsilon = 0$ or $\varepsilon \neq 0$ are real or complex coefficients. The harmonic function $\psi_h = \psi_{h0} + \psi_{h\varepsilon}$ is also biharmonic. Moreover, $x\psi_h$, $y\psi_h$ and $(x^2 + y^2)\psi_h$ functions are also biharmonic. Hence, the required current streamfunction is therefore:

$$\psi = (1 + x + y + x^2 + y^2)\psi_h \qquad [3.6]$$

For further details on this resolution, the reader is invited to refer to Hellou and Bach (2011).

The biharmonic function was applied to the zero-rate theoretical flow in a channel with parallel walls of infinite length, which corresponds to a flow with closed streamlines (Hellou and Coutanceau 1992; Pozrikidis 1992; Moreau and Bourot 1993). In this case, only the streamfunction $\psi_{h\varepsilon}$ is conserved. Two flow families can be distinguished:

– antisymmetric flow: $\psi(x, y) = \psi(x, -y)$;

– symmetric flow: $\psi(x, y) = -\psi(x, -y)$.

An arbitrary flow results from the superposition of these two basic flows (Moreau and Bourot 1993; Jeong 2001).

In each of the basic flows, writing the no-slip conditions at the walls parallel to the axis $\vec{x}$, for example, leads to the following equations:

$$\sin(2y_0\delta_n) + 2y_0\delta_n = 0 \text{ (Antisymmetric flow)}$$
$$\sin(2y_0\delta_n) - 2y_0\delta_n = 0 \text{ (Symmetric flow)} \quad [3.7]$$

where $2y_0$ represents the width of the channel. These equations are solved numerically by setting $\delta_n = \lambda_n + i\mu_n$. Values of λ_n and μ_n for n up to 50 can be found in Moreau and Bourot (1993), both for the antisymmetric series and for the symmetric series.

If we examine the flow with zero average rate, to each value of n corresponds a flow depending on whether it is antisymmetric or symmetric. The complexity of the flow structure increases when the value of n increases. Figure 3.1 shows the structure of the antisymmetric flow between two planes for $n = 1$ and $n = 2$. For each value of n, there is theoretically an infinity of successive cells that are established in the channel. Their damping, which is given by the relation $\kappa_n = \exp(\pi\lambda_n/\mu_n)$ also increases with n ($\kappa_1 = 358$, $\kappa_2 = 51{,}290$). Hence, taking into account the exponential increase of the damping, real flow, represented by the superposition of the set of individual flows, would be similar to a flow corresponding to $n = 1$.

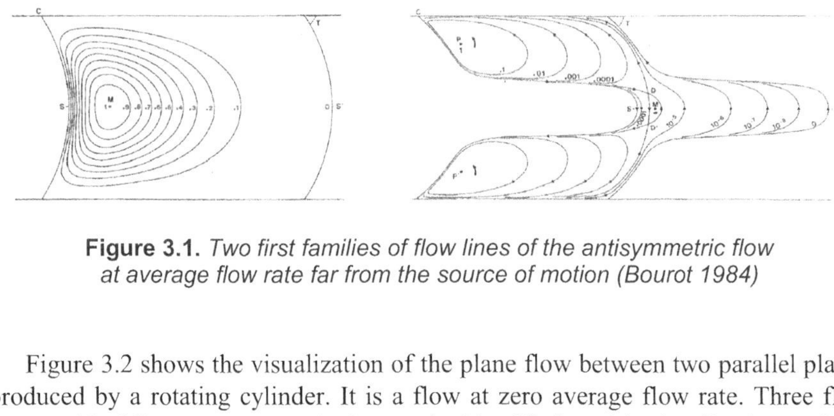

Figure 3.1. *Two first families of flow lines of the antisymmetric flow at average flow rate far from the source of motion (Bourot 1984)*

Figure 3.2 shows the visualization of the plane flow between two parallel planes produced by a rotating cylinder. It is a flow at zero average flow rate. Three flow zones with different average velocity can be identified on this visualization. Indeed, images are captured intermittently: 24 exposures of 10 s separated by an interval of 5 min. The flow around the cylinder is represented by continuous lines corresponding to the paths described by tracers in 10 s. The flow in the first cell is represented by discontinuous lines in the form of a series of points, each of which corresponds to an exposure of 10 s. The flow in the second cell is represented by dashes traveled by tracers for approximately 2 h. It is confirmed that a cell-structured flow is established similarly to the flow of the family $n = 1$ except near the source of motion, where the flow structure is affected by the nature of this source.

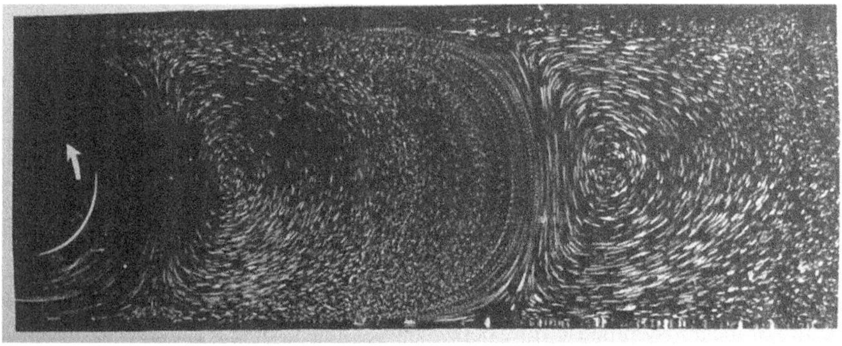

Figure 3.2. *Visualization of the cell-structured flow generated by a cylinder rotating between two parallel plane walls (Hellou and Coutanceau 1992)*

3.2.2. *Solution to the Stokes equation in polar coordinates*

As before, we use the variable separation method. Let us consider the streamfunction $\psi(r,\theta) = f(r)g(\theta)$. The solution to the biharmonic equation has the following form (Hellou and Coutanceau 1992):

$$\begin{aligned}\psi(r,\theta) = a_0 + b_0 Ln(r) + c_0 r^2 + d_0 r^2 Ln(r) + (a_1 r^{-1} + b_1 r Ln(r) + c_1 r \\ + d_1 r^3)\cos\theta + (a_1' r^{-1} + b_1' r Ln(r) + c_1' r \\ + d_1' r^3)\sin\theta + \sum_{n=2,3}^{\infty} (a_n r^{-n} + b_n r^{-n+2} + c_n r^n \\ + d_n r^{n+2} +)\cos(n\theta) + \sum_{n=2,3}^{\infty} (a_n' r^{-n} + b_n' r^{-n+2} + c_n' r^n \\ + d_n' r^{n+2} +)\sin(n\theta)\end{aligned} \quad [3.8]$$

This solution can be simplified in the particular cases of an antisymmetric flow such that $\psi(r,\theta) = \psi(r,\pi-\theta) = \psi(r,-\theta)$ or of a symmetric flow such that $\psi(r,\theta) = \psi(r,\pi-\theta) = -\psi(r,-\theta)$.

3.2.3. *Solution to the Stokes equation in elliptical coordinates*

Elliptical coordinates are related to Cartesian coordinates by the following relations:

$$\frac{y^2}{d_f^2 E^2} + \frac{x^2}{d_f^2 (E^2 - 1)} = 1$$
$$\frac{y^2}{d_f^2 E^2} + \frac{x^2}{d_f^2 (E^2 - 1)} = 1 \qquad [3.9]$$

E and X are the values of ellipses and hyperbolas that are orthogonal to them. $2d_f$ is the distance between the foci of an ellipse E (Figure 3.3).

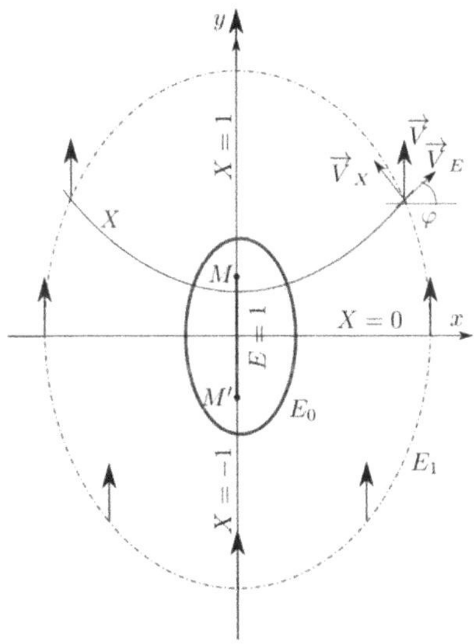

Figure 3.3. *Elliptical coordinates (E,X)*

The Laplace operator has the following expression in this system of coordinates:

$$\Delta = \frac{1}{E^2 - X^2}\left[(E^2 - 1)\frac{\partial^2}{\partial E^2} + (1 - X^2)\frac{\partial^2}{\partial X^2} + E\frac{\partial}{\partial E} - X\frac{\partial}{\partial X}\right] \qquad [3.10]$$

The resolution of the biharmonic equation leads to the following expression of the streamfunction (Hellou et al. 2021):

$$\frac{\psi(E,X)}{\sqrt{(1-X^2)(E^2-1)}} = a_0 + b_0 D_0 + a_1 C_2(E) + b_1 D_2(E) + \sum_{n=1}^{\infty} C_{2n}(X)[a_{2n-1}C_{2n-2} + b_{2n-1}D_{2n-2} + a_{2n}C_{2n} + b_{2n}D_{2n} + a_{2n+1}C_{2n+2} + b_{2n+1}D_{2n+2}] \quad [3.11]$$

where a_n and b_n are real coefficients, and C_n and D_n are Gegenbauer polynomials of first and second kind. The variables of these polynomials are X and E. The expressions and properties of polynomials can be found in Hellou et al. (2021).

3.3. Application to flows around obstacles

3.3.1. *Calculation of arbitrary coefficients of semi-analytical solutions*

The streamfunction solution to the biharmonic equation was determined for various systems of coordinates. Each solution depends on arbitrary coefficients of series that must be calculated.

When the boundaries of the flow domain are lines of coordinates in the chosen system of coordinates, these arbitrary coefficients of series can be calculated exactly after the application of boundary conditions. This is the case of the flow between two concentric circles. However, when boundary conditions must be met on a boundary that is not a line of coordinates, as a circle rotating in a rectangle for example, a part of the boundary conditions cannot be written exactly. In this case, we can resort to semi-analytical solution. The remaining coefficients can only be calculated by satisfying the remaining boundary conditions using approximation techniques or by numerical calculation.

In the following examples, we used the least squares method that consists of minimizing the quadratic deviation between the theoretical value of a quantity (boundary condition) and the value calculated on a set of points of the boundary. The method will be illustrated in more detail on one of the examples of flow presented below.

3.3.2. *Flows in a junction of orthogonal channels*

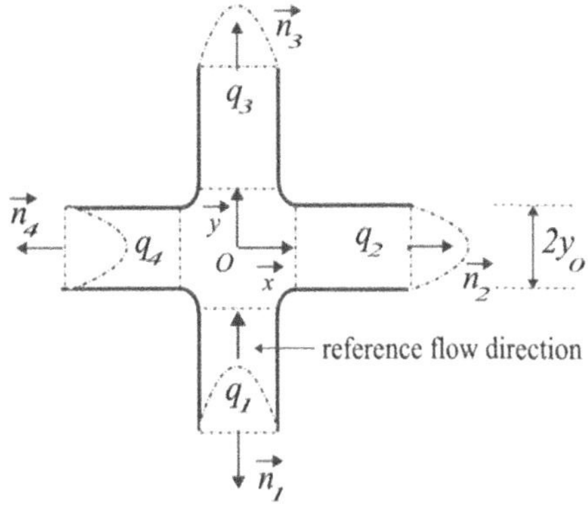

Figure 3.4. *Schematic representation of the four channel junction. The parabolic lines represent the velocity profile far from the junction in the case of the flow at nonzero average rate*

The streamfunction represented by equations [3.4] and [3.5] in each branch and the current function in the central domain of the junction can be used to solve this flow.

The junction being crossed by a flow rate, the biharmonic streamfunction is written in this case as follows:

$$\psi_0 = \sum_{m,p} A_{m,p}\, x^m y^p \qquad [3.12]$$

where $A_{m,p}$ are arbitrary coefficients, and m, p take integer values from 0 to 3.

For each branch, we use the current function obtained by the superposition of functions [3.4] and [3.5], which correspond to a flow at zero average flow rate, obviously by adapting the coordinates. In view to consider the flow rate that would be imposed to the entry of branch "i", a streamfunction corresponding to plane Poiseuille flow is superposed to the previous streamfunction. In the case of a channel with parallel walls of width $2y_0$, crossed by the flow rate q_i, this streamfunction obtained by using expression [3.12] is written as:

$$\psi_{pi} = A_i + \frac{3}{4} q_i \left(\frac{y}{y_0} - \frac{y^3}{y_0^3}\right) \qquad [3.13]$$

where q_i is the flow rate in the branch *i*.

We then physically connect the flow in each branch with the flow in the central domain of the junction. In the flow at zero average flow rate, we distinguish the antisymmetric flow that verifies $\psi_A(x, y) = \psi_A(x, -y)$ and the symmetric flow that verifies $\psi_S(x, y) = -\psi_S(x, -y)$. Their superposition represents the general flow whose streamfunction is: $\psi = \psi_A + \psi_S$.

If we consider semi-infinite branches, in this case only the combination $(1 + y + y^2)\psi_h$ is retained. This writing is valid for branches $i = 2, 4$ for which the $\vec{y}$ axis is perpendicular to the walls. For branches $= 1, 3$, the system axes should be rotated by $\pi/2$, or y should be replaced by x. If we adopt the configuration of the $\vec{y}$ axis perpendicular to the walls irrespective of the branch, the flow in a branch i will be represented by the streamfunction $(1 + y_i + {y_i}^2)\psi_{hi}$.

The matching conditions between the flow in each branch and the flow in the domain of the junction result from writing the equality between the components of the velocity and the stresses on the line that separates the branch and the junction. For the branch i, the following system is obtained:

Branches $i = 1, i = 3$	**Branches $i = 2, i = 4$**
$\begin{cases} u_0(x, \mp 1) = v_i(0, x) \\ v_0(x, \mp 1) = u_i(0, x) \\ p_0(x, \mp 1) = p_i(0, x) \\ \zeta_0(x, \mp 1) = \zeta_i(0, x) \end{cases}$	$\begin{cases} u_0(\mp 1, y) = u_i(0, y) \\ v_0(\mp 1, y) = v_i(0, y) \\ p_0(\mp 1, y) = p_i(0, y) \\ \zeta_0(\mp 1, y) = \zeta_i(0, y) \end{cases}$

Table 3.1. *Boundary conditions in each branch*

Index 0 corresponds to the central domain, and index i corresponds to the number of a branch. The coordinates are rendered dimensionless by dividing them by the half-width of each branch. To remove singular points, the walls were interconnected by curved lines that are given a shape. The central domain is thus modified as it can be seen in Figure 4.4. It is therefore recommended to add the no-slip conditions on these curved walls to be met by the components of velocity deduced from the streamfunction ψ_0 (equation [3.12]).

To satisfy the set of boundary conditions, we used the least squares method whose general methodology is summarized below.

Let us denote by g the quantity to be minimized to which we add the index from 0 to 4 depending on whether it is the central domain or one of the flow branches. We denote by S_1 the sum of quadratic errors on the conditions of connection between the central domain and each of the branches, and by S_2 the sum of quadratic errors on the boundary conditions in the junction: adhesion to curved walls. Thus, we have:

$$S_1 = \sum_{i=1}^{i=4} \sum_{m_i} (g_i - g_0)^2 \; S_2 = \sum_{m_0} g_0^2 \qquad [3.14]$$

where m_i is the number of minimization points distributed on the interface between the junction and the branch i, and m_0 is the number of points on the walls specific to the junction domain.

Let us write the expressions g_i and g_0 resulting from series [3.4], [3.5], [3.12], [3.13] in the following condensed forms:

$$g_i = q_i h_i(y) - \sum_{k=N_0+1}^{N} A_{k_i} f_{k_i}(x_i, y_i) \qquad [3.15]$$

$$g_0 = \sum_{k=1}^{N_0} A_{k_0} f_{k_0}(x, y) \qquad [3.16]$$

N_0 is the number of terms of equation [3.12] and $N = N_0 + N_{Ai} + N_{Si}$, where N_{Ai} and N_{Si} represent, respectively, the numbers of terms of antisymmetric and symmetric series.

The minimization of errors is expressed by:

$$\frac{\partial(S_1 + S_2)}{\partial A_j} = 0 \qquad [3.17]$$

Applying the quadratic minimization leads to the following algebraic system of equations:

$$\sum_{m} \sum_{k=1}^{N} A_k f_k f_j = \sum_{i=1}^{4} q_i h_i f_j \,, \quad j = 1, \ldots, N \qquad [3.18]$$

Figure 3.5 shows an example of flows in the case of the configuration of an inflow through branch 1 and outflows through other branches. We are only dealing with the case where the outflow through branch 2 is equal to the outflow through branch 4.

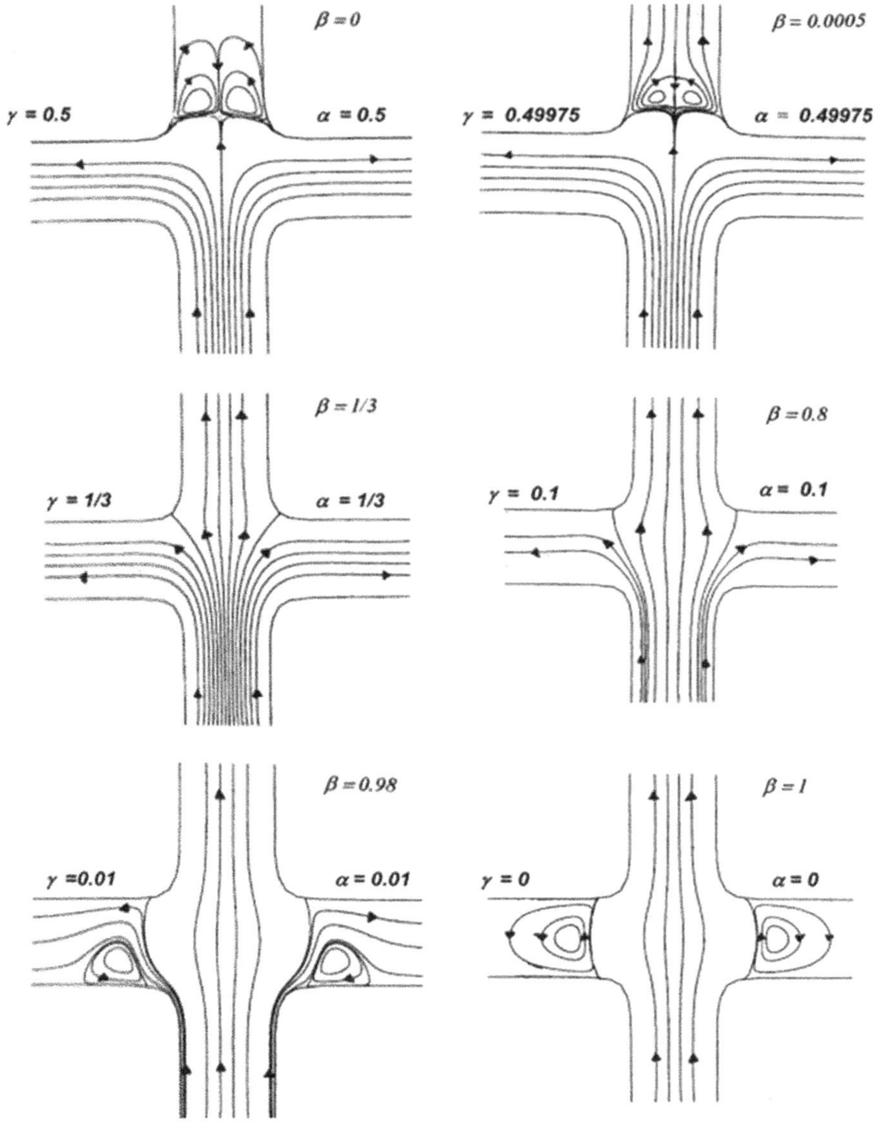

Figure 3.5. *Flow lines in a junction of four branches. Configuration: inflow through branch 1 and outflows through branches 2, 3 and 4 whose coefficients are α, β and γ. Case $\alpha = \gamma$. The reference flow rate is the inflow through branch 1 (Hellou and Bach 2011)*

Purely symmetric and purely antisymmetric flows can be identified for $\beta = 0$ and $\beta = 1$, respectively. These flows are characterized by zero average flow rate. They can also be found in other configurations: rotating cylinder in a channel (Hellou and Coutanceau 1992), translating cylinder in a channel (Moreau and Bourot 1993). They are formed of cells with closed streamlines that fill up the entire width of the channel. Their properties can be found in the above references. In the studied configuration, the flow rate is divided into three parts by separation lines. However, for β close to 0 or 1, recirculation cells whose size is smaller than the width of the channel are formed. Other configurations are presented in Hellou and Bach (2011).

As noted in the introduction, it is possible to control these cell-structured flows. A wide variety of structures can then be obtained. Figure 3.6 presents original configurations. Branches are closed by a wall at the entrance of each of them. The entire domain is closed. The flow is created only by the motion imposed on these walls. It is a flow with a zero-average flow rate. However, contrary to what has been previously presented, the branches have a length, and the solution to the equation of motion must therefore depend also on the variable in the direction of length. This type of solution obtained in the form of Fourier series is provided by Wang (1996). The solution is adapted to each of the branches. The conditions of connection with the flow in the central domain represented by the current function are still those indicated in Table 3.1. In addition to these conditions, we should consider those corresponding to motions imposed on end walls.

Figure 3.6 illustrates several original structures that can be exhibited by a Stokes flow. Fascinating structures can also be found in Pozrikidis (2000) and Guan et al. (2015).

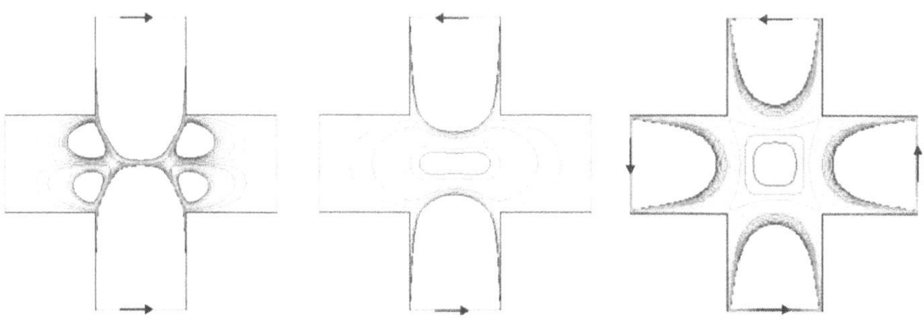

Figure 3.6. *Streamline patterns in a junction of four branches, each of which is closed by an end wall. The arrows indicate the wall motion direction. For a color version of this figure, see www.iste.co.uk/saouab/mechanics.zip*

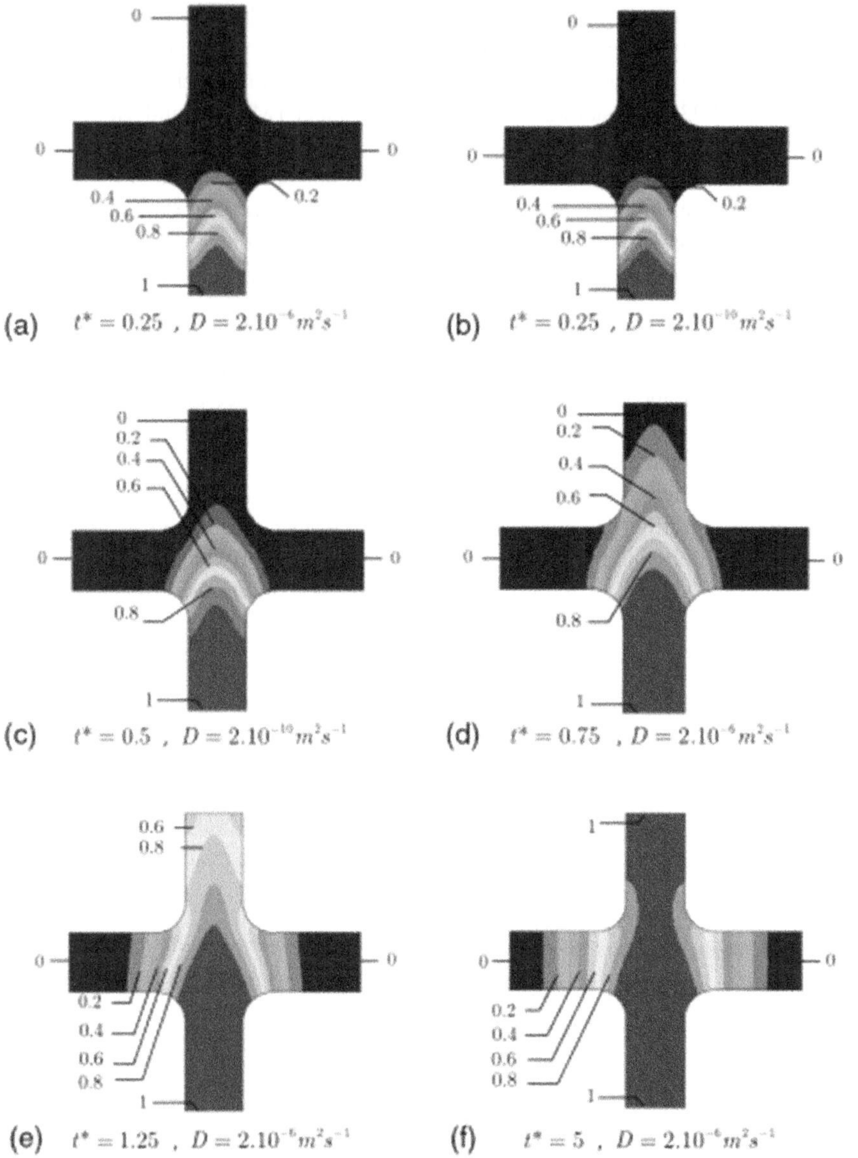

(a) $t^* = 0.25$, $D = 2.10^{-6} m^2s^{-1}$
(b) $t^* = 0.25$, $D = 2.10^{-10} m^2s^{-1}$
(c) $t^* = 0.5$, $D = 2.10^{-10} m^2s^{-1}$
(d) $t^* = 0.75$, $D = 2.10^{-6} m^2s^{-1}$
(e) $t^* = 1.25$, $D = 2.10^{-6} m^2s^{-1}$
(f) $t^* = 5$, $D = 2.10^{-6} m^2s^{-1}$

$P_1=10$

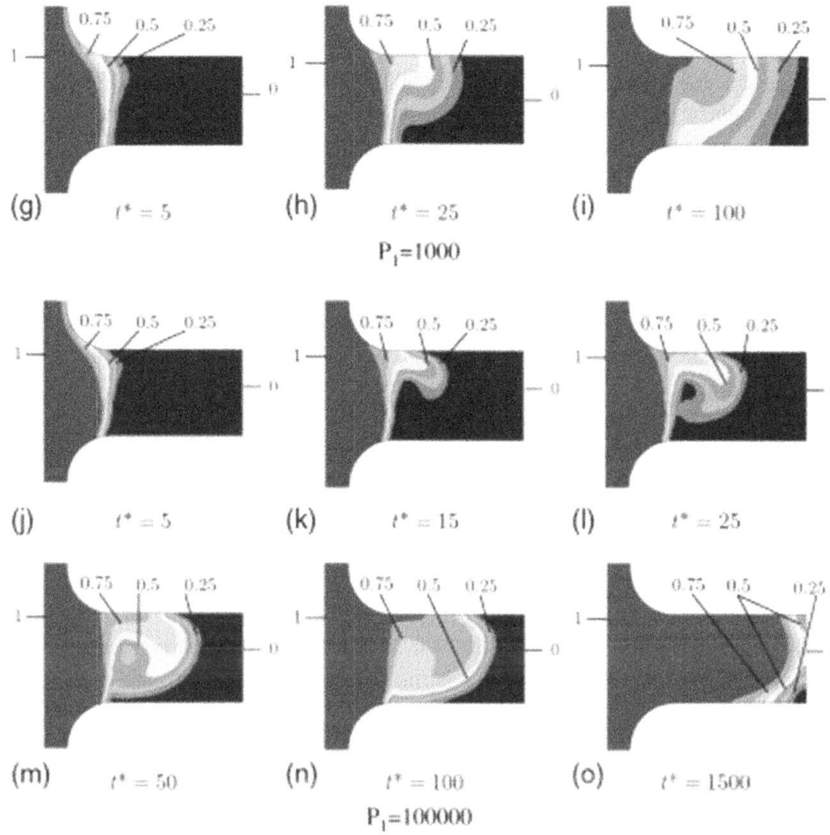

Figure 3.7. *Iso-concentration lines in a junction of channels for a ratio of flow rates* β *= 1. The indicated values correspond to C/C₀. The reference time is equal to 2 s, 0.02 s, 0.0002 s for* P_1 *equal to 10, 1,000 and 100,000, respectively (Hellou and Bour 2014). For a color version of this figure, see www.iste.co.uk/saouab/mechanics.zip*

Let us now examine how recirculation cells influence mass transport. We have also solved the advection-diffusion transport equation for various values of the Péclet number. We only discuss here the case of $\beta = 1$. In this case, the inflow through branch 1 goes out of the network entirely through the front branch, which is branch 3. The flow lines corresponding to this case are presented in Figure 3.5. The Péclet number is defined by $P_1 = U_1 y_0/D$. Index 1 is relative to branch 1. This number ranges from 10 to 100,000. A constant concentration is imposed at inlet 1. Figure 3.7 presents the iso-concentration lines for P_1 equal to 10, 1,000 and 100,000. In branches 1 and 3 (branches of the main flow), the concentration profile is similar to the velocity profile. However, the state remains diffusive in the adjacent branches in which velocity is low, and consequently the local Péclet number is smaller than that at the inlet, which is used as a reference. This state is illustrated by iso-concentration lines perpendicular to the walls of these branches. When *Pe* increases, the iso-concentration lines imitate the flow lines in the adjacent branches. Mass transfer tales place by the cellular movement as illustrated in Figures 3.7 from (g) to (o). The further the branch is penetrated, the more the iso-concentration lines tend toward a form of lines perpendicular to the walls, characteristic for a dominant diffusive transfer. This is due to velocity damping in the cells.

3.3.3. *Flows around networks of circular obstacles*

The use of the streamfunction in polar coordinates is adapted to this problem. In order to model flows in porous media, as will be seen further on, we choose assemblies of circles of radius *R* or a row of circles between parallel planes. The idea is to make a periodical assembly that will allow for the calculation in a periodic cell. As the Stokes flow is linear, we readily obtain the pressure drop in a sample composed of a certain number of cells.

Figure 3.8 shows examples of two-dimensional assembly: row of circles between two solid lines and arrangement of circles in a periodical grid in staggered rows. We choose a reference circle whose center is the origin of the study frame of reference $(O, \vec{e_r}, \vec{e_\theta})$. The domain in dotted lines in each figure represents the domain of calculation on whose border the boundary conditions are imposed.

Table 3.2 presents the set of boundary conditions that result from writing the adhesion to the walls, from the flow symmetry and from the connection of the calculation domain with the adjacent domains to meet the flow continuity.

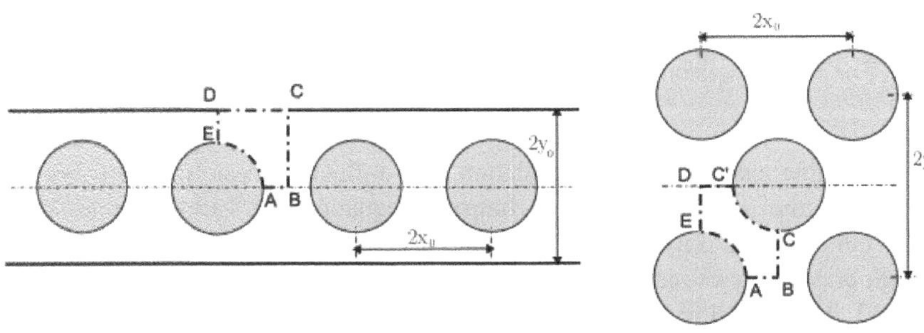

Figure 3.8. *Schemas of circle assemblies*

Row of circles	Staggered rows
$AB : v = 0,\ \zeta = 0$	$AB : v = 0,\ \zeta = 0$
$BC : v = 0, p = p_S$	$BC : v = 0,\ p = p_S$
$CD :\ u = 0, v = 0$	$CC' : u = 0, \quad v = 0$
$DE : v = 0,\ p = p_E$	$C'D : v = 0, \zeta = 0$
$EA : u = 0, v = 0$	$DE : v = 0, \quad p = p_E$
	$EA : u = 0, \quad v = 0$

Table 3.2. *Boundary conditions for the assembly of circles*

Here, u, v are the velocity components along $\vec{x}$ and $\vec{y}$, ζ represents the vorticity, and p_E and p_S are the constant pressure values at the inlet and outlet of the calculation domain, respectively. Some of these conditions are satisfied exactly by using the series [3.8]. Indeed, considering the symmetry, only the terms of the series containing $\sin\theta$ and $\sin(n\theta)$ are retained. In this case, the condition on AB is exactly met. As the circle is a line of coordinates, the adhesion on the reference circle, EA circle for example, is exactly satisfied. In this case, we obtain two equations that can be used to express half of the coefficients of the series as a function of the other half. The input and output pressure values are problem data, imposed according to the fluid nature and to the expected flow rate. However, it would be preferable to impose the value of the flow rate and deduce from it the resulting pressure drop. In this case, the conditions on pressure are replaced by conditions on the streamfunction on AB and CD or C'D lines. Indeed, the flow rate is proportional to $|\psi(x,0) - \psi(x,y0)|$. Hence by using the radius of a circle and the average exit velocity as reference quantities, the boundary conditions that are not analytically met by the series are presented in the following table:

Row of circles	Staggered rows
$\begin{cases} BC : v = 0 \\ CD : u = v = 0, \psi = Y_0 \end{cases}$	$\begin{cases} BC : v = 0 \\ CC' : u = v = 0, \\ C'D : v = 0, \psi = Y_0 \end{cases}$

Table 3.3. *Boundary conditions for the assembly of circles*

To meet these conditions, we use for example a least squares numerical calculation after having retained a sufficient number of coefficients.

Figure 3.9 presents the streamline patterns obtained for several chosen geometry cases.

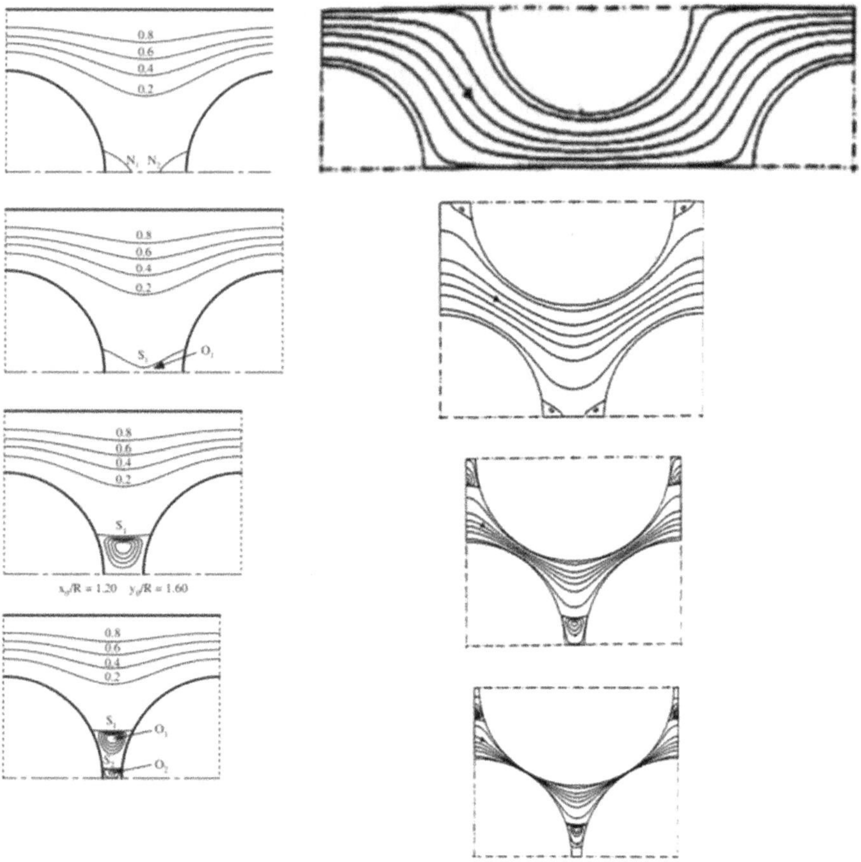

Figure 3.9. *Streamline patterns of the flow around an assembly of circles. Values of x0/R and y0/R. Circles in line: (1.41, 1.60); (1.40, 1.60); (1.20, 1.60); (1.10, 1.60). Circles in staggered rows: (2.61, 1.51); (1.30, 2.08); (1.10, 1.80); (1.05, 1.70)*

Recirculation motions can be established in both cases. For the row of circles, for a given width y_0/R, circles getting closer together lead to the appearance of cells, which coalesce as they continue to get closer. Cell multiplication can be observed. Ultimately, when the circles are in contact, the number of cells becomes infinite. For the circles in staggered rows, streamlines are even more curved, and all of the circles must be brought closer to obtain zones of significant recirculation.

3.3.4. *Stokes flow around an elliptical obstacle*

The biharmonic equation in elliptical coordinates has been solved, and its solution is presented. It will be applied to the simple problem of flow around an ellipse having one of its axes in the same direction as the average flow. This flow is imposed on a border that is also concentric elliptical and parallel to the obstacle. Unlike the case of the Cartesian or polar series, the elliptical series [3.11] does not lead to an exact expression of the streamfunction when the border is a coordinate line. Indeed, detailed calculations show that the coefficients of the series are interdependent. On the other hand, if the series is limited to the following expression:

$$\psi(E,X) = \sqrt{(1-X^2)(E^2-1)}\left[a_0 + b_0D_0(E) + a_1C_2(E) + BLn(E+\sqrt{(E^2-1)})\right] \qquad [3.19]$$

we obtain a simple approximate solution around an ellipse of equation E_0 surrounded by an elliptical boundary of equation E_l on which a uniform flow is imposed (Figure 3.3).

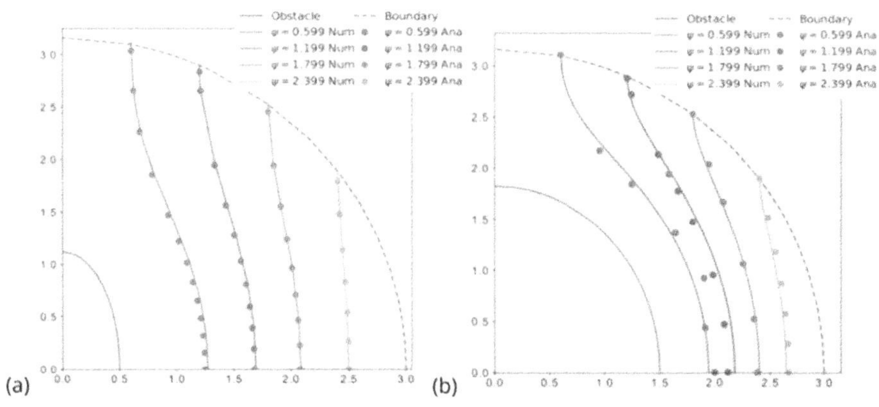

Figure 3.10. *Streamline patterns around an ellipse with uniform flow on an elliptical boundary. (a) d_f = 1, semi-minor axes of the internal and external ellipse λ_0 = 0.5, λ_1 = 3. (b) d_f = 1, λ_0 = 1.5, λ_1 = 3. For a color version of this figure, see www.iste.co.uk/saouab/mechanics.zip*

However, the latter is not sufficient for a proper representation of the flow when confinement increases. It is then necessary to further use terms of the semi-analytical solution (equation [3.11]) and proceed to a numerical calculation by least squares for example to satisfy the boundary conditions, uniform velocity on the border E_1 in our case. Figure 3.10 presents the flow lines for two cases of geometry with the same focal length. When the border is far enough away, the proposed analytical solution [3.19] is sufficient to calculate the flow. When the border is drawn closer, this solution deviates from the numerical results obtained by solution [3.11], even though in the presented case it remains sufficient.

3.4. Application to the calculation of the permeability of periodical porous media

These models of porous media are formed by assembling obstacles to the flow according to a periodic pattern. Solving Stokes' equation can be used to calculate the average pressure gradient in a grid per unit flow rate expressed for the entire porous medium by Darcy's law:

$$\frac{G}{Q} = \frac{\mu}{kS} \qquad [3.20]$$

where G is the pressure drop per unit length of the medium, S is the cross-sectional area crossed by the rate flow Q and k is the permeability in the flow direction.

3.4.1. *Two-dimensional flow*

Figure 3.11 presents the dimensionless permeability of the porous medium formed of cylindrical fibers arranged in staggered rows according to the pattern presented in Figure 3.8. We varied the porosity of the medium:

$$\phi = 1 - \frac{\pi R^2}{2x_0 y_0} \qquad [3.21]$$

by varying x_0 and y_0. The influence of y_0 is stronger than that of x_0 since y_0 is proportional to the fluid passage section. It is for this reason that the points are dispersed. This difference is great for low porosity values and nearly zero when porosity exceeds 0.6.

The hexagonal model represents an average behavior. We used practical formulas to approximate it:

$$\phi < 0.4 : \frac{k}{R^2} = \frac{\pi}{(1-n)}(-0.0004\phi^{2.5} + 0.032\phi^{3.5}) \qquad [3.22]$$

$$\phi \geq 0.4 : \frac{k}{R^2} = \frac{\pi}{1-\phi}(0.0406\phi^{2.5} - 0.2356\phi^{3.5} + 0.3422\phi^{4.5} + 0.5960\phi^{5.5} - 1.560\phi^{6.5} + 0.9092\phi^{7.5} \qquad [3.23]$$

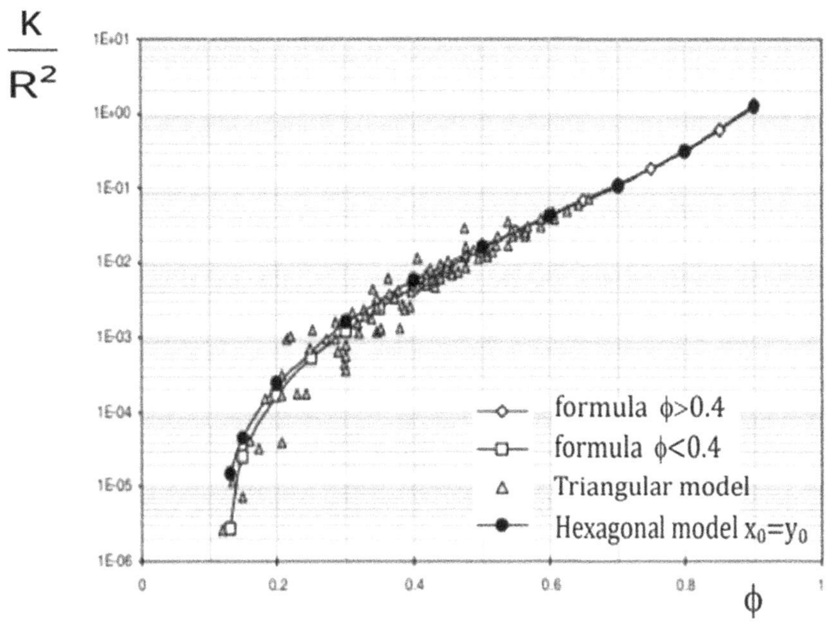

Figure 3.11. *Permeability as a function of the porosity of the medium consisting of an assembly of circles*

This behavior can also be reproduced by the Kozeny–Carman equation:

$$k = \frac{\phi^3}{C_k \Sigma^2} \qquad [3.24]$$

where Σ represents the specific surface and C_k is the Kozeny constant. In the case of staggered row model, we obtain:

$$\frac{k}{R^2} = \frac{\phi^3}{4C_k(1-\phi)^2} \qquad [3.25]$$

The value of constant C_k that gives satisfactory results is around 12. It should be recalled that in the case of an assembly of spheres, the constant C_k is 5.

In the case of the elliptical model, it can be shown that permeability is written as:

$$\frac{k}{\lambda_0 \gamma_0} = \frac{1}{4B(1-\phi)} \qquad [3.26]$$

where B is the coefficient of the streamfunction [3.19]. If we use the general streamfunction [3.11], we find that the force exerted on an obstacle is zero. It is then necessary to add a particular solution. In Hellou et al. (2021), we present a detailed calculation of this particular solution that can be written using the coefficient B as follows:

$$\psi = A + BLn(E + \sqrt{E^2 - 1}) \quad [3.27]$$

We also examined the assembly of fibers with elliptical section. Figure 3.12 presents an example of such assembly.

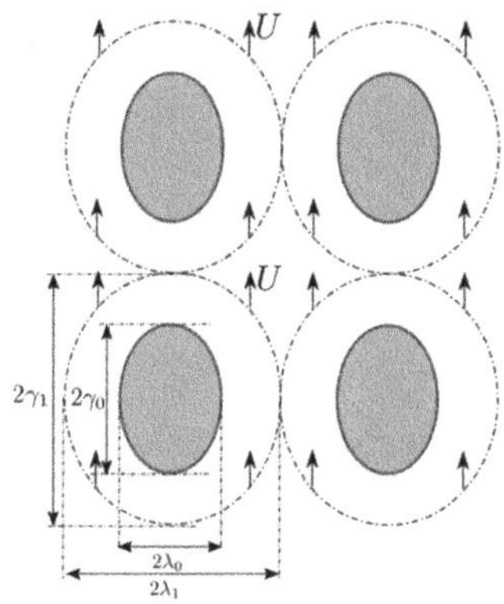

Figure 3.12. *Periodical porous medium formed of elliptical fibers*

We apply the semi-analytical solution with an elliptical boundary as previously presented. However, as noted in Figure 3.12, the set of elliptical boundaries does not cover the entire domain. Therefore, there is a space where the flow is not calculated. To solve this problem, we replaced the elliptical boundary with a rectangular one. In this case, the boundary conditions cannot be exactly expressed. COMSOL Multiphysics software was chosen for the computation.

Figure 3.13 shows the variation of permeability depending on porosity. We superposed the results obtained by the simplified analytical solution [3.19], the semi-analytical solution [3.11] and the numerical solution of COMSOL with elliptical boundary and rectangular boundary. When the medium is diluted, the results are superposed. However, the simplified solution diverges below the porosity 0.9. However, its results are similar to those of Raynor (2002) obtained for the flow around a fiber by using a simplified solution with four coefficients that take into account the fiber orientation. The model in Figure 3.12 and the semi-analytical solution [3.11] are adequate as long as the porosity exceeds 0.4.

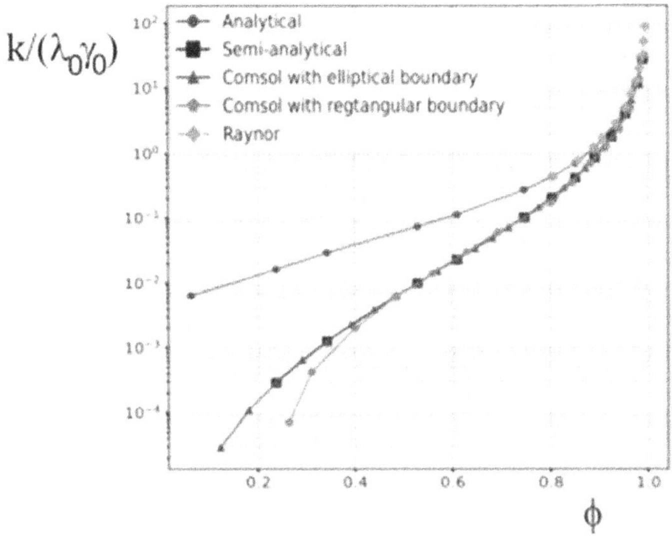

Figure 3.13. *Permeability as a function of the porosity of the medium consisting of an assembly of ellipses (Hellou et al. 2021). For a color version of this figure, see www.iste.co.uk/saouab/mechanics.zip*

The behavior is similar to that of the model with circular fibers. Permeability varies, however, significantly for low porosities, which indicates that medium structure has a significant influence on permeability when porosity is low.

3.4.2. *Three-dimensional porous media*

Figure 3.14 shows an example of structures of granular porous medium and fibrous porous medium consisting, respectively, of an assembly of cubes or rectilinear fibers of square cross-section. Tables 3.4 and 3.5 present the values of solid concentration ($C = 1 - \phi$) as a function of the ratio between the sides of the section of grain or fiber and of the grid. For an SC assembly, the square shape of solid obstacles makes it possible with to reach the maximal concentration value 1.

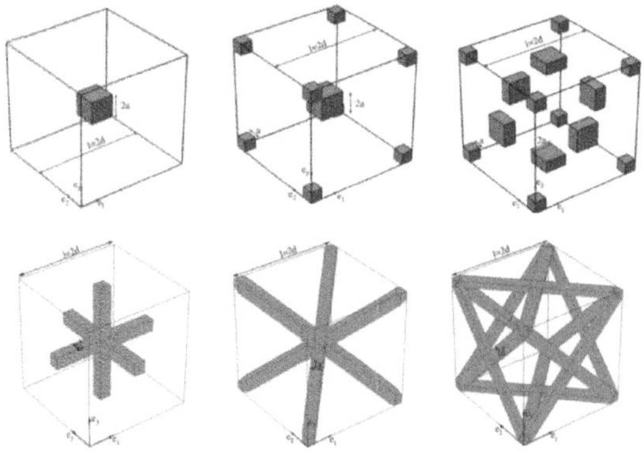

Figure 3.14. *Three-dimensional periodic porous medium: (a) granular; (b) fibrous. For a color version of this figure, see www.iste.co.uk/saouab/mechanics.zip*

Granular porous medium			
	SC (simple cubic)	BCC (body centered cubic)	FCC (face centered cubic)
	Cubes at vertices	Cubes at vertices and one cube at the center	Cubes at vertices and at the centers of faces
C	$\left(\frac{a}{d}\right)^3$	$2\left(\frac{a}{d}\right)^3$ if $\frac{a}{d} \leq 0.5$ $2\left(\frac{a}{d}\right)^3 - \left(2\frac{a}{d} - 1\right)^3$ if $0.5 \leq \frac{a}{d} \leq 1$	$4\left(\frac{a}{d}\right)^3$
(a/d) max	1	0.5 with overlapping 1 without overlapping	0.49
C max	1	1	0.47

Table 3.4. *Characteristics of the microstructure of the granular medium*

To complete the characteristics of these media, Figure 3.15 presents the specific surface as a function of porosity.

Fibrous porous medium			
	Fibers joining vertices	Fibers joining vertices and the center of the grid	Fibers joining vertices and the centers of the faces
C	$3\left(\frac{a}{d}\right)^2 - 2\left(\frac{a}{d}\right)^3$	$4\sqrt{3}\left(\frac{a}{d}\right)^2 - \frac{\sqrt{2}}{3}\left(10 + 3\sqrt{3}\right)\left(\frac{a}{d}\right)^3$	$6\sqrt{2}\left(\frac{a}{d}\right)^2 - 3\left(2 + \sqrt{2}\right)\left(\frac{a}{d}\right)^3$
(a/d) max	1	0.51	0.41
C max	1	0.85	0.72

Table 3.5. *Characteristics of the microstructure of the fibrous medium*

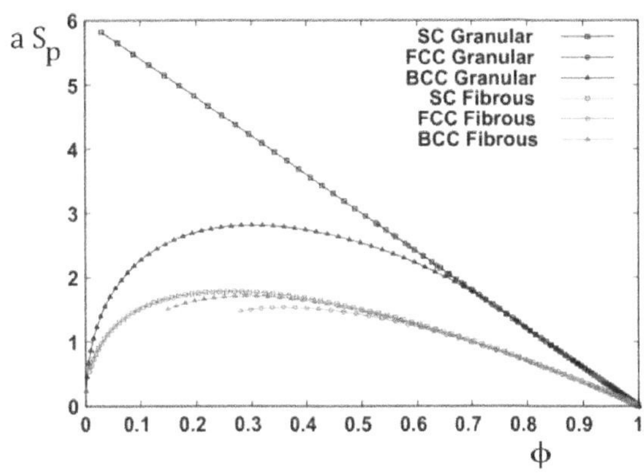

Figure 3.15. *Specific surface of the three-dimensional periodic porous medium as a function of porosity (Bourbatache et al. 2019). For a color version of this figure, see www.iste.co.uk/saouab/mechanics.zip*

Figures 3.16 present the dimensionless permeability as a function of porosity for the granular medium and for the fibrous medium, respectively. Permeability obviously increases with the increase in porosity. The values of permeability of the three assembly configurations and for the two media are close in the 0.75–0.99 porosity range. The influence of microstructure is significant below the porosity of 0.75. The FCC assembly is the less permeable in all cases. In the case of a granular medium, the permeability of SC and BCC assemblies are close. However, at the same porosity, the permeability of SC is lower than that of BCC because of the larger specific surface for SC. These results can be observed in the case of the fibrous medium. However, the BCC assembly of the fibrous medium features an anisotropy. Indeed, the projection of the solid phase in the $(\overrightarrow{e_1}, \overrightarrow{e_2})$ plane is not similar to the projection in $(\overrightarrow{e_1}, \overrightarrow{e_3})$ and $(\overrightarrow{e_2}, \overrightarrow{e_3})$ planes. Hence, the permeability in the $\overrightarrow{e_3}$ direction differs from the permeability in $\overrightarrow{e_1}$ direction. It is lower at the same porosity.

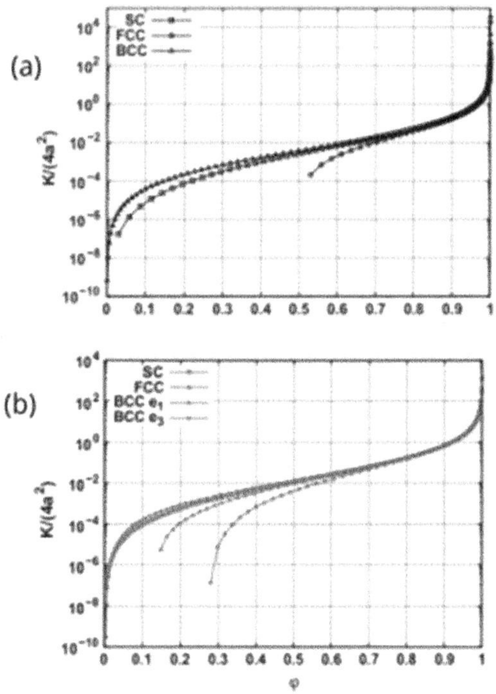

Figure 3.16. *Permeability of the three-dimensional periodical porous medium as a function of (a) granular medium and (b) fibrous medium (Bourbatache et al. 2019). For a color version of this figure, see www.iste.co.uk/saouab/mechanics.zip*

Figure 3.17 shows a superposition of the permeability: porosity results obtained for the two media compared with the results obtained for the spheres by Zick and Homsy (1982) and those of fibers of circular section obtained by Higdon and Ford (1996) for SC, BCC and FCC assemblies.

Overall, at the same porosity the fibrous medium is more permeable than the granular medium. This is largely explained by the fact that, at the same porosity, the specific surface is larger for the granular medium than for the fibrous medium. However, the specific surface is not the only parameter that would explain this difference. Indeed, tortuosity also influences permeability (Bourbatache et al. 2019).

Moreover, the shape of the solid obstacle has a smaller influence for the granular medium than for the fibrous medium.

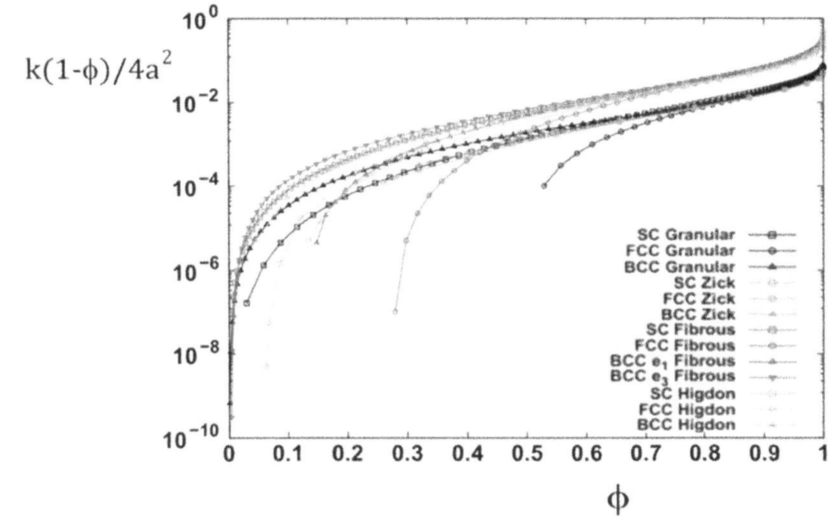

Figure 3.17. *Permeability of the three-dimensional periodical porous medium as a function of porosity – comparison with media formed of spheres, cylindrical, circular and elliptical fibers (Bourbatache et al. 2019). For a color version of this figure, see www.iste.co.uk/saouab/mechanics.zip*

3.5. Conclusion

This chapter presented a theoretical overview of Stokes' flows and their properties. Under certain conditions, these flows have closed-structure cellular motions most often attached to the walls of the flow domain or to the walls of the obstacles.

We applied the solutions of the biharmonic equation that represents the two-dimensional Stokes flow to the channel junction configuration. We have shown that depending on the conditions imposed in each channel, a cellular flow can be produced. Furthermore, the size of cells can be controlled by acting on the flow rate in each channel. Due to these properties, these cellular motions have numerous applications in microfluidics.

Flows in porous media are often at a low Reynolds number. Therefore, we used the pressure field obtained by solving the Stokes equation associated with a behavior according to Darcy's law to determine the relation between permeability and porosity for certain media formed of periodic arrangements of obstacles. We have thus shown that starting with a porosity threshold of around 0.7, permeability is influenced neither by the shape of obstacles nor by their arrangement type, but only by the level of porosity. Below this threshold, porosity shape, specific surface and also tortuosity, which characterize the microstructure, influence permeability. At the same porosity, permeability changes if the type of arrangement is modified. However, variations are less significant for the granular medium than for the fibrous medium.

3.6. References

Adler, P.M. and Thovert, J.F. (1999). *Fractures and Fracture Networks*. Kluwer Academic, The Netherlands.

Berkowitz, B., Naumann, C., Smith, L. (1994). Mass transfer at fracture intersections: An evaluation of mixing models. *Water Resour. Res.*, 30(6), 1765–1773.

Berre, I., Doster, E., Keilegavlen, F. (2019). Flow in fractured porous media: A review of conceptual models and discretization approaches. *Transp. Porous Med.*, 130(1), 215–236.

Bourbatache, M.K., Hellou, M., Lominé, F. (2019). Two-scale analysis of the permeability of 3D periodic granular and fibrous media. *Acta Mech.*, 230(10), 3703–372.

Bourot, J.M. (1984). Sur la structure cellulaire des écoulements plans de Stokes, à débit moyen nul, en canal indéfini à parois parallèles. *C. R. Acad. Sc. Paris, II*, 298(5), 161–164.

Brody, J.P., Yager, P., Goldstein, R.E., Austin, R.H. (1996). Biotechnology at low Reynolds numbers. *Biophys. J.*, 71(6), 3430–3441.

Bruderer, C. and Bernabé, Y. (2001). Network modeling of dispersion: Transition from Taylor dispersion in homogeneous networks to mechanical dispersion in very heterogeneous ones. *Water Resour. Res.*, 37(4), 897–908.

Cachile, M., Talon, L., Gomba, J.M., Hulin, J.P., Auradou, H. (2012). Stokes flow paths separation and recirculation cells in x-junctions of varying angle. *Phys. Fluids*, 24(2), 021704.

Champmartin, S., Ambari, A., Roussel, N. (2007). Flow around a confined rotating cylinder at small Reynolds number. *Phys. Fluids*, 19, 103101.

Charpentier, J.B., Bizet, L., Saouab, A. (2018). Gravity induced fluid fragmentation in an asymmetric microfluidic Y junction. *J. Colloid Interface Sci.*, 532, 300–308.

Corona P.T., Ruocco, N., Weigandt, K.M., Leal, L.G., Helgeson, M.G. (2018). Probing flow-induced nanostructure of complex fuids in arbitrary 2D flows using a fluidic four-roll mill (FFoRM). *Nature*, 8, 15559.

Costa, A. (2006). Permeability-porosity relationship: A reexamination of the Kozeny-Carman equation based on a fractal pore-space geometry assumption. *Geophys. Res. Lett.*, 33, 1–5.

Freund, J.B. (2014). Numerical simulation of flowing blood cells. *Annu. Rev. Fluid Mech.*, 46(1), 67–95.

Guan, J., Liu, J., Li, X., Tao, J., Wang, J. (2015). Stokes flow in a two-dimensional micro-device combined by a cross-slot and a microfluidic four-roll mill. *Z. Angew. Math. Phys.*, 66(1), 149–169.

Hajjam, M. (1998). Ecoulements cellulaires de Stokes générés par rotation dans un canal coaxial. *Z. Angew. Math. Phys.*, 49, 934–954.

Happel, J. and Brenner, H. (1973). *Low Reynolds Number Hydrodynamics. Mechanics of Fluids and Transport Processes.* Noordhoff International Publishing, The Netherlands.

Hellou, M. and Bach, T.D.P. (2011). Stokes flow in a junction of two dimensional orthogonal channels. *Z. Angew. Math. Phys.*, 62(1), 135–147.

Hellou, M. and Bour, O. (2014). Transient mass transport within Stokes eddies induced in a junction of orthogonal flow branches. *J. Hyd. Eng.*, 140(5), 040114007.

Hellou, M. and Coutanceau, M. (1992). Cellular Stokes flow induced by rotation of a cylinder in a closed channel. *J. Fluid Mech.*, 236, 557–577.

Hellou, M. and Lominé, F. (2018). Stokes flow within network of flow branches. *J. Fluids Eng.*, 140, 121110.

Hellou, M., Martinez, J., El Yazidi, M. (2004). Stokes flow through microstructural model of fibrous media. *Mech. Res. Commun.*, 61(3–4), 97–103.

Hellou, M., Lominé, F., Bourbatache, M.K. (2021). Solution of two-dimensional Stokes flow with elliptical coordinates and its application to permeability of porous media. *J. Appl. Mech.*, 88, 061009.

Higdon, J.J.L. and Ford, G.D. (1996). Permeability of three-dimensional models of fibrous porous media. *J. Fluid Mech*, 308, 341–361.

Hommel, J., Coltman, E., Class, H. (2018). Porosity-permeability relations for evolving pore space: A review with a focus on (bio-)geochemically altered porous media. *Transp. Porous Media*, 124, 1–41.

Huber, D., Oskooei, A., Casadevall i Solvas, X., deMello, A. (2018). Hydrodynamics in cell studies. *Chem. Rev.*, 118(4), 2042–2079.

Jeong, J.T. (2001). Slow viscous flow in a partitioned channel. *Phys. Fluids*, 13(6), 1577.

Li, G. (2002). Tracer mixing at fracture intersections. *Environ. Geol.*, 42(2–3), 137–144.

Mittal, R., Erath, B.D., Plesniak, M.W. (2013). Fluid dynamics of human phonation and speech. *Annu. Rev. Fluid Mech.*, 45(1), 437–467.

Moffatt H.K. (1964). Viscous and resistive eddies near a sharp corner. *J. Fluid Mech.*, 18, 1–18.

Moreau, F. and Bourot, J.M. (1993). Écoulements cellulaires de Stokes produits en canal plan illimité par la rotation de deux cylindres. *Z. Angew. Math. Phys.*, 44(5), 777–798.

Mourzenko, V.V., Yousefian, F., Kolbah, B., Thovert, J.F., Adler, P.M. (2002). Solute transport at fracture intersections. *Water Resour. Res.*, 38(1), 1–14.

O'Neill, M.E. (1983). On angles of separation in Stokes flows. *J. Fluid Mech.*, 133, 427–442.

Park, Y.J., Lee, K.K., Berkowitz, B. (2001). Effects of junction transfer characteristics on transport in fracture networks. *Water Resour. Res.*, 37(4), 909–923.

Pozrikidis, C. (1992). *Boundary Integral and Singularity Methods for Linearized Viscous Flow*. Cambridge University Press, Cambridge.

Pozrikidis, C. (2000). *Little Book of Streamlines*. Academic Press, San Diego.

Raynor, P.C. (2002). Flow field and drag for elliptical filter fibers. *Aerosol Sci. Technol.*, 36(12), 1118–1127.

Sangani, A.S. and Acrivos, A. (1982). Slow flow past periodic arrays of cylinders with application to heat transfer. *Int. J. Multiph. Flow*, 8(3), 193–206.

Shankar, P.N. (2007). *Slow Viscous Flows*. Imperial College Press, London.

Stone, H.A., Stroock, A.D., Ajdari, A. (2004). Engineering flows in small devices: Microfluidics toward a lab on a chip. *Annu. Rev. Fluid Mech.*, 36(1), 381–411.

Tamayol, A. and Bahrami, M. (2011). Transverse permeability of fibrous porous media. *Phys. Rev. E*, 83, 1–10.

Taneda, S. (1979). Visualization of separating stokes flows. *J. Phys. Soc. Jpn.*, 46(6), 1935–1942.

Toschi, F. and Sega, M. (2019). *Flowing Matter*. Springer, Berlin.

Valdés-Parada, F.J., Porter, M.L., Wood, B.D. (2011). The role of tortuosity in upscaling. *Transp. Porous Med.*, 88(1), 1–30.

Wang, C.Y. (1996). Stokes flow through an array of rectangular fibers. *Int. J. Multiph. Flow*, 22(1), 185–194.

Wang, C.Y. (2002). Slow viscous flow between hexagonal cylinders. *Transp. Porous Med.*, 47, 67–80.

Wang, C.Y. (2010). Flow through a finned channel filled with a porous medium. *Chem. Eng. Sci.*, 65(5), 1826–1831.

Yazdchi, K., Srivastava, S., Luding, S. (2011). Microstructural effects on the permeability of periodic fibrous porous media. *Int. J. Multiph. Flow*, 37, 956–966.

Zick, A.A. and Homsy, G.M. (1982). Stokes flow through periodic arrays of spheres. *J. Fluid Mech.*, 115, 13–26.

4

The Yield Stress Method for the Characterization of Pore Size Distribution of Porous Media: Progress Review

The current methods for the characterization of pore size distribution (PSD) of porous media have significant limitations. Mercury intrusion porosimetry (MIP), considered a reference technique, is polluting and toxic. Furthermore, the pore sizes obtained with this technique are not directly related to the distribution of hydraulic conductances of the flow paths inside the medium, essential information in many applications. The yield stress method (YSM) has been recently developed as an alternative. The YSM relies on the inversion of flow rate-effective pressure gradient measures during the single-phase flow of a yield stress fluid through a 1D sample of the studied medium. Indeed, a YSM can flow through a pore, provided there is a minimum pressure gradient, and its value depends on the hydraulic conductance. This allows the size-based discrimination of the flow paths in which the fluid is mobilized by the variation of the effective pressure gradient. This chapter reviews the literature presenting a mathematical model of YSM, its experimental validation based on measurements conducted on various porous media and results interpretation by comparison with MIP and X-ray micro-tomography.

4.1. Introduction

MIP is currently the most widespread method for determining PSD of porous media (Cieszko et al. 2019). The disadvantages of this technique are however well

Chapter written by Antonio RODRIGUEZ DE CASTRO, Azita AHMADI-SENICHAULT and Abdelaziz OMARI.

known. In addition to mercury toxicity, MIP overestimates the volume of small pores and underestimates that of large pores upstream of throats. Moreover, high external pressures applied to samples during measurements may lead to microstructure deformation. Besides MIP, 3D X-ray micro-tomography is also very widely used for non-destructive characterization of the microstructure of porous media (Saxena et al. 2018). Nevertheless, the required equipment is expensive and, despite recent progress, spatial resolution (~1 μm) conditions the characterization of small pores in the presence of dispersed PSD.

Ambari et al. (1990) proved that, theoretically, the PSD of a porous medium modeled as a bundle of parallel capillaries of equal length but of different radii can be derived from the measurement of flow rate as a function of pressure gradient during the injection of YSM. Taking advantage of these theoretical ideas, subsequent research has been dedicated to the development of an experimental porosimetry method by injection of YSMs, known as YSM (Rodríguez de Castro et al. 2014).

This chapter focuses on YSM. However, several groups of researchers have also contributed to the development of a porosimetry method based on the injection of complex fluids using various approaches. Their contributions are summarized in Table 4.1.

Authors	Contributions
Ambari et al. (1990)	Theoretical basis of a porosimetry method based on the injection of yield stress fluids. They showed that PSD of a bundle of rectilinear circular capillaries can be derived from measurements (Q, ∇P) upon the injection of a Bingham fluid, Q being the total flow rate through the analyzed porous sample and ∇P the average pressure gradient.
Chaplain et al. (1992)	Extension to a bundle of randomly oriented capillaries.
Oukhlef (2011) and Oukhlef et al. (2014)	Extension to Casson and Herschel–Bulkley fluids. Analytical inversion of numerically generated data (Q, ∇P) by calculating the partial derivatives of a polynomial data regression. This type of inversion is sensitive to experimental uncertainty and provides noisy results in the region of small size pores.
Malvault (2013)	Calculation of PSDs of porous media models based on flow experiments in laboratory using yield stress fluids. Alternative numerical inversion including data filtering by a polynomial function. The results obtained depend on the order of the polynomial used during filtering.

Rodríguez de Castro (2014) and Rodríguez de Castro et al. (2014)	Inversion algorithm without filtering adapted to experimental data: YSM method. Selection and characterization of concentrated solutions of xanthan gum as yield stress fluid for laboratory tests. Experimental characterization of artificially consolidated porous media with YSM. The characterized length scale was not analyzed.
Rodríguez de Castro et al. (2016)	Use of YSM to characterize a set of natural and artificial consolidated porous media. Identification and evaluation of the main experimental stakes related to the use of xanthan gum solutions. The characterized length scale was not analyzed.
Abou Najm and Atallah (2016)	Development of an alternative porosimetry method based on the injection of non-Newtonian fluids without yield stress: ANA method. The obtained solutions are not unique, and the injection of several fluids is required.
Malvault et al. (2017)	The authors proved that the precision of results provided by YSM could be significantly improved by considering the axial variation of the pores cross-section.
Rodríguez de Castro et al. (2018)	Effects of polymer concentration on the precision of YSM tests and sensitivity of the method to the errors of rheological characterization of the fluid.
Rodríguez de Castro et al. (2019)	Experimental characterization of packs of spherical beads using YSM. Comparison with MIP and with the pore network model extracted from a 3D image of a porous medium.
Atallah and Abou Najm (2019)	Experimental validation of ANA method. The uniqueness of the solution is not guaranteed in all cases.
Rodríguez de Castro et al. (2020a)	YSM extension to rock fractures with rough walls. Modeling, numerical experimentations and experimental validation.
Rodríguez de Castro et al. (2020b)	Application of YSM to the results of direct numerical simulations using as geometry a digital image of the porous medium. Good agreement with MIP and with the results obtained through pore network modeling.
Mackaya et al. (2021)	Improvement of the model of bundle of capillaries used in YSM by including triangular and rectangular pores.
Rodríguez de Castro et al. (2021)	Analysis of the length scale characterized by YSM in the case of consolidated porous media by comparing the results of this technique with the PSDs obtained by MIP and by X-ray micro-tomography. It can be concluded that YSM can be reliably used to replace MIP in the case of the studied consolidated porous media, given the general agreement observed between these methods.

Table 4.1. *Summary of the contributions to the development of a porosimetry method based on the injection of complex fluids*

4.2. Yield stress method

According to the YSM, the porous medium is considered homogenous, isotropic and well represented by a bundle of parallel capillaries of various radii. The number of pores whose radii are within a range of sizes is given by a probability distribution $p(r)$, according to the schematic representation in Figure 4.1. Let us denote by Q the total flow rate of the fluid crossing the porous sample and by ∇P the average pressure gradient in the sample (total pressure loss in the sample divided by its length). Considering the bundle model of N capillaries of size distribution $p(r)$, the total flow rate through the porous medium $Q(\nabla P)$ and the average flow rate in the pores $\overline{q(\nabla P)}$ are, respectively, given by:

$$Q(\nabla P) = N\int_0^{\infty} q(\nabla P, r)p(r)\mathrm{d}r\,;\ \overline{q(\nabla P)} = \frac{Q(\nabla P)}{N} \qquad [4.1]$$

For an incompressible fluid flowing through a capillary, assuming a fully developed 1D flow, the balance of momentum at equilibrium leads to a linear dependence of the shear stress τ with respect to the radial coordinate r for a given pressure gradient ∇P:

$$\tau = \frac{\nabla P r}{2} \qquad [4.2]$$

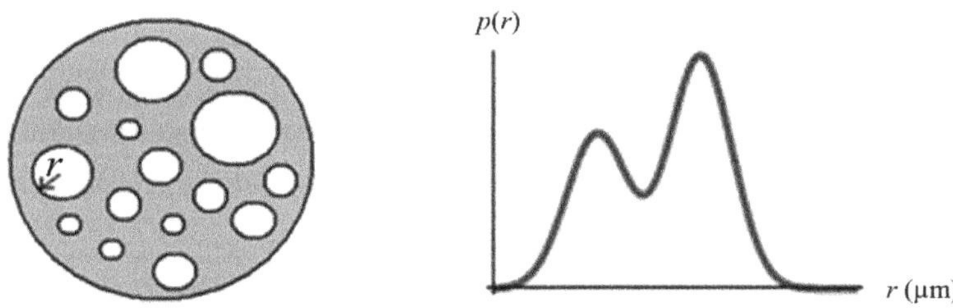

Figure 4.1. *Representation of the capillary bundle model and its PSD. For a color version of this figure, see www.iste.co.uk/saouab/mechanics.zip*

A large number of complex fluids used in the industry have a yield stress. The fundamental characteristic of YSMs is that they do not flow unless they are subjected to a shear stress exceeding a critical threshold value τ_0. Otherwise, they are deformed in a finite manner as elastic solids. According to equation [4.2], a YSM only flows through the pores whose radius exceeds $r_0 = 2\tau_0/\nabla P$ as long as there is no slip at the pore walls.

The empirical law of Herschel–Bulkley is often used in YSM to describe the rheological behavior under simple shear of YSMs. This law can be written as follows:

$$\begin{cases} \tau = \tau_0 + k\dot{\gamma}^n \text{ for } \tau \geq \tau_0 \\ \dot{\gamma} = 0 \text{ for } \tau \leq \tau_0 \end{cases} \qquad [4.3]$$

k being the consistence and n the flow index. k and n are generally obtained by adjusting the data resulted from the measurement of the shear rate $\dot{\gamma}$ as a function of the applied shear stress τ using a rheometer. The flow rate of a Herschel–Bulkley fluid through an individual capillary represented by a pore of radius r of the bundle of capillaries, denoted by $q(\nabla P, r)$, is given by the following expression (Skelland 1967):

$$q(\nabla P,r) = \begin{cases} \left[1-2\left[\frac{\left(1-\frac{2\tau_0}{r\nabla P}\right)^2}{\frac{1}{n}+3}+\frac{\frac{2\tau_0}{r\nabla P}\left(1-\frac{2\tau_0}{r\nabla P}\right)}{\frac{1}{n}+2}\right]\right] \times \left[\frac{r}{\frac{1}{n}+1}\left(\frac{r\nabla P}{2k}\right)^{\frac{1}{n}}\left(1-\frac{2\tau_0}{r\nabla P}\right)^{\frac{1}{n}+1}\right]\pi r^2 \text{ if } \nabla P \geq \frac{2\tau_0}{r} \\ 0 \text{ if } \nabla P \leq \frac{2\tau_0}{r} \end{cases} \qquad [4.4]$$

Oukhlef et al. (2014) proposed an analytical inversion of equation [4.1] for a Hershel–Bulkley fluid:

$$p(r) = \left[\left(\frac{1}{n}+4\right)\frac{\partial^{\frac{1}{n}+1}\overline{q(\nabla P)}}{\partial \nabla P^{\frac{1}{n}+1}} + \nabla P\frac{\partial^{\frac{1}{n}+2}\overline{q(\nabla P)}}{\partial \nabla P^{\frac{1}{n}+2}}\right]\frac{2^{\frac{1}{n}+4}k^{\frac{1}{n}}\nabla P}{16\left(\frac{1}{n}\right)!\,\pi r_0^{\frac{1}{n}+4}} \qquad [4.5]$$

Using a polynomial calibration of numerically obtained results (Q, ∇P), the latter authors determined $p(r)$ using equation [4.5] and obtained good results for a Bingham fluid ($n = 1$). However, when $1/n$ is a real number, the determination of $p(r)$ from equation [4.5] requires the calculation of fractional derivatives, complicating the calculation of $p(r)$ and providing noisy PSD in the region of small pores.

Other inversion methods have been considered. Based on the (Q, ∇P) measurements conducted during the injection of the YSM in a sample of the porous medium to be characterized, Rodriguez de Castro et al. (2014) proposed an inversion algorithm consisting of isolating the incremental increase of Q resulting from the flow of fluid in the pores of a given size. Indeed, the fluid flows in increasingly smaller pores as the applied ∇P increases, as can be seen in Figure 4.2. In this respect, Bauer et al. (2019) and Chaparian and Tammisola (2021) showed the existence of three distinct flow states: (1) a state of the lowest flow rates in which the relation between Q and ∇P follows a power law, as the YSM flows only through the highest conductance pores; (2) an intermediate region, in which the YSM is gradually mobilized in increasingly more pores as the flow rate increases; and (3) a power law of ∇P with an exponent of $1/n$ at the highest flow rates, the YSM flowing through all channels.

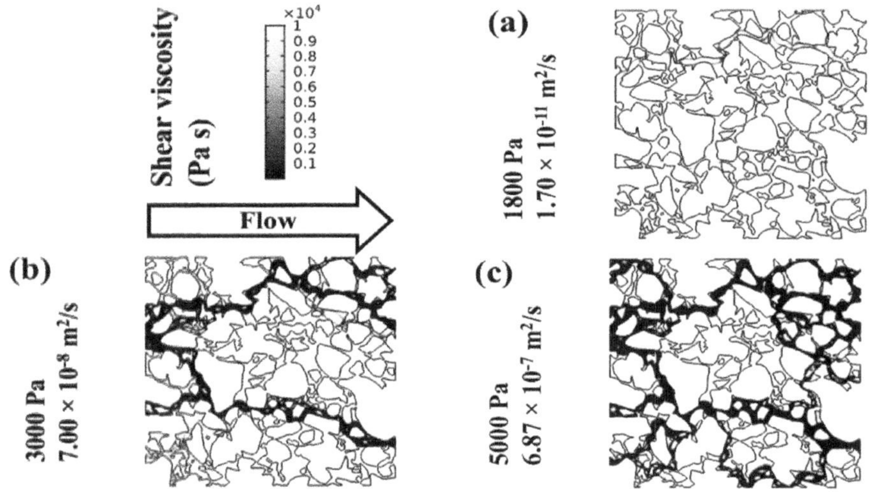

Figure 4.2. *Shear viscosity maps provided by numerical simulations of the injection of a yield stress fluid in a sand pack (Rodríguez de Castro et al. 2020b). The extent of the non-stagnant zone represented in black is a function of the applied injection pressure. At the lowest pressure (a), there is only one flow channel, whereas with increasing pressure (b, c), fluid progressively flows through additional channels. For a color version of this figure, see www.iste.co.uk/saouab/mechanics.zip*

The inversion algorithm relies on taking into consideration the additional increment of Q when ∇P increases at a higher value, as a consequence of the flow incorporating pores of smaller radius. For a set of data $(Q_i, \nabla P_i)$ collected through a standard YSM experiment, each ∇P_i defines a class of pores of representative radius $r_i = \alpha \frac{2\tau_0}{\nabla P_i}$, where α is a numerical factor greater than unity $(r_1 > r_2 > \cdots > r_N)$. The fundamental reason for $\alpha > 1$ is that for a given ∇P_i, the resulting Q_i is due to the pores whose radii are strictly above $\frac{2\tau_0}{\nabla P_i}$. In other terms, ∇P_i represents only the setting off of the contribution to flow of the pores of radius $\frac{2\tau_0}{\nabla P_i}$. In the practice used to calculate the number of pores of a targeted class of pores n_i (i = 1, … , N), and starting by n_1, successive and increasing positive values are scanned until $Q_s < \sum_{g=1}^{i} n_g\, q(\nabla P_s, r)$ for a given s. The maximal value is then assigned to n_i. In the scanning procedure for calculating each n_i, and as the large pores have a very significant contribution to the total flow rate, scanning must be finer when the radius of the considered pore is large. The scanning step used in the calculation of each n_i is:

$$\text{Step}_i = \frac{\zeta q(\nabla P_{N+1}, r_N)}{q(\nabla P_{N+1}, r_i)} \tag{4.6}$$

where ζ is a parameter that determines the scanning refinement. According to equation [4.6], the minimal contribution to the global flow rate for a given pore size r_i at the highest pressure gradient ∇P_{N+1} is the same for all r_i.

Furthermore, a filtering stage is necessary. Indeed, considering all the classes of pore radius, we exclude the pores of class r_j for which the number n_j is only limited by the pair of experimental data $(Q_{N+1}, \nabla P_{N+1})$. The reason is that these classes of pores are delimited by $(Q_{N+1}, \nabla P_{N+1})$ only because no higher flow rate or pressure gradient was imposed during the experiments, so no significant information on the number of these pores can be obtained. Consequently:

$$\begin{aligned} &\text{If } Q_{N+1} > \sum_{g=1}^{s-1} n_g q(\nabla P_{N+1}, r_g) \\ &\text{and } Q_{N+1} = \sum_{g=1}^{s} n_g q(\nabla P_{N+1}, r_g) \end{aligned} \tag{4.7}$$

then n_s is unknown.

Once n_i ($i = 1, \dots, \mathrm{N}$) is calculated, the probabilities in relative frequency $p(r_i) = \frac{n_i}{\sum_{j=1}^{N} n_j}$ for $i = 1 \cdots \mathrm{N}$ and in terms of relative volume $p_v(r_i) = \frac{n_i \pi r_i^2}{\sum_{j=1}^{N} n_j \pi r_j^2}$ for $i = 1 \cdots N$ are also obtained. The total flow rate Q as a function of the pressure gradient ∇P corresponding to the obtained PSD is then calculated as follows:

$$Q(\nabla P) = \sum_{i=1}^{N} n_i q(\nabla P, r_i) \tag{4.8}$$

To determine the retained value of α, denoted by α^*, we use successive increasing values to calculate PSD, from $\alpha = 1$ to $\alpha = 2$ and by using a step of δ. α^* is the value of α that minimizes the relative difference between calculated data (via equation [4.8]) and experimental data $(Q_i, \nabla P_i)$ in the sense of the least squares method.

The robustness of the proposed inversion algorithm and its sensitivity to input rheological parameters was numerically and experimentally studied in the work of Rodríguez de Castro et al. (2018). It was thus possible to evaluate its applicability to running real experiments in the laboratory. The results of these experiments showed that the precision of these PSD is mainly sensitive to errors in n, flow index (equation [4.1]), while the effects of errors in τ_0 and k are moderate. It was also noted that the effects of errors in n are weaker in the case of narrow PSD with only one peak of probability.

4.3. Experimental validation of YSM method

4.3.1. *Aqueous solutions of xanthan gum as injection fluid in YSM method*

YSM relies on the injection of a YSM with negligible levels of thixotropy and viscoelasticity and also a well-defined yield stress. Among a certain number of candidate fluids, aqueous solutions of xanthan gum have been selected (Rodríguez de Castro 2014) due to their negligible thixotropy, their limited viscoelasticity (Jones and Walters 1989; Sorbie 1991), to the rigidity of xanthan gum macromolecules and to their plane monolayer adsorption on the pore walls (Chauveteau 1982; Sorbie 1991; Dario et al. 2011).

Despite the above-mentioned advantages, the shear viscosity of xanthan gum solutions at low shear rates is finite, as described by the Carreau model (Carreau 1972). Consequently, xanthan gum solutions are in reality pseudo YSMs. Due to the rigidity of polymer macromolecules, semi-diluted aqueous solutions of xanthan develop a high viscosity level at low shear rates and a very marked shear-thinning behavior. Consequently, it has often been considered that xanthan gum solutions develop an apparent yield stress (Withcomb and Macosko 1978; Carnali 1991; Khodja 2008; Benmouffok-Benbelkacem et al. 2010), and several works have shown that the shear rheology at equilibrium of these solutions is properly described by the Herschel–Bulkley model (Rodríguez de Castro et al. 2014, 2018; Song et al. 2006).

According to equation [4.2], the fluid is subjected to higher shear stresses in the largest pores. Consequently, the concentrated solutions of xanthan gum flowing through a porous medium at constant pressure gradient ∇P are subjected to low shear stresses in small pores, which corresponds to the first Newtonian plateau of Carreau model (high viscosity), but are subjected to higher shear stresses in large pores (outside of this plateau).

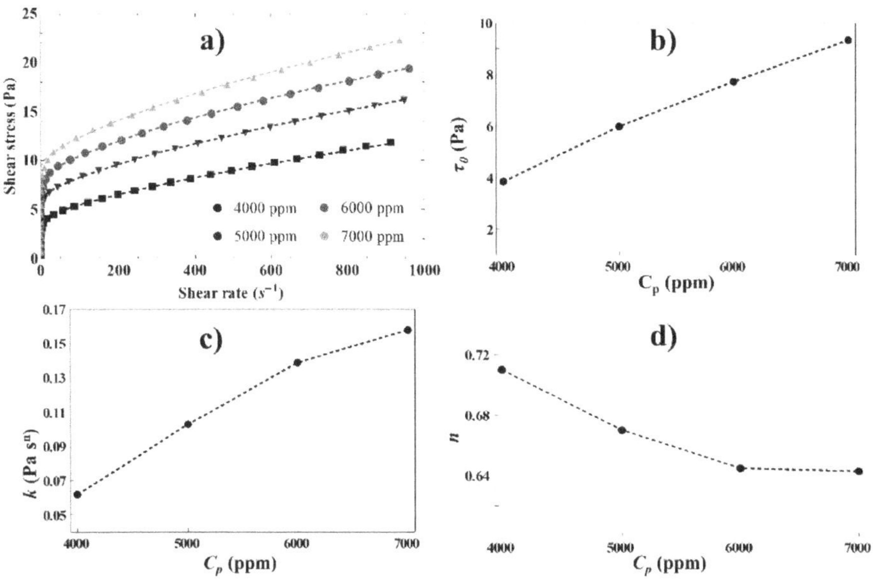

Figure 4.3. *(a) Rheograms and (b–d) Herschel–Bulkley parameters for polymer concentrations, Cp, between 4,000 and 7,000 ppm. For a color version of this figure, see www.iste.co.uk/saouab/mechanics.zip*

It is known that the relations between shear viscosity and shear stress are highly dependent on polymer concentration C_p. This is the reason why various xanthan solutions with various C_p have been used and characterized in order to evaluate the effect of C_p on the precision of PSD provided by YSM. Figure 4.3 shows the rheograms and parameters of the Herschel–Bulkley model for C_p ranging between 4,000 and 7,000 ppm (Rodríguez de Castro et al. 2018).

Rodríguez de Castro et al. (2016) studied the importance of adsorption, retention and mechanical degradation of polymer molecules during their flow through a consolidated porous sample. This was done through a comparison of measurements of density and viscosity of ingoing and outgoing fluids in a set of porous media.

Though no sign of significant adsorption or mechanical degradation was observed, polymer retention proved to be potentially significant at the lowest flow rates. At these low flow rates, macromolecules are not oriented in the flow direction and may block small pores. This is why other YSMs must be identified for the characterization of porous media, whose pore size is typically below 1 µm.

Considering the previous aspects, xanthan solutions have allowed a satisfactory characterization of PSD of a set of heterogeneous porous media: sandstones, sintered silicates and packs of spheres (Rodríguez de Castro et al. 2016, 2018, 2019). The method has also been extended and used to characterize rock fractures with aperture distribution.

4.3.2. *Experimental set-up and procedure*

A simple experimental configuration can be used to conduct YSM injection tests. The sample is first of all prepared differently, according to its nature. The classical preparation for consolidated porous media consists of dressing a cylindrical core of non-wetting epoxy resin, then coating it with glass fiber also dressed with epoxy resin to ensure its lateral impermeability and its mechanical strength (Rodríguez de Castro et al. 2014). In the case of consolidated media with low permeability (typically < 1 darcy), the core is mounted in a Hassler cell (Rodríguez de Castro et al. 2016). In this configuration, the sample is inserted in a rubber sleeve and two metallic grids are placed at both ends to prevent grain detachment. Then the sleeve and the grids are mounted in the Hassler sample holder and confined by the injection of pressurized brine surrounding the sleeve to prevent lateral leakages. In the case of unconsolidated porous media, packs of solid grains are prepared in a 1D Plexiglas column under vibration to reach the expected compactness level (Rodríguez de Castro et al. 2019). A layer of textile material is placed at each end of the pack to prevent solid grains from penetrating into effluent injection and recovery pipes. Whether for consolidated or unconsolidated samples, injectors fitted with flow distributors are installed at the ends of cores or columns. The preparation of samples for the characterization of rough fractures must follow a more specific procedure, which is detailed in previous works (Rodríguez de Castro et al. 2020a). A constant displacement pump is used to inject the YSM, and the loss of charge over the length of the porous sample L is measured using a differential pressure sensor. The injection circuit is open, and the effluent is collected using a container.

During each experiment, a set of different rate flows Q_j (j = 1…N + 1) is imposed, according to the sequence as follows:

$$Q_j = Q_{\min}\left(\frac{Q_{\max}}{Q_{\min}}\right)^{\left(\frac{N+1-i}{N}\right)} \qquad [4.9]$$

where $Q_{\min}$ is typically of the order of 0.01 mL/h for samples used in the cited works. This low flow rate must ensure that the YSM remains stagnant in all pores excepting the largest ones. $Q_{\max}$ is chosen so that the evolution of ∇P with Q follows a power law at the highest flow rates, which reflects that all pores contribute to the flow (this occurs when $\nabla P \geq \frac{2\tau_0}{r_{min}}$, r_{min} being the representative radius of the smallest pores). The temperature of the experiment room must be recorded throughout the test duration.

4.3.3. *Unconsolidated porous media: packs of spheres*

An essential step in YSM validation consists of the accurate characterization of model porous media with well-known morphologies and a microstructural topology. From this perspective, four packs of mono-disperse spherical glass beads were characterized by Rodríguez de Castro et al. (2019) using various porosimetry methods: experimental YSM, numerical simulation of MIP and pore network extraction from a 3D digital image. After saturating the columns of spheres with water, a xanthan solution with C_p = 7,000 ppm was injected at various flow rates and ∇P was measured for each flow rate. Figure 4.4 presents the raw measurements of YSM tests. $u = Q/A$, is Darcy velocity and A is the cross-sectional area of the columns. The error bars represented in the mentioned figure correspond to a confidence interval at 95% of the uncertainty related to the measurement repeatability and to the accuracy of the differential pressure sensors. From these measurements, it can be deduced that the relation between u and ∇P is nonlinear. Moreover, as expected, more significant pressure losses are obtained in the columns filled with small spheres, considering the lower permeability of these media. A decrease of one order of magnitude in ∇P leads to smaller values of u by 3–4 orders of magnitude. Consequently, it is confirmed that the injected fluid has a strong shear-thinning behavior in the porous medium, particularly at the lowest injection velocities, which is characteristic for YSMs.

The procedure presented in section 4.3.2 was used to obtain the PSD of the four columns of packed beads and the results were compared to those obtained by numerical simulation of MIP tests. To make such a comparison, PSDs were first normalized by the radius of spherical balls r_s. Rodríguez de Castro et al. (2019) noted that the PSDs obtained by YSM for all the studied packs of beads overlap, independently of r_s. The results obtained with YSM are in very good agreement with those provided by the numerical MIP. The size distributions obtained by the two techniques are represented in Figure 4.5(a), showing that the probability peaks coincide (r_i/r_s = 0.27) and the range of pore size is quasi-identical. Moreover, the average values $E[r_i/r_s]$ of PSDs obtained are also very close, with $E[r_i/r_s]$ = 0.30 for YSM and $E[r_i/r_s]$ = 0.27 for MIP. The PSDs provided by YSM were also compared to those numerically determined by Nolan and Kavanagh (1994) and the results also show a satisfactory agreement. These authors calculated the PSD of a pack of spherical particles, based on the sizes of the largest spheres that could be inserted in the interstices using compact random networks of spheres of equal sizes.

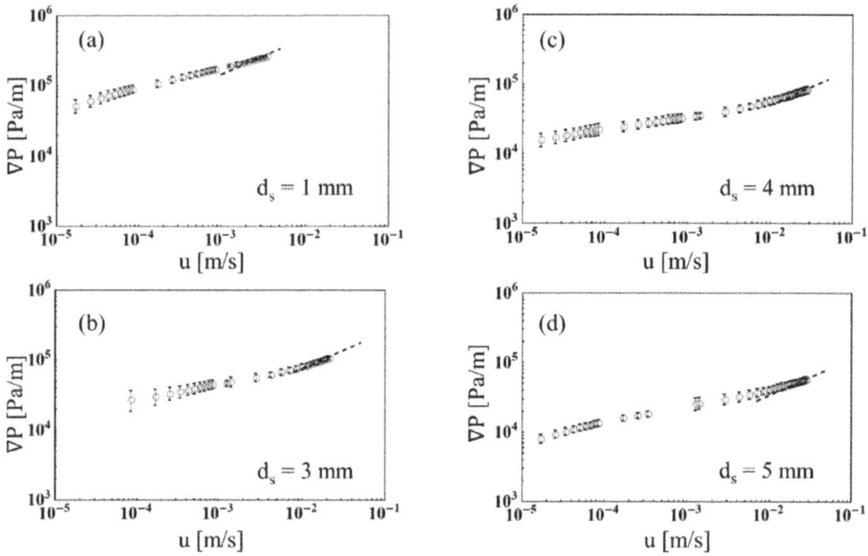

Figure 4.4. *Raw u-∇P measurements obtained from YSM experiments for the characterization of packs of spherical beads. These measurements are the input data of the YSM algorithm allowing the determination of PSD. (a–d) Various diameters of spherical beads (d_S). The dotted lines represent a power law of ∇P with an exponent of $1/n$ adjusted to the measurements made at the four highest flow rates*

4.3.4. *Consolidated porous media: experiment on the core*

Rodríguez de Castro et al. (2016, 2021) experimentally characterized 1D cylindrical samples (cores) of various consolidated porous media. These materials are listed in Table 4.2, which indicates their porosities ε and macroscopic permeability K, their lengths L and their diameters D.

Figures 4.5(b–d) shows the PSDs obtained using MIP and YSM for three of these materials. The pore sizes provided by YSM are not always similar to those characterized by MIP in this type of consolidated porous media and the reasons will be analyzed in section 4.4. However, the distributions obtained by the two methods have a similar shape.

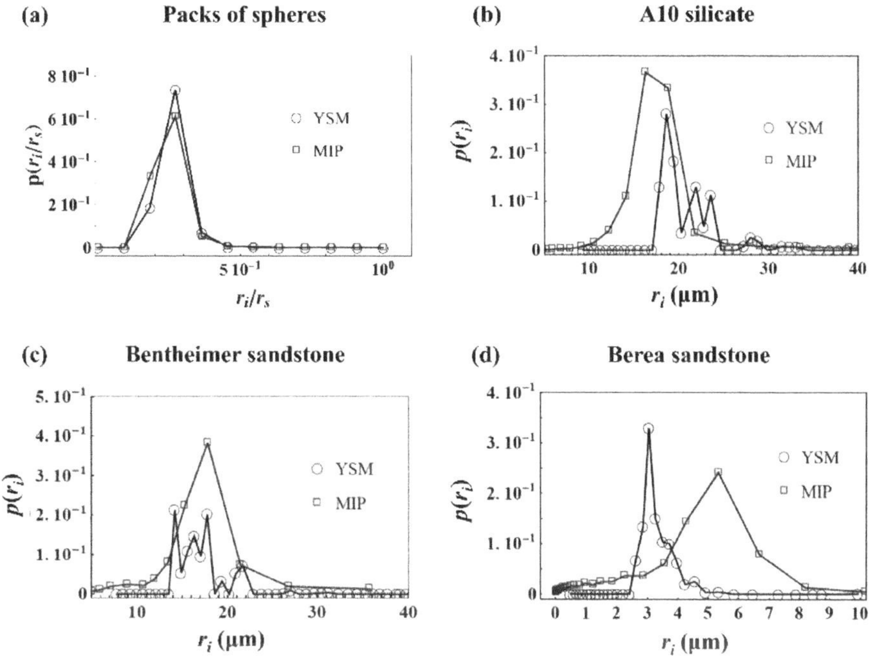

Figure 4.5. *Comparison between experimentally obtained PSD using YSM (black) and MIP method (blue). (a) For packs of spherical beads of radius r_S (Rodríguez de Castro et al. 2019). Abscissas represent the pore radius normalized by the radius of the spheres r_i/r_S. $p(r_i/r_S)$ is the relative probability of a given class r_i/r_S . (b–d) The consolidated samples characterized in the work of Rodríguez de Castro et al. (2021). For a color version of this figure, see www.iste.co.uk/saouab/mechanics.zip*

Sample	Material	K (mD)	ε	L (cm)	D (cm)
1	Aerolith®10 (PALL Corporation, USA)	9,000	0.45	10	5
2	Aerolith®10 (PALL Corporation, USA)	2,400	0.40	5	3.8
3	Bentheimer sandstone (Kocurek Industries, Inc., USA)	2,200	0.25	5	3.8
4	Castlegate sandstone (Kocurek Industries, Inc., USA)	750	0.25	5	3.8
5	Berea sandstone (Cleveland Quarries, USA)	38	0.22	5	3.8

Table 4.2. *Permeability, porosity and sizes of samples of consolidated porous media characterized by Rodríguez de Castro et al. (2016, 2021)*

4.4. Physical significance of the pore size characterized by YSM

While the results of YSM have been compared to those provided by MIP in the previous sections, the relations between the determined PSD and the actual geometry of pores have not yet been approached. It is however essential to examine the nature of pore dimension actually provided by YSM. This matter is particularly interesting for geoscientists involved in the search for relevant information based on porous sample characterization operations. To this end, the present section focuses on the evaluation of the agreement between PSDs characterized by YSM, the pore-opening size distributions provided by MIP tests, and the throat size distribution (TSD) and the pore body size distribution (BSD) obtained by computerized X-ray microtomography. With this objective, a set of artificial and natural porous samples with permeability values extending over six orders of magnitude were characterized by Rodríguez de Castro et al. (2019, 2021) using YSM and MIP laboratory tests at the same time. Then, the results were compared to the geometries of pore network models extracted from digital images of the actual microstructure.

4.4.1. *Pore network modeling*

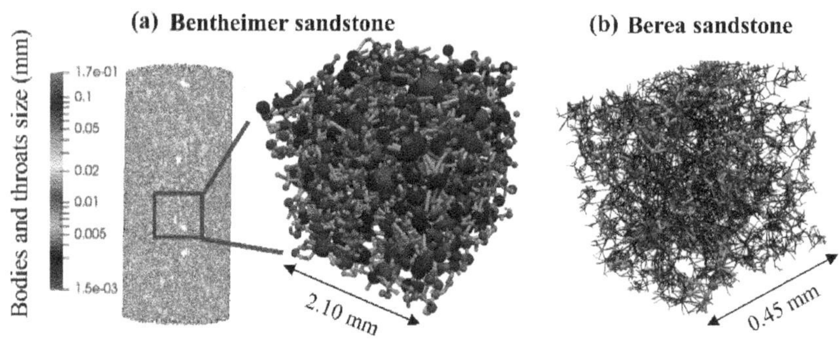

Figure 4.6. *Pore network models extracted from two porous media studied by Rodríguez de Castro et al. (2021): (a) Bentheimer sandstone and (b) Berea sandstone. For a color version of this figure, see www.iste.co.uk/saouab/mechanics.zip*

Digital images of the studied media can be used as input information for the extraction of equivalent pore networks. These equivalent networks consist of spherical pores bodies connected by cylindrical throats. In order to extract such a simplified geometry, the sub-network of the over-segmented watershed algorithm (SNOW) presented by Gostick (2017) was used by Rodríguez de Castro et al. (2019, 2021). Some examples of resulting pore networks are shown in Figure 4.6. SNOW algorithm starts by extracting the map of distances in the voids, which is then filtered by the exclusion of peaks on saddles and plateaus. Then, the peaks that are very close to one another are merged, thus generating a set of markers. These markers are used as inputs for a marker-based watershed segmentation algorithm. Then, this algorithm finds the basins of the distance map, producing the segmentation of each pore region. The SNOW algorithm is included in PoreSpy (Gostick et al. 2019), an open source toolbox for the quantitative post-processing of digital images of porous media.

4.4.2. Comparison of PSD provided by YSM, MIP and by imaging

Figure 4.7 compares the results provided by YSM and MIP for the packs of identical spheres, A10 silicate and Bentheimer and Berea sandstones studied in the previously mentioned works. It is first noted that a strong agreement is found between the results provided by MIP and YSM, both in terms of ranges of pore radii and average pore radii. A remarkable characteristic of this comparison is that, while YSM yields pore sizes slightly higher than MIP for A10 silicate and Bentheimer sandstone, PSDs are shifted toward small pores in the case of Berea sandstone. This is consistent with the results of numerical experiments conducted by Rodríguez de Castro et al. (2020b) on unconsolidated porous media. In the previous works, it has been noted that YSM yielded pore sizes slightly larger than MIP for porous media displaying wide PSDs due to the existence of preferential flow paths for the YSM preventing the smallest throats. These numerical results have been confirmed by the experimental observations provided by YSM tests conducted on packs of spherical beads (Rodríguez de Castro et al. 2019), for which slightly larger pore sizes have also been obtained by YSM, compared to MIP. The same is expected for A10 sintered silicate and Bentheimer sandstone, which present a higher statistical dispersion of their PSD than Berea sandstone.

Figures 4.7(a–d) present the BSDs and the TSDs extracted from digital images of the media studied by Rodríguez de Castro et al. (2019, 2021). These distributions are compared with the results of YSM tests. These results clearly show that the TSD and BSD of Berea sandstone are very similar and nearly overlap (in good agreement with the results of Peksa et al. 2015). There is a greater difference between TSD and BSD in the case of A10 silicate and Bentheimer sandstone, which may lead to the generation of preferential flow paths. The same figure shows that YSM results are within a range in which TSD and BSD overlap for A10 silicate and Bentheimer sandstone. In the case of Berea sandstone, YSM yields PSDs in the range of TSD and BSD. In all cases, PSDs obtained by YSM are closer to the size of throats than to that pores, given the strong influence of throats on the generation of pressure losses throughout the medium. In this respect, it should be recalled that the pressure loss is the quantity measured in YSM tests. The differences between the TSD obtained from pore network modeling and the PSD provided by YSM and MIP are summarized in Table 4.3. It can be concluded from this table that YSM is reasonably close to TSD for these porous media, and a particularly good agreement is obtained for Berea sandstone in terms of average pore size. Moreover, the PSDs provided by the three porosimetry methods are compared in Figures 4.7(e–h), showing that the range of pore sizes provided by MIP and YSM is within the TSD range. The differences between the average pore sizes provided by the three methods studied are below 28% for A10 silicate and for the samples of Bentheimer sandstone. As Table 4.3 indicates, the most significant difference in relative terms is observed between the average size of pores obtained with MIP for Berea sandstones and those obtained with the other two methods. The fact that smaller pore sizes are obtained with YSM compared to MIP for Berea sandstones may result from its narrower PSD with respect to A10 and Bentheimer, as mentioned above.

It should be kept in mind that the three porosimetry methods consider simplified geometries of pores. This is understandable considering the continuous and irregular nature of the connected porous space. Such a continuous porous space must be discretized when defining a PSD, in which a single characteristic length is used to describe pore dimensions. Moreover, the use of a single dimension is possible only by the decomposition of complex microscopic geometries in a set of elements of regular shape, which is just a useful idealization. For example, MIP assumes a model of bundle of capillaries in which the actual sizes of pores are assimilated to the size of pore openings (pore throats giving access to a part of the pore volume). When the injection pressure is higher than the threshold capillary pressure corresponding to the pore opening, the pores are invaded by mercury. Consequently, the injection pressure in MIP is equivalent to the capillary pressure and the capillary radius is considered as the pore radius (Giesche 2006; Jiao et al. 2020). As explained in section 4.2, a model of bundle of capillaries is also used in YSM. The pore radii provided by YSM are closely related to the throat size of pores, as shown by the numerical results of Rodríguez de Castro et al. (2020b) and the experimental results presented in Figure 4.7. In the case of microtomography, pore networks composed of a set of spherical pores interconnected by cylindrical throats are extracted from digital images. In the cited works, TSD and BSD provided by pore network modeling correspond to the distributions of inscribed diameters (or radii) of pore bodies and throats, respectively.

It should be highlighted that, while YSM and MIP assume that the porous space is a bundle of capillary tubes, the extraction of the pore network from digital images does not share the same hypothesis. Consequently, a direct comparison is not simple. In this sense, a recent numerical work used pore network modeling to simulate the flow of shear-thinning fluids in model porous media (Rodríguez de Castro and Goyeau 2021). The results of this work have shown that the pressure loss at Darcy scale, which is the quantity measured in YSM, is mainly sensitive to the fluid viscosity in the active flow channels, meaning the flow channels in which flow rate is not negligible. During the injection of a YSM, as in the YSM tests, a channel is active (or not) under a given pressure gradient, depending on the size of pore throats.

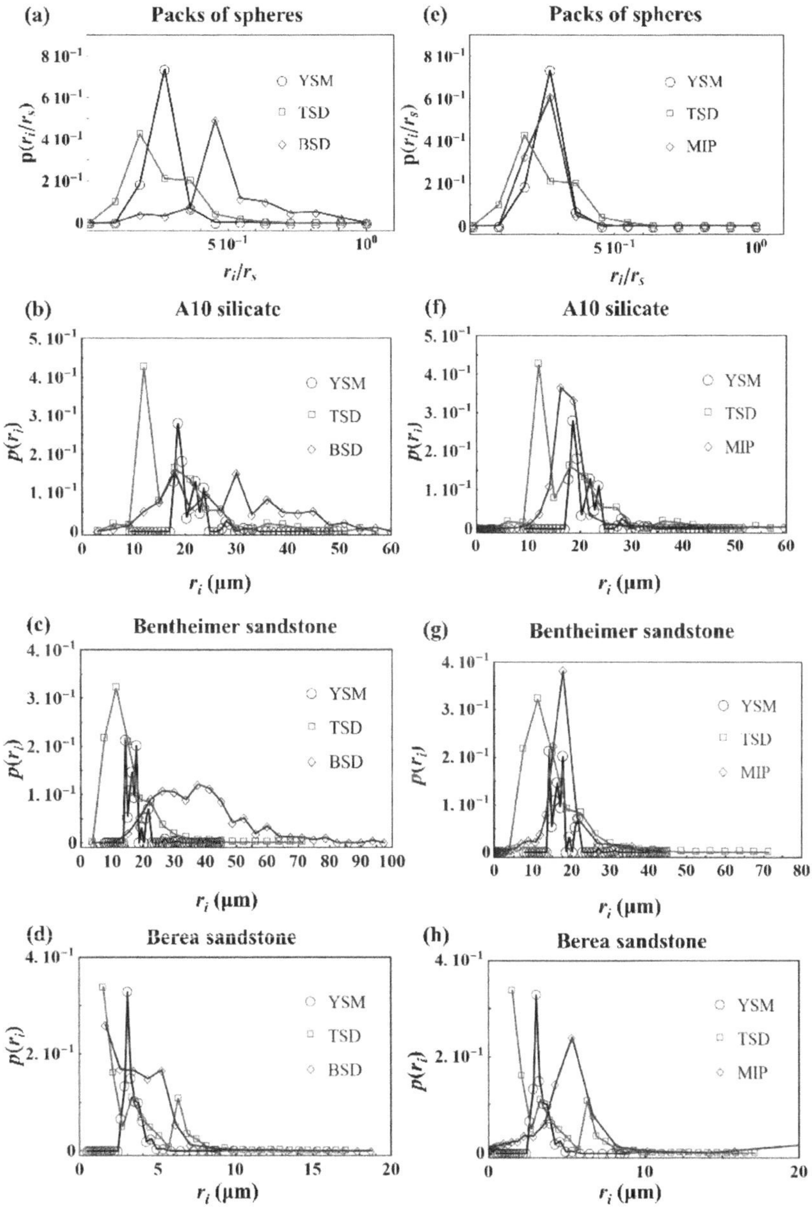

Figure 4.7. *(a–d) PSDs of various porous media obtained by YSM and the resulting TSD and BSD of pore networks extracted from μ-tomographic images. (e–h) PSDs of various porous media obtained by YSM and MIP compared to TSD obtained from pore networks. For a color version of this figure, see www.iste.co.uk/saouab/mechanics.zip*

It should be recalled that the result of MIP represents the volume fraction of pores accessible by pore openings of a given size. MIP is therefore sensitive to the largest openings to the bodies of pores. The size of the largest pore openings is not exactly identical to the throat size. On the other hand, YSM provides the radii of cylindrical capillaries in a bundle of capillaries with the same length and producing the same pressure loss as the actual porous medium irrespective of the flow rate. Consequently, the dimension provided by YSM is more closely related to the hydrodynamic radius than to the throat radius. The differences between pore throat radius, hydrodynamic radius and pore opening radius have been discussed in the past by Chauveteau et al. (1996). They studied the relation between the equivalent hydrodynamic diameter of a pore d_h and the diameters of pore openings derived from MIP tests d_{Hg} in sandstones. d_h is the diameter of a cylindrical capillary having the same length and giving the same pressure loss as the actual pore for a considered flow rate. Independently from the refinement of the analysis process, mercury intrusion remains a non-directional static process, as mercury invades porous bodies through the largest throats. On the other hand, the directional character of fluid flow involves the preferential use of throats located upstream and downstream. Hence, the equality between the dimensions of pores determined by mercury injection and the flow experiments would be obtained only for unrealistic porous media, where all pore throats would have the same size.

Sample	Average (MIP)	Average (TSD)	Average (YSM)	Diff YSM-MIP (%)	Diff YSM-TSD (%)	Diff MIP-TSD (%)
A10	16.28	17.29	20.78	27.64	20.18	–5.84
Bentheimer	15.14	14.44	17.05	12.62	18.07	4.85
Berea	5.17	3.33	3.27	–36.75	–1.80	55.26

Table 4.3. *Average sizes of pores provided by the various porosimetry methods: YSM, MIP and by extraction of pore network from micro-tomographic images (TSD). The differences between YSM and MIP (Diff YSM-MIP), between YSM and TSD (Diff YSM-TSD) and between MIP and TSD (Diff MIP-TSD) are summarized*

4.5. Extension of YSM to rough fractures

Fractures are preferential paths for the flow of a fluid though a geological porous medium, their permeability being far higher than that of rock matrices in which they are formed (Oladyshkin et al. 2007). Taking into account the flows through factures is therefore essential, among other applications, for successful remediation operations of polluted soils. YSM was therefore extended to the characterization of the hydraulic aperture distribution of rock fractures with rough walls

(Rodríguez de Castro et al. 2020a). There is a significant interest in such an extension, given that the distinct characteristics of rock fractures with respect to the matrix represent a burden for other traditional porosimetry techniques. In the particular case of X-ray microtomography, a time-consuming calibration is often necessary and serious difficulties arise due to the hardening of the bundle and image reconstruction artifacts.

Neuzil and Tracy (1981), Tsang and Tsang (1987) and Silliman (1989) proposed simplified fracture models with a variability of 1D aperture in which the total flow rate of a Newtonian fluid through a fracture with rough walls is considered as being the sum of a set of elementary flow rates through smooth channels of various apertures. This type of model has then been extended to flows of Ostwald fluids (Di Federico 1998) and more recently to flows of fluids of the "truncated power law" type (Felisa et al. 2018). Using a similar approach, Auradou et al. (2006) developed a simple model for shear-displaced fractures presenting self-affine geometries. In their model, the field of apertures is replaced by a set of parallel channels, each of constant aperture perpendicular to the direction of shear displacement. These latter authors have shown that the streamlines are only slightly tortuous for the flows parallel to these channels, and the velocity variations along the flow paths are small compared to the velocity contrasts from one channel to another. For each channel, fluid velocity is related to the pressure gradient by Darcy equation. In a subsequent work, Auradou et al. (2008) studied the geometry of the fluid–fluid interface generated during the displacement of a transparent shear-thinning polymer solution by a tinted solution of similar composition. Their results showed a very weak transversal mixture between the flow channels of shear-thinning fluids, which suggests that the fracture can be considered as a set of parallel pipes of constant hydraulic apertures.

Inspired by this research, Rodríguez de Castro et al. (2020a) proposed a model of YSM flow through fractures in which the void space between rough walls is represented by a bundle of Z parallel and rectilinear channels of rectangular section of widths w_i and apertures h_i (with i ranging from 1 to Z). A schematic representation of this model is shown in Figure 4.8. The transversal flow between channels is considered negligible (Auradou et al. 2008). Consequently, the flow is one-dimensional in each channel by neglecting the effects of walls in the direction of width and assuming a sliding condition between neighboring flow channels, as shown in Figure 4.8(c). This model is consistent with the numerical works of Lavrov (2013), who reported the generation of a strong flow channeling during the injection of shear-thinning fluids through fractures with rough walls. This means that the velocity components in the main flow direction are far more significant than those in the perpendicular direction, given the high values of viscosity developed by this type of fluids when the pressure gradient is low.

It is worth highlighting that the dimensions w_i and h_i should not be expected to accurately represent the actual geometry and topology of the fracture, but only that they determine the generated pressure losses. Considering the previous observations, the total flow rate Q through the fracture can be written as follows:

$$Q(\nabla P) = \sum_{i=1}^{Z} w_i q(\nabla P, h_i) \qquad [4.10]$$

where $\nabla P = \Delta P/L$ is the absolute value of the pressure gradient, ΔP is the absolute value of the pressure loss between the inlet and the outlet of the fracture, L is the length of channels and $q(\nabla P, h_i)$ is the individual flow rate per unit width in a rectangular channel of aperture h_i, when a pressure gradient ∇P is applied.

The hydraulic aperture of a rough channel is defined as the aperture of a Hele–Shaw cell producing the pressure loss as the actual rough channel during the flow of a Newtonian fluid, whatever the flow rate. It is known that, in the case of a Newtonian fluid, the values of h_i to be used in equation [4.10] to describe the relation between q and ∇P are equal to those of hydraulic apertures of the different channels. This equivalence is however not obvious in the case of YSMs: should the value of the hydraulic aperture be assigned to h_i in equation [4.10] in order to predict the flow of a YSM through a rough channel? Numerical results on a set of real rock fracture geometries (Rodríguez de Castro et al. 2020b) have proved that the answer is yes.

Consequently, the data (w_i,h_i) correspond to the distribution of hydraulic apertures of the flow channels of the studied fracture, and $q(\nabla P, h_i)$ is the individual flow rate per unit width in a rectangular channel of hydraulic aperture h_i, when a pressure gradient ∇P is applied.

In the case of the one-dimensional Stokes flow of a Newtonian fluid of shear viscosity μ, the expression of $q(\nabla P, h_i)$ to be used in equation [4.10] is given by Hele–Shaw equation:

$$q(\nabla P, h_i) = \frac{h_i^{\,3}}{12\mu} \frac{\nabla P}{T} \qquad [4.11]$$

Here, T is the hydraulic tortuosity. Shear rheology of YSMs at the steady state and under simple shear is often described by Herschel–Bulkley law (equation [4.1]). For a Herschel–Bulkley fluid, $q(\nabla P, h_i)$ is expressed as follows (Skelland 1967; Chhabra and Richardson 2008; Ferrás et al. 2012):

$$q(\nabla P, h_i) = \begin{cases} 2\left[\frac{n}{n+1}\left(\frac{\nabla P}{Tk}\right)^{1/n}\left[\frac{h_i h}{2}\left(\frac{h_i}{2}-\frac{h_0}{2}\right)^{1+1/n} - \frac{n\left(\frac{h_i}{2}-\frac{h_0}{2}\right)^{2+\frac{1}{n}}}{1+2n}\right]\right] \text{if } h_i \geq h_0 \\ 0 \text{ if } h_i \leq h_0 \end{cases} \quad [4.12]$$

where $h_0 = 2\tau_0 T/\nabla P$ is the critical aperture below which the fluid remains stagnant for the considered value of ∇P. Obviously, equation [4.12] is identical to equation [4.11] for $\tau_0 = 0$, $n = 1$ and $k = \mu$.

Furthermore, rectangular flow channels have been considered as perfectly straight in the current model ($T = 1$ in equation [4.11]). This is consistent with laboratory measurements (Shojaei et al. 2019) proving the low tortuosity of streamlines during the flow of YSMs in a rock fracture replica. In order to characterize the hydraulic aperture distribution of a rock fracture with rough wall with YSM, an experiment must be conducted for collecting a set of data (Q_i, ∇P_i), Q_i being the total flow rate through the fracture. These steady-state measurements are conducted during the injection of a Herschel–Bulkley fluid through the studied fracture. The number of collected data pairs is $Z + 1$ ($i = 1, \ldots, Z + 1$). From $Z + 1$ measurements, a set of Z apertures, named h_j, representatives of flow channels are defined as follows:

$$h_j = \alpha \frac{2\tau_0}{\nabla P_j} \; (j = 1, \ldots, Z) \quad [4.13]$$

It should be reminded that the flow under a given pressure gradient ∇P_j occurs only in channels of aperture strictly above $\frac{2\tau_0}{\nabla P_j}$. Consequently, the value of the numerical factor α in equation [4.13] must be above the unit, as mentioned in section 4.2.

In order to extend YSM to rock fractures with rough walls, we first assume that the bundle of Z straight parallel channels with rectangular cross-sections presented in Figure 4.8 can accurately predict the relation between the flow rate and the generated pressure loss. Then, the width w_i of the flow channel of aperture h_i is calculated as being the maximal positive value of w_j for which the following condition is satisfied:

$$Q_s \geq \sum_{g=1}^{j} w_g \, q(\nabla \mathrm{P_s}, \mathrm{h_g}) \text{ for } s \geq j+1 \quad [4.14]$$

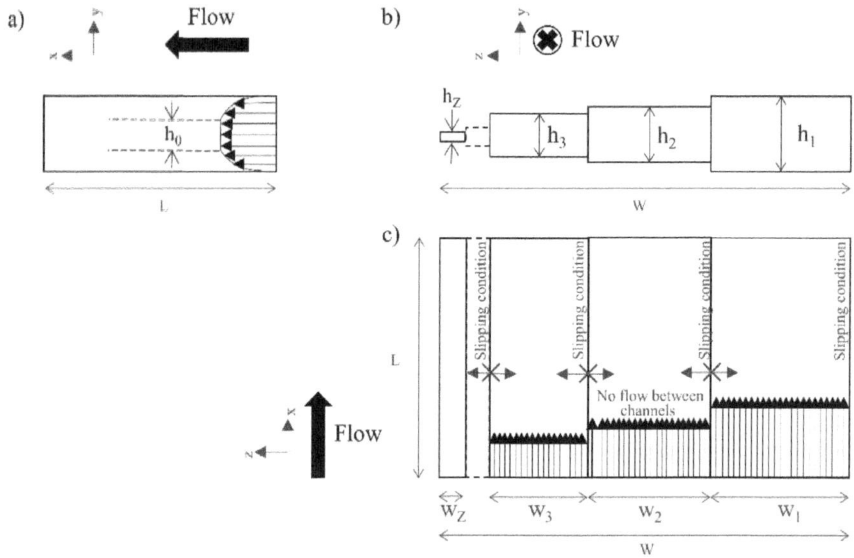

Figure 4.8. *Schematic representation of the model of bundle of rectangular channels used to represent the flow of a yield stress fluid through a fracture (Rodríguez de Castro et al. 2020a): (a) view of the vertical plane parallel to the main flow direction; (b) view of the cross-section perpendicular to the flow direction displaying various apertures of rectangular channels (h_1...h_Z); and (c) view of the horizontal plane parallel to the main flow direction displaying various widths of rectangular channels (w_1...w_Z). The black arrows in (a) and (c) represent the velocity profiles in each channel. The aperture class h_Z is considered as being below $h_0 = 2\tau_0/\nabla P$, which means no flow is produced in channels of aperture h_Z. For a color version of this figure, see www.iste.co.uk/saouab/mechanics.zip*

where $q(\nabla P_s, h_g)$ is given by equation [4.12]. The width, w_1, of the flow channel with the largest aperture h_1 is first determined by solving the system of inequations [4.12] for $s \geq 2$. Then, w_2 is calculated for $s \geq 3$, then the width of flow channels of increasingly small apertures is calculated until w_Z (the width of the flow channel with the smallest aperture) is calculated. At the end of this process, a set of data (h_j, w_j) is obtained. Based on these results, the relative frequency $p(h_j)$ of each aperture class of the flow channel h_j is given by:

$$p(h_j) = \frac{h_j w_j}{\sum_{i=1}^{Z} h_i w_i} \qquad [4.15]$$

The appropriate value of α, denoted by α^*, is calculated as follows: first, $\alpha = 1$ is used to calculate the data (h_j, w_j) and the total width of flow channels $\sum_{i=1}^{Z} w_i$ is calculated. Then, the same procedure is repeated by successively increasing the value of α with a scanning step of δ ($\delta = 0.1$ has been used in this chapter) until the data (h_j, w_j) corresponding to $\alpha = 2$ have been calculated. α^* was chosen as being the value of α minimizing the absolute value of the difference between the actual width of the fracture W and the one calculated, that is, $|W - \sum_{i=1}^{Z} w_i|$. Based on the calculated data (h_j, w_j), the macroscopic relation between the total flow rate through the fracture Q and the pressure gradient ∇P can be obtained using equation [4.10].

Using the YSM, two replicas of rough rocks (granite and Vosges sandstone) were characterized by Rodríguez de Castro et al. (2020a). The aperture distributions of these fractures compared to the PSDs obtained by YSM are shown in Figure 4.9. A good agreement is obtained.

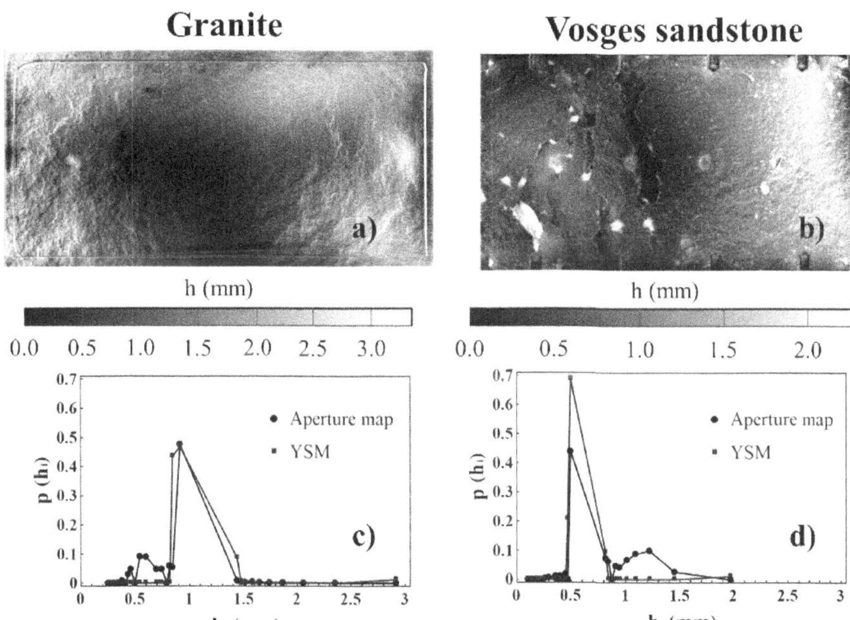

Figure 4.9. *Aperture distribution of replicas of (a) granite and (b) Vosges sandstone studied by Rodríguez de Castro et al. (2020a). PSD of (c) granite and (d) Vosges sandstone obtained by YSM and comparison to those obtained by Nowamooz et al. (2013) for the same fractures by an image processing procedure based on the attenuation of a source of light according to the Beer–Lambert law. YSM results are represented in blue and imaging-based results are represented in black. For a color version of this figure, see www.iste.co.uk/saouab/mechanics.zip*

4.6. Conclusion

The porosimetry method based on the injection of YSMs was developed on the basis of theoretical concepts presented by Ambari et al. (1990). This technique can be used to characterize the PSD of a porous sample thanks to the inversion of raw flow rate–pressure gradient data obtained during simple and non-toxic single-phase flow experiments. The research articles cited in this literature review proved that YSM provides PSDs representative for the microscopic heterogeneity in a variety of consolidated and unconsolidated samples, and also in rock fractures. The PSDs obtained are very close to those characterized by MIP. Thanks to the pore network modeling of the microstructure of characterized samples, it was observed that the pore dimension characterized by YSM is closely linked to the hydraulic conductance of the flow paths in the analyzed samples. Despite these promising results, the research on YSMs allowing the characterization of porous media with very low permeability should be continued. These fluids should consist of particles whose size is sufficiently small to avoid plugging the pores. Moreover, the model of capillary bundle used in YSM should be progressively enriched to take into account the effect of other topological and geometric properties of the microstructure of porous media on the relation between flow rate and pressure gradient measured in the tests. Finally, rheological models that are more apt to describe the rheology of the fluid employed (e.g. the laws of Casson or Carreau) could be implemented in the inversion algorithm to improve the accuracy of the obtained PSDs.

4.7. References

Abou Najm, M.R. and Atallah, N.M. (2016). Non-Newtonian fluids in action: Revisiting hydraulic conductivity and pore size distribution of porous media. *Vadose Zone Journal*, 15(9), 1539–1663.

Ambari, A., Benhamou, M., Roux, S., Guyon, E. (1990). Distribution des tailles des pores d'un milieu poreux déterminée par l'écoulement d`un fluide à seuil. *C.R. Acad. Sci. Paris*, 311(II), 1291–1295.

Atallah, N.M. and Abou Najm M.R. (2019). Characterization of synthetic porous media using non-Newtonian fluids: Experimental device. *European Journal of Soil Science*, 70, 257–267.

Auradou, H., Drazer, G., Boschan, A., Hulin, J.P., Koplik, J. (2006). Flow channeling in a single fracture induced by shear displacement. *Geothermics*, 35, 576–588.

Auradou, H., Boschan, A., Chertcoff, R., Gabbanelli, S., Hulin, J.P., Ippolito, I. (2008). Enhancement of velocity contrasts by shear-thinning solutions flowing in a rough fracture. *Journal of Non-Newtonian Fluid Mechanics*, 153, 53–61.

Bauer, D., Talon, L., Peysson, Y., Ly, H.B., Batôt, G., Chevalier, T., Fleury, M. (2019). Experimental and numerical determination of Darcy's law for yield stress fluids in porous media. *Phys. Rev. Fluids*, 4, 063301.

Benmouffok-Benbelkacem, G., Caton, F., Baravian, C., Skali-Lami, S. (2010). Non-linear viscoelasticity and temporal behavior of typical yield stress fluids: Carbopol, xanthan and ketchup. *Rheol Acta*, 49, 305–314.

Carnali, J.O. (1991). A dispersed anisotropic phase as the origin of the weak-gel properties of aqueous xanthan gum. *Journal of Applied Polymer Science*, 43(5), 929–941.

Carreau, P.J. (1972). Rheological equations from molecular network theories. *Transactions of the Society of Rheology*, 16, 99–127.

Chaparian, E. and Tammisola, O. (2021). Sliding flows of yield-stress fluids. *Journal of Fluid Mechanics*, 911, A17.

Chaplain, V., Mills, P., Guiffant, G., Cerasi, P. (1992). Model for the flow of a yield fluid through a porous medium. *Journal de Physique II France*, 2, 2145–2158.

Chauveteau, G. (1982). Rodlike polymer solution flow through fine pores: Influence of pore size on rheological behavior. *Journal of Rheology*, 26, 111.

Chauveteau, G., Nabzar, L., El Attar, L., Jacquin, C. (1996). Pore structure and hydrodynamics in sandstones. SCA Conference Paper Number 9607.

Chhabra, R.P. and Richardson, J.F. (2008). *Non-Newtonian Flow and Applied Rheology: Engineering Applications*. Butterworth-Heinemann/Elsevier, Amsterdam. doi: 10.1016/B978-0-7506-8532-0.X0001-7.

Cieszko, M., Kempinski, M., Czerwinski, T. (2019). Limit models of pore space structure of porous materials for determination of limit pore size distributions based on mercury intrusion data. *Transport in Porous Media*, 127, 433–458.

Dario, A.F., Hortencio, L.M.A., Sierakowski, M.R., Neto, J.C.Q., Petri, D.F.S. (2011). The effect of calcium salts on the viscosity and adsorption behavior of xanthan. *Carbohydrate Polymers*, 84, 669–676.

Di Federico, V. (1998). Non-Newtonian flow in a variable aperture fracture. *Transport in Porous Media*, 30(1), 75–86.

Felisa, G., Lenci, A., Lauriola, I., Longo, S., Di Federico, V. (2018). Flow of truncated power law fluid in fracture channels of variable aperture. *Advances in Water Resources*, 122, 317–327.

Ferrás, L.L., Nóbrega, J.M., Pinho, F.T. (2012). Analytical solutions for Newtonian and inelastic non-Newtonian flows with wall slip. *Journal of Non-Newtonian Fluid Mechanics*, 175, 76–88.

Giesche, H. (2006). Mercury porosimetry: A general (practical) overview. *Particle & Particle Systems Characterization*, 23, 1–11.

Gostick, J.T. (2017). Versatile and efficient pore network extraction method using marker-based watershed segmentation. *Physical Review E*, 96, 023307.

Gostick, J.T., Khan, Z.A., Tranter, T.G., Kok, M.D., Agnaou, M., Sadeghi, M., Jervis, R. (2019). PoreSpy: A Python toolkit for quantitative analysis of porous media images. *Journal of Open Source Software*, 4, 1296.

Jiao, L., Andersen, P.Ø., Zhou, J., Cai, J. (2020). Applications of mercury intrusion capillary pressure for pore structures: A review. *Capillarity*, 3, 62–74.

Jones, D.M. and Walters, K. (2021). The behavior of polymer solutions in extension-dominated flows with applications to enhanced oil recovery. *Rheol Acta*, 28, 482–498.

Khodja, M. (2008). Les fluides de forage: étude des performances et considerations environnementales. PhD Thesis, Institut National Polytechnique de Toulouse, Toulouse.

Lavrov, A. (2013). Redirection and channelization of power-law fluid flow in a rough-walled fracture. *Chemical Engineering Science*, 99, 81–88.

Mackaya, T.E., Ahmadi, A., Omari, A., Rodríguez de Castro, A. (2021). Empirical flow rate/pressure drop relationships for non-circular capillaries used in yield stress fluid porosimetry. *Transport in Porous Media*, 136, 587–605.

Malvault, G. (2013). Détermination expérimentale de la distribution de taille de pores d'un milieu poreux par l'injection d'un fluide à seuil ou par analyse fréquentielle. PhD Thesis, Arts et Métiers ParisTech, Paris.

Malvault, G., Ahmadi, A., Omari, A. (2017). Numerical simulation of yield stress fluid flow in capillary bundles: Influence of the form and the axial variation in the cross section. *Transport in Porous Media*, 120, 255–270.

Neuzil, C.E. and Tracy, J.V. (1981). Flow through fractures. *Water Resources Research*, 17(1), 191–199.

Nolan, G.T. and Kavanagh, P.E. (1994). The size distribution of interstices in random packings of spheres. *Powder Technology*, 78, 231–238.

Nowamooz, A., Radilla, G., Fourar, M., Berkowitz, B. (2013). Non-Fickian transport in transparent replicas of rough-walled rock fractures. *Transport in Porous Media*, 98(3), 651–682.

Oladyshkin, S., Skachkov, S., Panfilova, I., Panfilov, M. (2007). Upscaling fractured media and streamline HT-splitting in compositional reservoir simulation. *Oil & Gas Science and Technology – Rev. IFP*, 62, 137–146.

Oukhlef, A. (2011). Détermination de la distribution de tailles de pores d'un milieu poreux. PhD Thesis, Arts et Métiers ParisTech, Paris.

Oukhlef, A., Champmartin, S., Ambari, A. (2014). Yield stress fluids method to determine the pore size distribution of a porous medium. *Journal of Non-Newtonian Fluid Mechanics*, 204, 87–93.

Peksa, A.E., Wolf, K.-H.A.A., Zitha, P.L.J. (2015). Bentheimer sandstone revisited for experimental purposes. *Mar. Pet. Geol.*, 67, 701–719.

Rodríguez de Castro, A. (2014). Flow experiments of yield stress fluids in porous media as a new porosimetry method. PhD Thesis, Arts et Métiers ParisTech, Paris.

Rodríguez de Castro, A. and Goyeau, B. (2021). A pore network modelling approach to investigate the interplay between local and Darcy viscosities during the flow of shear-thinning fluids in porous media. *Journal of Colloid and Interface Science*, 590, 446–457.

Rodríguez de Castro, A., Omari, A., Ahmadi-Sénichault, A., Bruneau, D. (2014). Toward a new method of porosimetry: Principles and experiments. *Transport in Porous Media*, 101, 349–364.

Rodríguez de Castro, A., Ahmadi-Sénichault, A., Omari, A., Savin, S., Madariaga, L.-F. (2016). Characterizing porous media with the yield stress fluids porosimetry method. *Transport in Porous Media*, 114, 213–233.

Rodríguez de Castro, A., Ahmadi-Sénichault, A., Omari, A. (2018). Using xanthan gum solutions to characterize porous media with the yield stress fluid porosimetry method: Robustness of the method and effects of polymer concentration. *Transport in Porous Media*, 122(2), 357–374.

Rodríguez de Castro, A., Agnaou, M., Ahmadi-Sénichault, A., Omari, A. (2019). Application of non-toxic yield stress fluids porosimetry method and pore-network modelling to characterize the pore size distribution of packs of spherical beads. *Transport in Porous Media*, 130(3), 799–818.

Rodríguez de Castro, A., Ahmadi-Sénichault, A., Omari, A. (2020a). Determination of the aperture distribution of rough-walled rock fractures with the non-toxic yield stress fluids porosimetry method. *Advances in Water Resources*, 146, 103794.

Rodríguez de Castro, A., Agnaou, M., Ahmadi-Sénichault, A., Omari, A. (2020b). Numerical porosimetry: Evaluation and comparison of yield stress fluids method, mercury intrusion porosimetry and pore network modelling approaches. *Computers & Chemical Engineering*, 133, 106662.

Rodríguez de Castro, A., Ahmadi-Sénichault, A., Omari, A. (2021). Analysis of the length scale characterized by the yield stress fluids porosimetry method for consolidated media: Comparison with pore network models and mercury intrusion porosimetry. *Hydrogeology Journal*, 29, 2853–2866.

Saxena, N., Hows, A., Hofmann, R., Alpak, F.O., Freeman, J., Hunter, S., Appel, M. (2018). Imaging and computational considerations for image computed permeability: Operating envelope of digital rock physics. *Advances in Water Resources*, 116, 127–144.

Shojaei, M.J., Rodríguez de Castro, A., Méheust, Y., Shokri, N. (2019). Dynamics of foam flow in a rock fracture: Effects of aperture variation on apparent shear viscosity and bubble morphology. *Journal of Colloid and Interface Science*, 552, 464–475.

Silliman, S.E. (1989). An interpretation of the difference between aperture estimates derived from hydraulic and tracer tests in a single fracture. *Water Resources Research*, 25, 2275–2283.

Skelland, A.H.P. (1967). *Non-Newtonian Flow and Heat Transfer*. Wiley, New York.

Song, K.-W., Kim, Y.-S., Chang, G.S. (2006). Rheology of concentrated xanthan gum solutions: Steady shear flow behavior. *Fibers Polym.*, 7(2), 129–138.

Sorbie, K.S. (1991). Rheological and transport effects in the flow of low-concentration xanthan solution through porous media. *Journal of Colloid and Interface Science*, 145(1), 74–89.

Tsang, Y.W. and Tsang, C.F. (1987). Channel model of flow through porous media. *Water Resources Research*, 23(3), 467–479.

Withcomb, P.J. and Macosko, C.W. (1978). Rheology of xanthan gum. *Journal of Rheology*, 22, 493.

5

Instabilities and Transition to Turbulence in the Couette–Taylor Flow of a Shear-thinning Viscoelastic Liquid

This chapter focuses on the study of the first instability modes and their transition to turbulence in the Couette–Taylor viscoelastic flow when the inner cylinder rotates while the outer cylinder is fixed. The flow is driven by two control parameters: the Taylor number Ta that measures the ratio of the centrifugal force to viscous forces and the elastic number E that reflects the degree of elasticity of the viscoelastic solution. For moderate values of $E \in [0.01; 0.5]$ and when Ta reaches a critical value Ta_c, the circular Couette flow becomes unstable only near cylinder ends. This instability appears in the form of two spirals localized near the ends of cylinders and which coexist with the laminar Couette flow in the central part of the system. These localized spirals, known as *wall modes*, are modeled using the complex Ginzburg–Landau equations without coupling term. The area occupied by spirals increases with the control parameter Ta at the expense of the laminar zone until saturation, when the spiral pattern occupies the entire system. The spirals transit toward turbulence with the increase of Ta via three scenarios: defects turbulence in spirals, spatiotemporal intermittency (turbulent spots in the presence of spirals) for $0.01 < E < 0.05$ and flame patterns for $0.05 < E < 0.5$.

5.1. Introduction

In the industry, hydrodynamic instabilities and turbulence are often sought after to increase the mixing capacity (e.g. in a chemical reactor) or dreaded when they limit

---

Chapter written by Noureddine LATRACHE, Fayçal KELAI, Olivier CRUMEYROLLE and Innocent MUTABAZI.

the efficiency of a process. The addition of a polymer of high molar mass makes it possible to control these instabilities and the transition toward turbulence in certain flows. In a turbulent flow, the addition of a very small quantity of polymer may lead to drag reduction (de Gennes 1984; Gyr and Bewersdorff 1995). Drag reduction by adding polymer has several industrial applications such as the increase in the reach of fire hose nozzles, the increase in sewer flow rate, the increase in lift of navigating objects, hydraulic transportation, central heating systems and rotating flows such as centrifuges. When the added polymer quantity is moderate, elastic and inertial forces are comparable and the instabilities and turbulence are elasto-inertial (Latrache et al. 2008, 2012, 2016, 2021; Kelai 2010). When the quantity of added polymer exceeds a critical value, the elastic force may destabilize the flow which transits directly into elastic turbulence, even at very low Reynolds number *Re* (Larson et al. 1990; Groisman and Steinberg 1996a, 1998; Baumert and Muller 1999).

This chapter addresses the study of viscoelastic instabilities in the Couette–Taylor flow composed of two coaxial cylinders. The inner cylinder of radius a rotates at an angular frequency Ω, while the outer cylinder of radius b is fixed (Figure 5.1). The flow zone is the gap of width d and length L (Figure 5.1). The flow shear rate is given by $\dot{\gamma} = \Omega a/d$ (Figure 5.1).

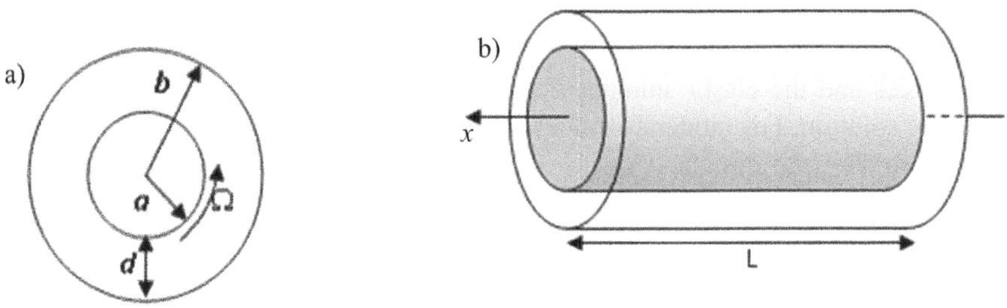

Figure 5.1. *Schematic representation of the Couette–Taylor system: (a) Cross section (x = const); (b) three-dimensional view of the cylindrical annulus with horizontal axis x*

A Newtonian solvent in which a certain quantity of polymers of high molar mass is dissolved can be used to obtain a viscoelastic solution whose viscosity η is higher than that of the solvent η_s (Bird et al. 1987). The difference $\eta_p = \eta - \eta_s$ is the polymer contribution to viscosity. The viscosity of a viscoelastic solution is determined by rheological measurements. The rheological behavior of the viscoelastic solution depends on the concentration of the polymer dissolved in the solution. The viscosity of the highly diluted viscoelastic solution (i.e. whose polymer concentration c is significantly below that of the overlap concentration c^*, i.e. $c \ll c^*$) does not vary with the shear rate $\dot{\gamma}$. For diluted solutions ($c < c^*$), semi-diluted solutions (c ~ c^*) or concentrated solutions (c > c^*), viscosity varies with $\dot{\gamma}$. There is a distinction between shear-thinning viscoelastic liquids whose total viscosity η decreases with $\dot{\gamma}$ and shear-thickening liquids whose viscosity increases with $\dot{\gamma}$. Viscoelastic liquids, for which viscosity remains constant with respect to $\dot{\gamma}$, are known as *Boger liquids* (Boger 1977).

Viscoelastic liquids are also characterized by the existence of normal stress differences denoted by N_1 and N_2. In the viscoelastic Couette–Taylor flow, they are given by $N_1 = \tau_{\theta\theta} - \tau_{rr}$; $N_2 = \tau_{zz} - \tau_{rr}$. These stresses intervene in the destabilization mechanism of the viscoelastic Couette–Taylor flow. Bai et al. (2015, 2021, 2022) have shown that the radial gradient of N_1 modifies the Rayleigh discriminant by introducing the possibility of instability induced by elastic forces even when the centrifugal force would be stabilizing.

Another characteristic of viscoelastic solutions is the existence of relaxation time λ, which determines the characteristic relaxation of the energy stored in macromolecules when they are sheared or stirred in the flow. The definition of λ is not unambiguous and depends on the chosen (molecular or macroscopic) model and on the nature of the stress (elongation, shear) applied on the polymer (Bird et al. 1987).

The viscosity of a shear-thinning viscoelastic solution (Figure 5.2) presents a Newtonian plateau with $\eta = \eta_0$ for small values of $\dot{\gamma}$ and a decreasing zone described by the Carreau law $\eta = \eta_0[1 + (\lambda_C\dot{\gamma})^2]^{-k}$, where k is the shear-thinning index and λ_C is the characteristic time of Carreau that will be used as characteristic scale of relaxation time in our work, as it is experimentally accessible.

To describe viscoelastic flows, we use the Weissenberg number $Wi = \dot{\gamma}\lambda$ that is the analogue of the Reynolds number $Re = \dot{\gamma}\tau_\nu$. The ratio between Wi and Re defines the elasticity number $E = Wi/Re = \lambda/\tau_\nu = \lambda\nu/d^2$ where ν is the kinematic viscosity of the viscoelastic solution. The elastic number is a characteristic property of the viscoelastic solution in a geometry of characteristic scale d. For the same solution, the elasticity of the solution will have more effects in a confined geometry than in an extended one. For very low values of the elasticity number E, the elastic effects are negligible compared to the inertia effects which dominate the flow dynamics and conversely for high values of E, the inertia effects will be negligible compared to the elastic effects which generate elastic instabilities and elastic turbulence (Larson et al. 1989, 1992). For intermediate values of E, inertial and elastic effects are competing and generate elasto-inertial instabilities (Groisman and Steinberg 1996a; Crumeyrolle et al. 2002; Latrache et al. 2012, 2016; Lacassagne et al. 2020; Boulefentis et al. 2023; Moazzen et al. 2023).

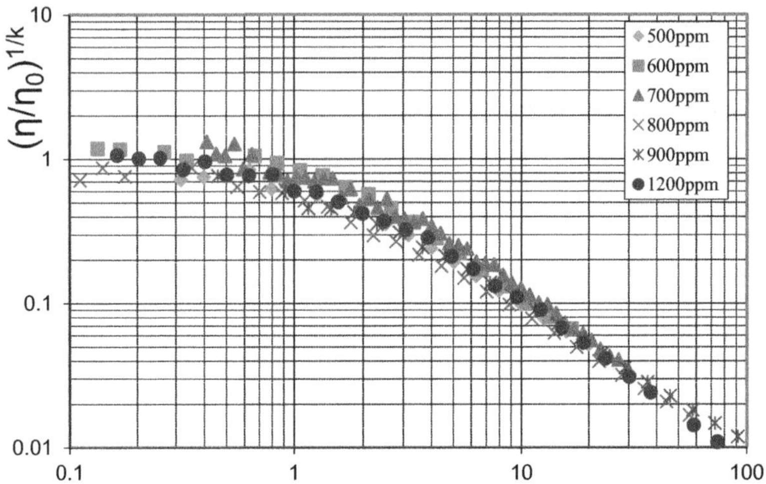

Figure 5.2. *Dynamic viscosity as a function of shear rate. For a color version of this figure, see www.iste.co.uk/saouab/mechanics.zip*

The following section examines the instabilities of the flows of shear-thinning viscoelastic liquids in the Couette–Taylor system. The Couette–Taylor flow, between two coaxial cylinders in differential rotation, is a model hydrodynamic system for the study of instabilities and the transition to turbulence with viscoelastic liquids. Viscoelastic flows in the Couette–Taylor system can be controlled by the Taylor number $Ta = Re\left(\frac{d}{a}\right)^{1/2}$ for inertia and curvature effects, by the elasticity number E and the viscosity ratio $S = \eta_p/\eta$ for viscoelastic effects. For very low values of the elasticity number E, elastic effects are weak compared to inertia effects. In this case, experiments with various polymer solutions (Denn and Roisman 1969; Crumeyrolle et al. 2002, 2003, 2005) revealed a stabilization of the circular Couette flow with the increase in the threshold of primary instability, which manifests itself through the appearance of stationary Taylor vortex flows (TVFs) and secondary instability in form of wavy vortex flows (WVFs). For moderate values of elasticity, inertial effects remain significant, but the elasticity of the solution modifies the nature of the instability and the critical state is produced through a supercritical Hopf bifurcation in the form of spiral waves (Baumert and Muller 1995, 1997, 1999; Groisman and Steinberg 1996a, 1998; Crumeyrolle et al. 2002, 2003, 2005; Latrache et al. 2008, 2012, 2016, 2021). For high values of elasticity, elastic effects dominate inertia effects and instability appears in the form of large disordered vortices even at very low values of Ta (Larson et al. 1990; Groisman and Steinberg 1998) or in the case of a stable Newtonian flow, for example, when the inner cylinder is fixed and only the outer cylinder rotates (Larson et al. 1990; Bai et al. 2023).

After this introduction, we briefly describe the experiment, data processing techniques and also the physicochemical characterization of the polymer (rheology and light diffusion). We will describe various scenarios of the transition to turbulence in the Couette–Taylor system with a viscoelastic solution based on polyoxyethylene (POE) for values of the elastic number ranging between 0.01 and 0.5. We will describe the space-time dynamics of the spirals observed near the endplates (wall modes) using the Ginzburg–Landau (GL) theory.

5.2. Experimental set-up and physicochemical characterization of the polymer

Figure 5.3. *Experimental set-up of the Couette flow. For a color version of this figure, see www.iste.co.uk/saouab/mechanics.zip*

The experimental set-up consists of a horizontal Couette cell formed of two coaxial cylinders. The inner cylinder is made of black Delrin with a radius $a = 4.46$ cm, and the outer cylinder is made of plexiglas with a radius $b = 5.05$ cm. The size of the gap is $d = b - a = 0.59$ cm and the useful length of the gap is $L = 27$ cm. The radius ratio and aspect ratio are respectively $a/b = 0.883$ and $\Gamma = L/d = 46.6$. The two cylinders are driven by direct current servomotors

(Figure 5.3). In the experiment described here, the outer cylinder is fixed, while the inner cylinder rotates at an angular frequency Ω.

The polymer solutions have been prepared by dissolving a solution of POE of high molecular mass (M ~ 10^6 g/mol) in 40 mL of isopropyl alcohol then in 760 mL of water. According to the rheological study, POE solutions of concentrations c between 500 and 1200 ppm have a shear-thinning behavior in which the shear viscosity decreases as a function of shear rate according to Carreau law (Figure 5.2). Light diffusion has shown that the POE macromolecule has an average radius of gyration $< R_g >= 165$ nm and an average molar mass $< M_w >= 4 \times 10^6$g/mol. The radius of gyration of POE varies with the power of the molar mass ($R_g \sim M_w^x$) with a Flory exponent $x \approx 0.6$. This means that the POE macromolecule is dissolved in a good solvent and takes the form of a coil (Brochard and de Gennes 1977). The relaxation time of the macromolecule is calculated using the Flory formula $\lambda_M = \frac{\eta_s R_g^3}{k_B T} = 1$ ms, where k_B is the Boltzmann constant and T is the temperature. The intrinsic viscosity $[\eta_0] = 4.72 \times 10^{-3}\text{ppm}^{-1}$ is calculated using Fox–Flory formula $[\eta_0] = \phi R_g^3/M_w$ and the measured radius of gyration, where $\phi = 4.2 \times 10^{24}$g/mol (Crumeyrolle 2002). The overlap concentration is calculated using the Graessley formula $c^* = 0.77/[\eta_0]$, which yields $c^* = 164$ ppm. Therefore, the POE solutions of concentrations between 500 ppm and 900 ppm used in our study are semi-diluted solutions.

c (ppm)	η_0 (mPa.s)	λ_C (s)	k	λ (s)	S0 = η_{p0}/η	λ/λ_M
500	5.13	0.12	0.078	0.088	0.83	88
600	7.40	0.16	0.098	0.122	0.88	122
700	10.21	0.29	0.105	0.233	0.92	233
800	15.03	0.47	0.116	0.63	0.93	630
900	20.5	0.77	0.126	1.09	0.95	1,090
1,200	26.23	0.87	0.115	1.16	0.96	1,160

Table 5.1. *Coefficients of the Carreau model of POE solutions*

The two largest scales of flow time are the relaxation time of the solution λ (Table 5.1) and the viscous diffusion time $\tau_v(\dot{\gamma}) = \rho d^2/\eta(\dot{\gamma}) \in [3.6\text{ s}; 7.8\text{ s}]$. To visualize the flows, we added to the polymer solutions 2% Kalliroscope AQ 1000, which is a strongly anisotropic suspension of reflective platelets. No significant change in the viscosity of solutions due to these platelets was observed. A linear CCD camera of 1,024 pixels was used to record the reflected light intensity $I(x)$ from one line (parallel to the axis of the cylinder) with a sampling of 8 bits. The lines recorded at regular intervals (0.2 s) are superposed to produce space-time diagrams $I(x,t)$ along a line of the vortex pattern. In most experiments, the data recording time is $T_{rec} = 18{,}618\ \lambda$ over a length $L_{rec} = 41$ d.

5.3. Results

5.3.1. *Stability diagram*

The states observed in the experiment with POE solutions between 0 and 900 ppm are represented in the stability diagram (E, Ta) in Figure 5.4. Three regimes have been identified:

i) Inertial regime $0 < E < 0.01$: here, the elasticity forces are negligible compared to the inertia forces and the two first critical modes are the same as for a Newtonian liquid, that is, TVF and WVF (Figure 5.5). In this regime, the polymer increases the viscosity of the solution. It plays a stabilizing role that is manifested in a small increase in critical values Ta_c of the two first modes with respect to the Newtonian case. The thresholds of these two modes do not vary with the elasticity number E (Figure 5.4).

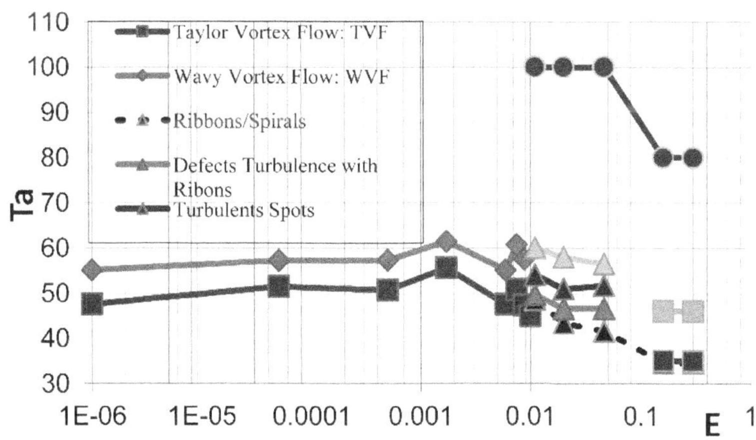

Figure 5.4. *Stability diagram in the (E, Ta) plane of the flows observed in the viscoelastic Couette–Taylor system. For a color version of this figure, see www.iste.co.uk/saouab/mechanics.zip*

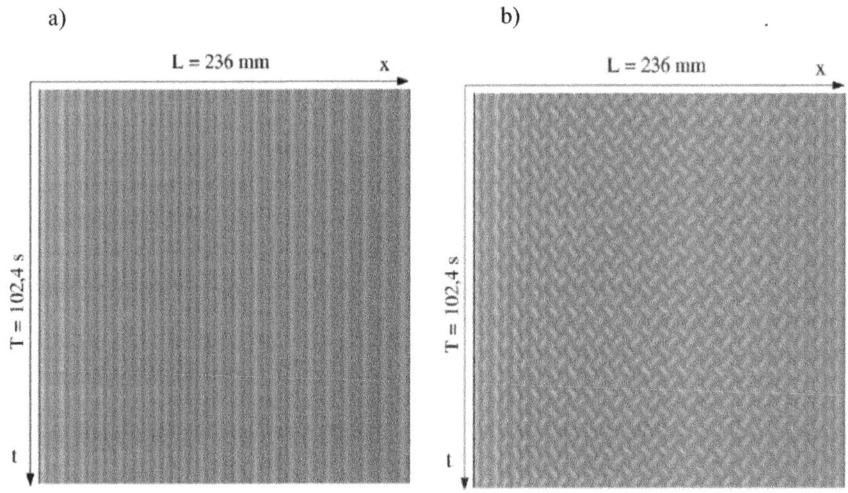

Figure 5.5. *Space-time diagrams observed in a POE solution of concentration c = 400 ppm corresponding to E = 0.009: (a) Taylor vortex flows (TVF) at Ta = 47.4; (b) wavy vortex flows (WVF) at Ta = 57.5*

ii) The elasto-inertial regime ($0.01 \leq E < 0.05$) in which the inertia and elasticity forces are comparable: for Ta > Ta_c, the first instability modes are axisymmetric and unsteady and occur in the form of spirals or ribbons (superposition of spirals of opposite helicity) (Figure 5.6(a)). A slight increase in the Taylor number ($Ta > Ta_d$) generates space-time defects in the ribbon pattern (Figure 5.6(b)) due to a secondary Benjamin–Feir instability, characterized by the appearance of a band of frequency peaks located around the fundamental frequency (Latrache et al. 2016).

iii) Elastic regime ($0.05 \leq E < 0.5$): here, the circular Couette flow branches off to a pattern of spirals at Ta = Ta_S = 34.6. A small increase in Taylor number (Ta_F = 35) leads to the sudden appearance of very strong inflows (from the outer cylinder toward the inner cylinder) in the form of solitons observed in the pattern of spiral waves. The resulting pattern is known as flame pattern (Figure 5.7(a)). The flame pattern rapidly pertains to a turbulent flow dominated by solitons. The transition of the flame pattern to turbulent flow has a sub-critical behavior as in the case of the transition of diwhirls to elastic turbulence observed by Groisman and Steinberg (1996b, 1998).

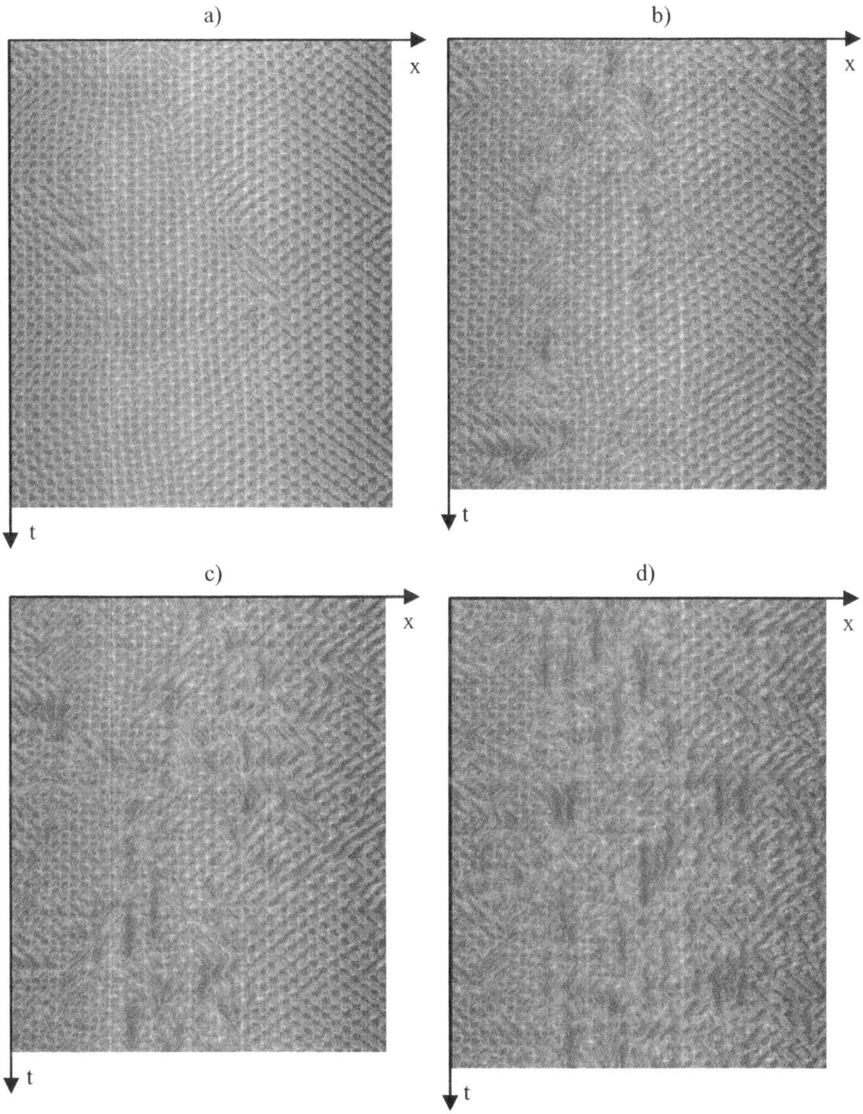

Figure 5.6. *Space-time diagrams observed with PEO E = 0.046: (a) spirals with defects Ta = 46; (b) intermittent regime: spirals with spots Ta = 52, (c–d) elasto-inertial turbulence regime Ta = 62 and Ta = 82*

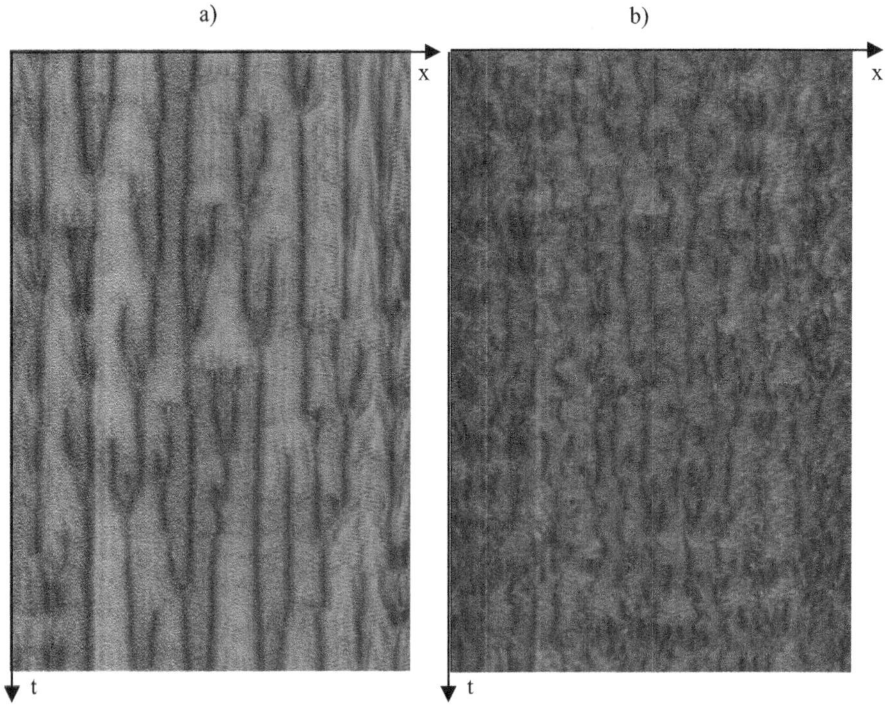

Figure 5.7. *Space-time diagrams observed with PEO solution with E = 0.30: (a) flame patterns at Ta = 38; (b) regime of inertio-elastic turbulence of flames at Ta = 50*

5.3.2. Characterization of defects

The radial and axial components of the ribbon pattern without and with defects have been measured using particle image velocimetry (PIV) in a solution with an elasticity E = 0.137 (Latrache et al. 2016). For the ribbons without defects, the velocities of right and left spirals are comparable (Latrache et al. 2016). For the ribbons with defects, the velocity field of the right spiral shows that the radial velocity vanishes at the core of the defect ($u'_r = 0$). The axial velocity is also zero at the core of the defect ($u'_z = 0$). According to the condition of incompressibility, the azimuthal component of the disturbance also vanishes ($u'_\theta = 0$). This result confirms that the amplitude of defects is A = 0, both in theory and in our visualization experiments using Kalliroscope. In this regime of defect turbulence ($Ta > Ta_d$), the number of defects increases according to Arrhenius law (Latrache et al. 2016). By increasing the Taylor number, the pattern with defects bifurcates to turbulent spots when the time between two consecutive defects T_{lam} is equivalent to the lifetime of a defect T_{def} ($T_{lam} \approx T_{def}$). These turbulent spots represent zones of strong flow from the outer cylinder to the inner cylinder (Latrache et al. 2016). The coexistence of turbulent spots and ribbons forms a spatiotemporal intermittent flow where ribbons are considered as passive (laminar) zones, while turbulent spots are active zones (Figure 5.6(b)). The increase in Ta generates more and more turbulent spots leading the flow to inertio-elastic turbulence (Figures 5.6(c) and (d)) with a 50% rate of turbulent spots for high values of Ta $\geq$ Ta_t = 56.5 (Figure 5.2). The threshold of this inertio-elastic turbulence (Ta_t = 56.5) is characterized by the divergence of the diffusion velocity of the pattern and the local Weissenberg number (Latrache et al. 2016).

5.3.3. *Characterization of the flame pattern*

In the regime of flames in the presence of counter-propagating spirals, the pattern is organized in domains separated by walls fluctuating in time and space (Figure 5.8). These "walls" are zones where the flow is strongly oriented to the inner cylinder (inflow), proving the presence of elastic vortices that are superposed to the inertio-elastic waves (Figures 5.8(b)–(d)). The domains are formed of disordered spiral patterns (Figure 5.8(b)). The average profile of these elastic vortices is the signature of the soliton behavior (Figure 5.8(c)).

PIV measurements (Figures 5.8(e)–(g)) confirm the inflow nature near the walls. Figures 5.8(e)–(g) support our comparison between inflow zones (black zones in Figures 5.8(a)–(d)) and the blue zones of radial velocity measured with PIV. Figure 5.8(e) shows the existence of two vortices.

The maximal radial velocity is equal to 9.7 mm/s and it is comparable to the ratio of the gap and the relaxation time of the solution measured at shear-thinning $\left(u_{r,\max} \approx \frac{d}{\lambda}\right)$. This time is far longer than the molecular relaxation time ($\lambda = 0.99$ s $>> \lambda_M = 0.001$ s), this suggests an elastic instability which is controlled by the elasticity of the solution.

In the flame regime, the number of solitons with strong inflows increases with the control parameter Ta, while in the inertio-elastic turbulence regime, the number of solitons remains constant (Figure 5.7(b)). We have characterized this transition from flames to inertio-elastic turbulence by using a statistical model (Latrache and Mutabazi 2021) to obtain the stability curve A_c as a function of Ta: $A_c \sim (Ta - Ta_F)^{-1}$. The value of the exponent is comparable to those obtained in sheared flows cited in Latrache and Mutabazi (2021).

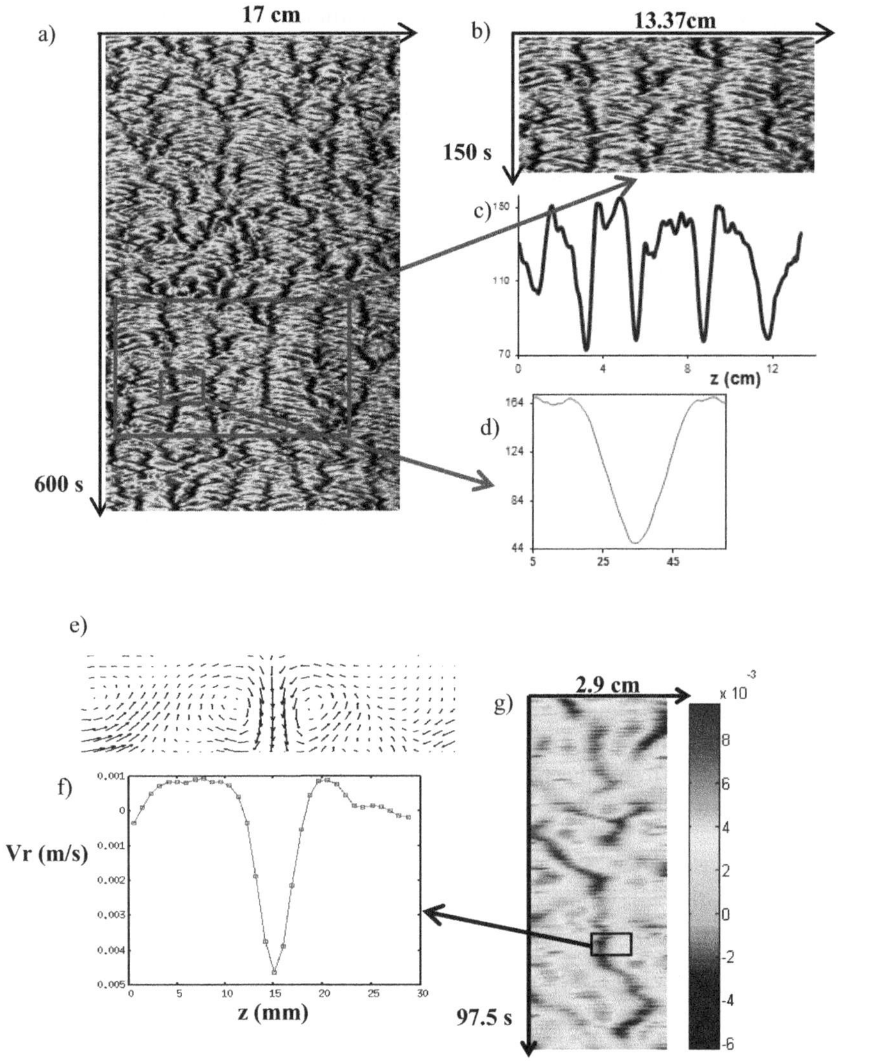

Figure 5.8. *(a) Space-time diagram I(x,t) of a flow of flames at 1,200 ppm for ε = 0.122; (b) zoom of the framed part of a); (c) time average of the zoom; (d) profile of the light intensity in the core of a wall; (e) instantaneous velocity field; (f) velocity profile extracted near a soliton (flame); (g) space-time diagram of radial velocity u'_r. For a color version of this figure, see www.iste.co.uk/saouab/mechanics.zip*

5.3.4. *Spiral pattern at the threshold near the walls*

This section focuses on wall modes that are patterns of spiral waves observed at the threshold of instability near cylinder ends.

For concentrations between 500 and 700 ppm ($0.01 \leq E < 0.05$), we progressively increase the rotation velocity of the inner cylinder with a step of 1 mHz. At a critical rotational frequency Ω_{ic} ($Ta_c = 35$), the basic laminar flow undergoes a supercritical Hopf bifurcation near the endplates in the form of spiral waves (Figure 5.9). These spirals are left and right waves propagating along and around the axis of the cylinder. The pattern of spirals presents a light superposition of right and left waves near the two edges.

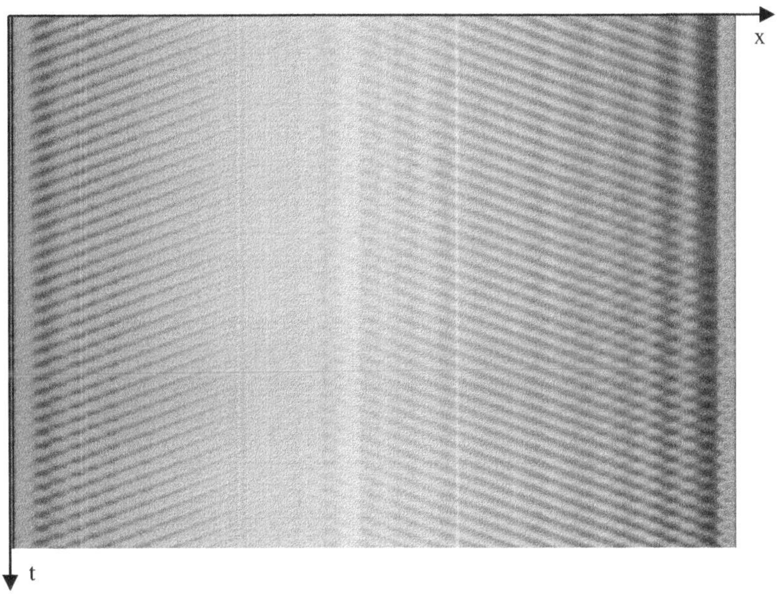

Figure 5.9. *Space-time diagram of spirals near the threshold for a solution of 700 ppm (E = 0.046); Ta = 35.2; ε = 4.18 × 10*$^{-3}$*; Δt = 204.8 s and Δx = 23.5 cm*

One-dimensional (1D) and two-dimensional (2D) spectral analysis shows that the frequencies and the wavenumbers of the two right and left spirals (Figure 5.10) are nearly equal ($f_{A0} \approx f_{B0}$; $k_{A0} \approx k_{B0}$). Hence, the pattern of spiral waves can be represented as follows:

$$I(x,t) = Re\{A(x,t)\,exp[i(k_A x - \omega_A t)] + B(x,t)\,exp[i(k_B x - \omega_B t)]\} + c.c \quad [5.1]$$

where A, B, $\omega_A = 2\pi f_A$ and $\omega_B = 2\pi f_B$ are, respectively, the amplitude and the frequencies of right and left spiral waves.

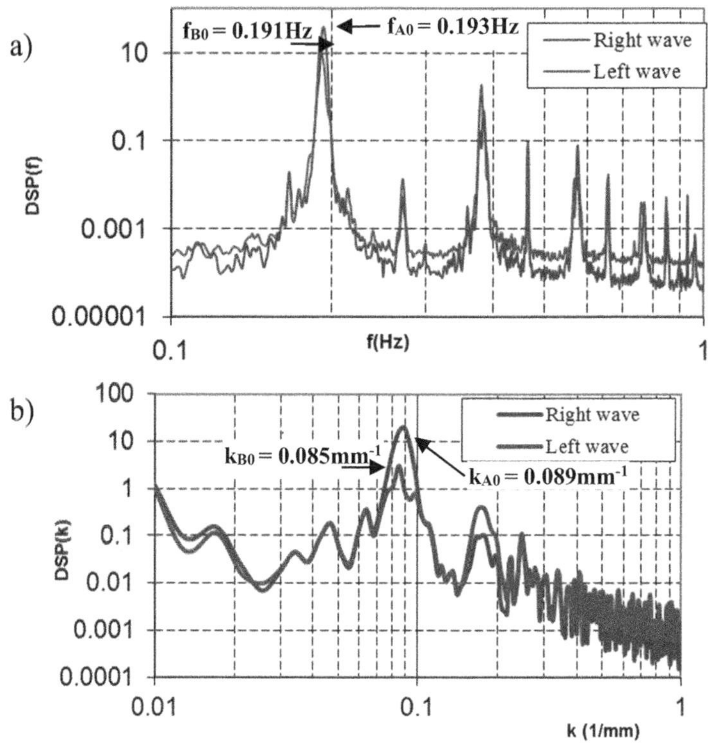

Figure 5.10. *Frequency power spectra (a) and wavenumber power spectra (b) of critical spiral patterns for a solution of 700 ppm (E = 0.046); Ta = 35.2; ε = 4.18 × 10^{-3}. For a color version of this figure, see www.iste.co.uk/saouab/mechanics.zip*

The space profile of the time-averaged amplitude (Figure 5.11) shows that the right half of the system is essentially occupied by the primary right wave in the presence of a low-amplitude reflected left wave. Conversely, the left half of the system is essentially occupied by the primary left wave in the presence of a low-amplitude reflected right wave. The distance between the maxima of the two amplitudes is quite significant ($\delta X/L$ = 0.78), which makes it possible for the primary right and left waves to propagate without interaction. This explains why there is no separation front (source or sinks) between the right and left primary waves. Thus, we have a pattern of two spatially decoupled counter-propagating waves. This pattern is known as wall modes and has been observed in several hydrodynamic systems (Tobias 1998; Lepiller et al. 2008).

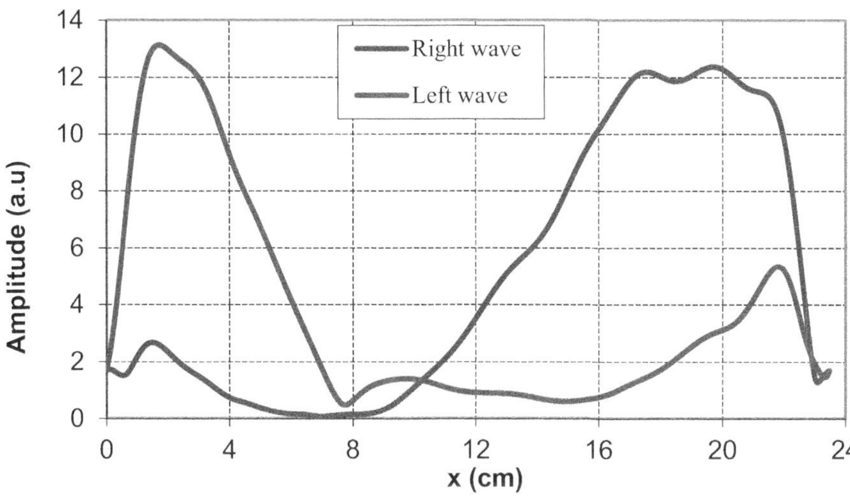

Figure 5.11. *Space profile of time-averaged amplitude in Figure 5.9 for a solution of 700 ppm (E = 0.046); Ta_c = 35.8; ε = 4.18 × 10^{-3}. For a color version of this figure, see www.iste.co.uk/saouab/mechanics.zip*

The space distribution of the time-averaged frequency (Figure 5.12(a)) shows that the frequency of the right part of the system $<f^*_{A0}>$ is slightly higher than that of left part of the system $<f^*_{B0}>$. In the right half, the right wave and the reflected left wave have identical frequencies $<f^*_{A0}>$. Similarly, in the right half of the system, the frequency of the left wave is identical to that of the reflected right wave (Figure 5.12(a)).

The spatial distribution of the time-averaged wavenumber (Figure 5.12(b)) shows that the wavenumber of primary or secondary waves has a quasi-parabolic distribution (Figure 5.12(b)). The areas with small wavenumbers correspond to areas of high amplitude for each mode, while the areas of high wavenumbers correspond to low amplitudes (Figures 5.12(a) and (b)). The minimal value of the wavenumber for the right and left spirals corresponds to its fundamental value $q_{A0} = q_{B0} = q_0 = 3.12$ (Figure 5.12). Introducing the perturbation $q'_i(x)$ of each wavenumber $q_i, i \in \{A, B\}$, the wavenumber can be written as follows:

$q_A(x) = q_0 + q'_A(x)$ for the right spiral wave

$q_B(x) = q_0 + q'_B(x)$ for the left spiral wave

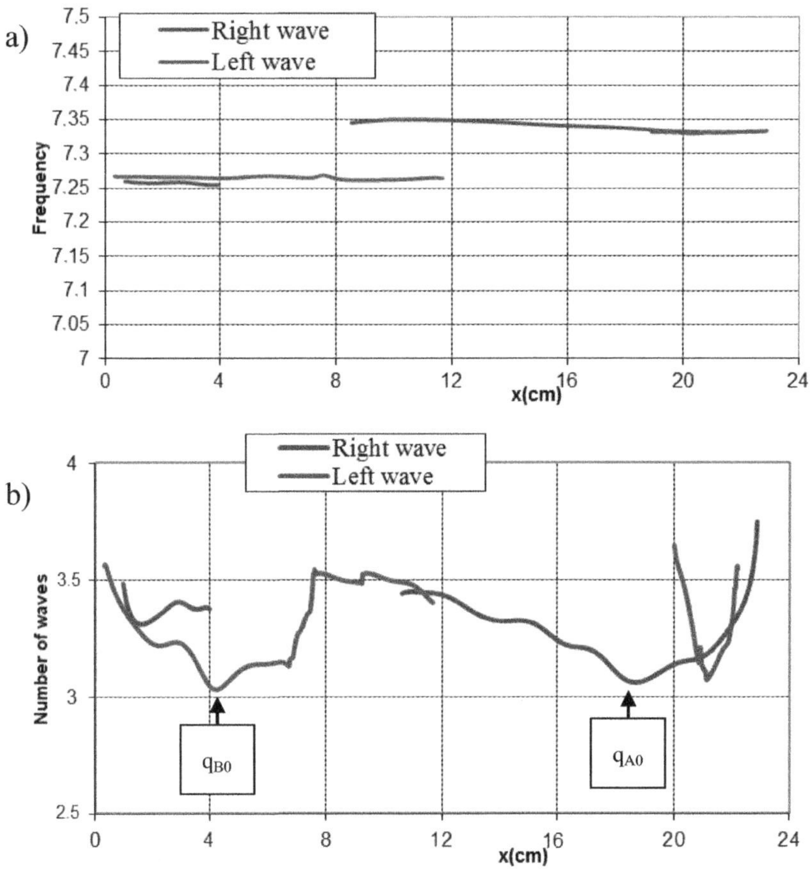

Figure 5.12. *Spatial profile of frequency and wavenumber of Figure 5.6 for a 700 ppm solution (E = 0.046); Ta = 35.2; ε = 4.18 × 10*$^{-3}$*. For a color version of this figure, see www.iste.co.uk/saouab/mechanics.zip*

The spatial distribution of the wavenumber is explained by the mechanism of the wavenumber selection by the slowdown of spiral amplitudes by the edges (Zaleski et al. 1985).

For a better understanding of the spiral amplitude slowdown, we need information on group velocity, which is responsible for energy transportation for a wave system (van Hecke et al. 1999; Ezersky et al. 2004). We have measured the group velocity following the method developed in the reference (Ezersky et al. 2004). The spatial variation of the group velocity can be used to locate the areas of wave amplitude slowdown (Figure 5.11). The linear group velocity corresponding to the linear variation of amplitude is $V_{g,A*} = 0.03$ (4.67×10^{-4} cm/s) for the right wave and $|V_{g,B*}| = 0.014$ (2.18×10^{-4} cm/s) for the left wave. The nonlinear group velocity corresponding to the high amplitude area is $V^{*}_{gNL,A} = 0.14$ (2.25×10^{-3} cm/s) for the right wave and $|V^{*}_{gNL,B}| = 0.18$ (2.8×10^{-3} cm/s) for the left wave. As shown in Figure 5.11, it can be noted that the group velocity of spirals is low in low amplitude areas (linear area near the ends of the system) and it is high in high-amplitude areas. This confirms that ends effects are responsible for the decrease in the group velocity near cylinder ends. Wave slowdown near the ends can be the result of the presence of Eckman layers (Zaleski et al. 1985). We observe that group velocities have the same signs as phase velocities ($V_g \times V_{ph} > 0$), therefore the spirals have normal dispersion.

5.3.5. *Discussion of wall modes using complex Ginzburg–Landau equations*

The spiral pattern resulting from the first supercritical Hopf bifurcation of the viscoelastic Couette flow with a POE concentration ranging from 500 to 700 ppm (Figure 5.9) is described by the coupled complex Ginzburg–Landau equations [5.2]:

$$\tau_0(\frac{\partial A}{\partial t} + V_g \frac{\partial A}{\partial x}) = \varepsilon A + \xi_0^2(1 + ic_1)\frac{\partial^2 A}{\partial x^2} - g(1 + ic_2)|A|^2 A + \delta(1 + ic_3)|B|^2 A \quad [5.2a]$$

$$\tau_0(\frac{\partial B}{\partial t} - V_g \frac{\partial B}{\partial x}) = \varepsilon B + \xi_0^2(1 + ic_1)\frac{\partial^2 B}{\partial x^2} - g(1 + ic_2)|B|^2 B + \delta(1 + ic_3)|A|^2 B \quad [5.2b]$$

where $A(t,x)$ and $B(t,x)$ represent the amplitudes of left and right spiral respectively, τ_0 is the characteristic relaxation time of disturbances in the system, V_g is the linear group velocity of spirals, $\varepsilon = (Ta - Ta_c)/Ta_c$ is the criticality, that is, the relative distance to threshold, ξ_0 is the coherence length of disturbances, c_l and $c_{2,3}$ are, respectively, the linear and nonlinear dispersion coefficients that are used for frequency corrections, g is Landau saturation constant and δ expresses the interaction of the two modes. The spirals observed in our experiment are supercritical and stable and, according to the nonlinear stability theory (Demay and Iooss 1985; Chossat and Iooss 1994), the saturation coefficient must be positive ($g > 0$). Moreover, in the vicinity of the threshold, there is no interaction between wall modes, which suggests that the coupling coefficient is negligible, compared to

g ($\delta << g$). Therefore, the wall modes in Figure 5.8 can be described by two equations in which we consider $\delta = 0$.

Let us assume that the basic states of right and left spirals are, respectively, disturbed by disturbances of the following form:

$$a(X,T) = |a|\, exp[i(\Omega_a T - q'_a X)] \qquad [5.3a]$$

$$b(X,T) = |b|\, exp[i(\Omega_b T + q'_b X)] \qquad [5.3b]$$

where X = x/L-1/2, and $T = t\nu/d^2$ represent the dimensionless variables of space and time.

Let us assume that the disturbances of spirals are slightly nonlinear waves. For this reason, the nonlinear term is temporarily neglected. The amplitudes of disturbances $a(X,T)$ and $b(X,T)$ are therefore described by the following equations:

$$\frac{\partial a}{\partial T} + V^*_{g,A}\frac{\partial a}{\partial X} = \mu_A a + \lambda_a \frac{\partial^2 a}{\partial X^2} \qquad [5.4a]$$

$$\frac{\partial b}{\partial T} - V^*_{g,B}\frac{\partial b}{\partial X} = \mu_B b + \lambda_B \frac{\partial^2 b}{\partial X^2} \qquad [5.4b]$$

with $\mu_{A,B} = \sigma_{0,A,B}\varepsilon = \frac{\varepsilon}{\tau^*_{0\,A,B}}$ and $= \sigma_{0,A,B}\xi^{*2}_{0,A,B}(1 + ic_{1,A,B})$ and $-0.5 \leq X \leq 0.5$.

Substituting the solutions of disturbances [5.3a] and [5.3b] in equations [5.4a] and [5.4b] yields the following dispersion equations:

$$\lambda_A {q'_A}^2 - iV^*_{g,A} q'_A + i\Omega_A - \mu_A = 0 \qquad [5.5a]$$

$$\lambda_B {q'_B}^2 - iV^*_{g,B} q'_B + i\Omega_B - \mu_B = 0 \qquad [5.5b]$$

The theory of absolute instability (Kuramoto 1984; Tagg et al. 1990, 1984) shows that the instability threshold is associated with the appearance of a double root of the dispersion relation [5.5] in the complex plane (the determinant of the system of equations [5.5] is zero). This makes it possible to search the wavenumber disturbances in the following complex form:

$$q'_A = q'^r_A + iq'^i_A \text{ et } q'_B = q'^r_B + iq'^i_B$$

The real parts q'^{r}_{A}, q'^{r}_{B} represent the wavenumber disturbances, while the imaginary parts q'^{i}_{A}, q'^{i}_{B} represent the spatial growth rate. Solving equations [5.5a] and [5.5b] yields:

$$q'_A = q'_B = \frac{c_1 V_g^* + i V_g^*}{2\sigma_0 \xi_0^{*2}(1 + c_1^2)} \qquad [5.6]$$

The separation of real and imaginary parts yields the following expressions:

$$q'^{r}_{A} = q'^{r}_{B} = \frac{c_1 \tau_0 V_g^*}{2\xi_0^{*2}(1 + c_1^2)} \quad [5.7a] \quad \text{and} \quad q'^{i}_{A} = q'^{i}_{B} = \frac{\tau_0^* V_g^*}{2\xi_0^{*2}(1 + c_1^2)} \qquad [5.7b]$$

$$\mu_A = \frac{\tau_0^* V_g^{*2}}{4\xi_0^{*2}(1 + c_1^2)} \quad [5.7c] \quad \text{and} \quad \Omega_{a,b} = \frac{\tau_0^* c_1 V_{g,A,B}^{*2}}{4\xi_0^{*2}(1 + c_1^2)} \qquad [5.7d]$$

where it was admitted that $\sigma_{0,A} = \sigma_{0,B} \equiv \sigma_0$ and $\xi_{0,A} = \xi_{0,B} \equiv \xi_0$.

The wavenumber disturbances of the two spirals are extracted in the linear zones of the wavenumber variation as a function of space coordinate X (Figure 5.11): $q'^{r}_{A} \approx q'^{r}_{B} \approx 0.45$. In the presence of reflective endplates, the boundary conditions on the amplitudes of disturbances in $X = -1/2$ and $X = 1/2$ are obtained from the amplitude profile represented in Figure 5.8, where $|A|(X = 1/2) \approx |B|(X = -1/2) \approx 2$.

With these conditions on edges, the disturbances of the two right and left spirals take the following forms:

$$a(X,T) = |a| \exp i\,(\Omega_A T - q'^{r}_{A} X) \qquad [5.8a]$$

$$b(X,T) = |b| \exp i\,(\Omega_B T + q'^{r}_{B} X) \qquad [5.8b]$$

where the amplitudes of disturbances of right and left spirals are given by:

$$|a|(X) = a_0 \cos(\pi X - \varphi_a) \exp(q'^{i}_{A}.X) \qquad [5.9a]$$

$$|b|(X) = b_0 \cos(\pi X + \varphi_b) \exp(q'^{i}_{B}.X) \qquad [5.9b]$$

where φ_a and φ_b are phases at the origin.

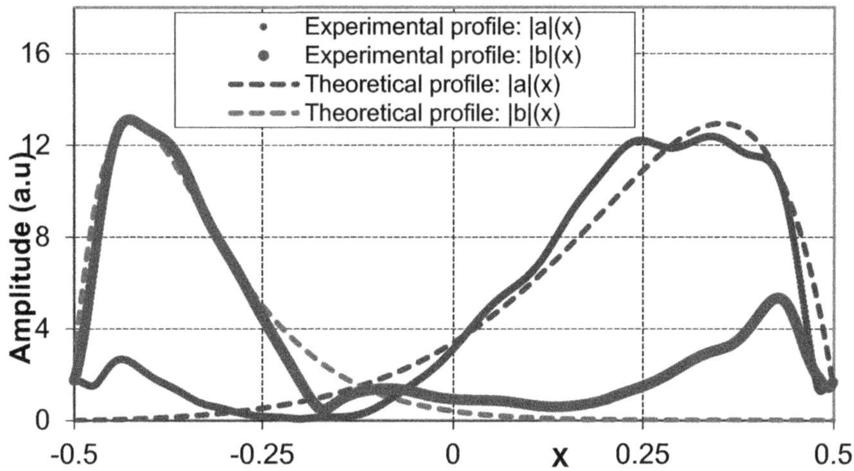

Figure 5.13. *Spatial profile of experimental and theoretical amplitudes of right and left spirals of Figure 3.6 for a 700 ppm solution (E = 0.046); Ta = 35.8; ε = 4.18 × 10*$^{-3}$*. For a color version of this figure, see www.iste.co.uk/saouab/mechanics.zip*

The fitting of the spatial distribution of amplitudes given by equations [5.9] with those experimentally obtained (Figure 5.13) is represented in Figure 5.14 and it leads to the coefficients given in Table 5.2.

E	a_0	b_0	$\varphi_a = \varphi_b$	$\tilde{q}_A^i$	$\tilde{q}_B^i$
0.046	3.70 ± 0.05	0.40 ± 0.03	0.02	5.97 ± 0.04	11.4 ± 0.2

Table 5.2. *Measured values of coefficients of equations [5.9a] and [5.9b]*

Relations [5.7a] and [5.7b] are linked by the following formula:

$$c_{1,A,B}\tilde{q}^r_{A,B} = \tilde{q}^i_{A,B} \quad [5.9c]$$

Using the numerical values found by the fitting of the spatial distribution of amplitudes with formula [5.9c], we deduce the numerical value of coefficient c_1 for each wave: $c_{1,A} = 7.54 \times 10^{-3}$ for the right spiral and $c_{1,B} = 3.94 \times 10^{-3}$ for the left spiral.

In the zone of exponential growth (right wave) or spatial decrease (left wave), the amplitudes of right and left spirals can be described by the following equations:

$$V^*_{g,A}\frac{\partial |a|}{\partial X} = \mu|a| \qquad [5.10a] \quad \text{and} \quad -V^*_{g,B}\frac{\partial |b|}{\partial X} = \mu|b| \qquad [5.10b]$$

which have the following solutions:

$$|a| = A_1 \exp(\frac{\mu_A}{V^*_{g,A}}X) \qquad [5.11a] \quad \text{and} \quad |b| = B_1 \exp(-\frac{\mu_B}{V^*_{g,B}}X) \qquad [5.11b]$$

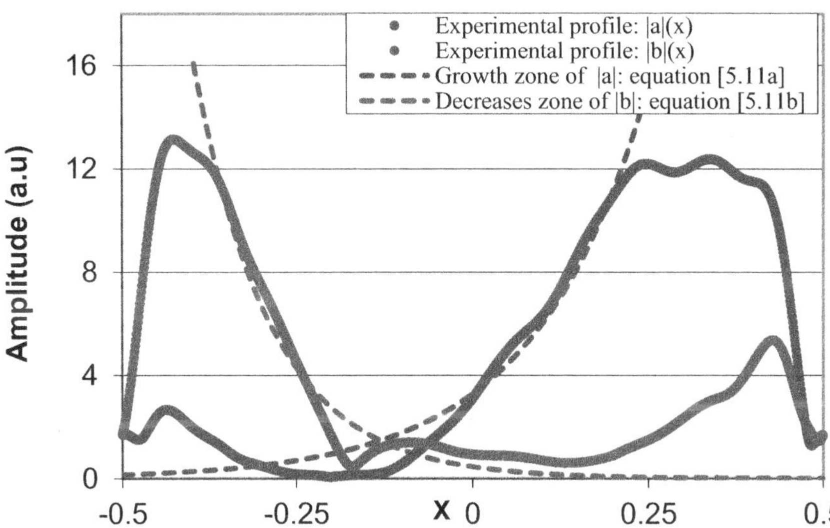

Figure 5.14. *Spatial profile of experimental and theoretical amplitudes in the growth zone (right wave: equation [5.11a]) and decay zone (left wave: equation [5.11b]) of right and left spirals for a 700 ppm solution (E = 0.046), Ta = 35, ε = 4.18 × 10^{-3}. For a color version of this figure, see www.iste.co.uk/saouab/mechanics.zip*

The fitting of the spatial distribution of theoretically obtained amplitudes (equations [5.11]) with those experimentally obtained in the growth and decay zones represented in Figure 5.14 yields the following coefficients:

$$A_1 = 3.28 \pm 0.04\,;\ B_1 = 0.46 \pm 0.02 \text{ and}$$

$$\frac{\mu_A}{V^*_{g,A}} = \frac{\varepsilon}{\tau^*_{0,B}V^*_{g,A}} = 5.12 \pm 0.07\,;\ \frac{\mu_B}{V^*_{g,B}} = \frac{\varepsilon}{\tau^*_{0,B}V^*_{g,B}} = 8.94 \pm 0.02.$$

Using the average values found for the linear group velocity, we deduce the characteristic time values for each wave: $\tau^*_{0,A} = 2.23 \times 10^{-2}$ for the right spiral and $\tau^*_{0,B} = 3.94 \times 10^{-2}$ for the left spiral.

Using relation [5.7b], we deduce the numerical value of the coherence length of disturbances for each wave: $\xi_{0,A} = 1.84 \times 10^{-2}$ for the right spiral and $\xi_{0,B} = 1.53 \times 10^{-2}$ for the left spiral. For the maximum values of the two right and left amplitudes, we have $\frac{\partial A}{\partial X} = \frac{\partial^2 A}{\partial X^2} = 0$ for the right wave and $\frac{\partial B}{\partial X} = \frac{\partial^2 B}{\partial X^2} = 0$ for the left wave. The amplitudes of disturbances of the right and left spirals can thus be written in the following form:

$$(\varepsilon - g_A |A_{max}|^2)a = 0 \qquad [5.12a]$$

$$(\varepsilon - g_B |B_{max}|^2)b = 0 \qquad [5.12b]$$

We can therefore deduce the experimental value of the Landau coefficient for the two spirals:

$$g_A = \frac{\varepsilon}{|A_{max}|^2} = 2.52 \times 10^{-5} \qquad [5.13a]$$

$$g_B = \frac{\varepsilon}{|B_{max}|^2} = 2.44 \times 10^{-5} \qquad [5.13b]$$

The results of the measurements of Ginzburg–Landau coefficients of equation [5.5] are summarized in Table 5.3. The resulting difference between the numerical values of the left and right wave is the signature of the dissymmetry between these two spiral waves above the threshold of this instability.

	V^*_g	$V^*_{g,NL}$	τ^*_0	ξ^*_0	c_1	g
Right spiral	7.54×10^{-4}	3.52×10^{-3}	1.1	8.33×10^{-3}	7.54×10^{-2}	2.52×10^{-5}
Left spiral	3.52×10^{-4}	4.53×10^{-3}	0.75	3.39×10^{-3}	3.94×10^{-2}	2.44×10^{-5}

Table 5.3. *Measured values of Ginzburg–Landau coefficients for a 700 ppm solution (E = 0.046); Ta_c = 35.2; $\varepsilon = 4.18 \times 10^{-3}$*

5.3.6. *Evolution of the size occupied by spirals as a function of Ta*

We have noted that, for low values of the control parameter, the pattern of spirals does not appear along the whole length of the cylinders (Figure 5.15). The extent l of the pattern of spirals has been measured as a function of the control parameter

Ta: it grows as a linear function of the criticality ε before invading the entire system (Figure 5.16):

$$\frac{l(Ta)}{L} = \begin{cases} 9.1\varepsilon + 0.45, & 0 < \varepsilon < 0.06 \\ 1 & 0.06 < \varepsilon \end{cases}$$

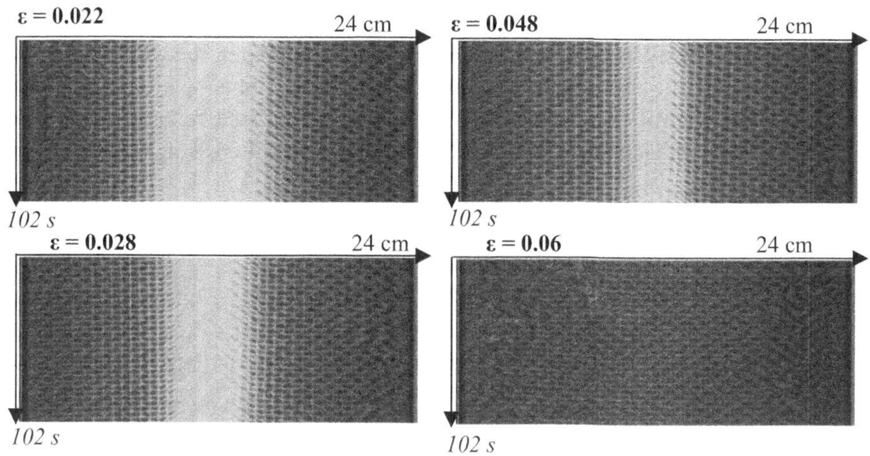

Figure 5.15. *Space-time diagrams of spirals in a viscoelastic solution (E = 0.046) for various values of the criticality ε*

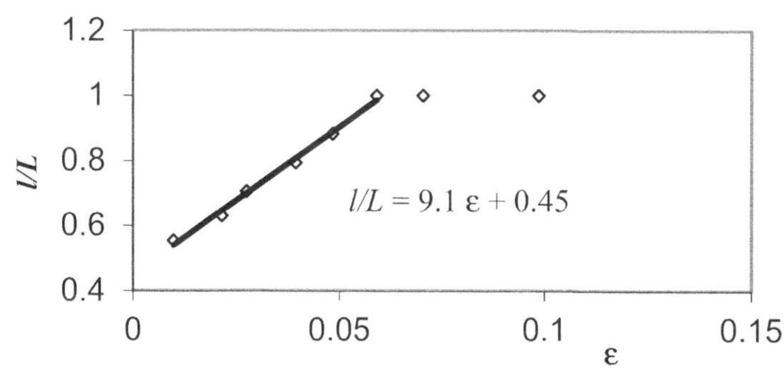

Figure 5.16. *Evolution of the size of spirals as a function of the criticality ε*

5.4. Conclusion

We have investigated the scenarios of the transition to turbulence of the viscoelastic Couette–Taylor flow with PEO solutions with concentrations ranging between 500 and 700 ppm. These concentrations yield moderate values of elasticity ($0.01 \leq E \leq 0.50$). In our experiments, the outer cylinder was fixed, and only the inner cylinder was rotating. With the increase in the rotation velocity of the inner cylinder, the stationary viscoelastic Couette flow bifurcates to inertio-elastic spirals in form of wall modes. The size of these wall modes increases with the control parameter before invading the entire system. The wall modes are described by the complex Ginzburg–Landau equations (without coupling term) whose coefficients have been measured. With the increase of the control parameter, the wall modes gain energy by increasing the control parameter and form ribbon pattern. A further increase in the control parameter leads to turbulent flow states classified in three different regimes: defects turbulence spatiotemporal intermittence and inertio-elastic turbulence.

5.5. References

Bai, Y., Crumeyrolle, O., Mutabazi, I. (2015). Viscoelastic Taylor-Couette instability as analog of the magnetorotational instability. *Phys. Rev. E*, 92, 031001(R).

Bai, Y., Vieu, T., Crumeyrolle, O., Mutabazi, I. (2021). Viscoelastic Taylor-Couette instability in the Keplerian regime. *Geophys. Astrophys. Fluid Dyn.*, 115(3), 322344.

Bai, Y., Latrache, N., Kelai, F., Crumeyrolle, O., Mutabazi, I. (2023). Viscoelastic instabilities of Taylor-Couette flows with different rotation regimes. *Phil. Trans. R. Soc. A*, 223, 289–343.

Baumert, B. and Muller, S. (1995). Flow visualization of the elastic Taylor-Couette in stability in Boger fluids. *Rheol. Acta*, 34, 147–159.

Baumert, B. and Muller, S. (1997). Flow regimes in model viscoelastic fluids in a circular Couette system with independently rotating cylinders. *Phys. Fluids*, 9, 566–586.

Baumert, B. and Muller, S. (1999). Axisymmetric and non-axisymmetric elastic and inertio-elastic instabilities in Taylor–Couette flow. *J. Non-Newton. Fluid Mech.*, 83, 33–69.

Bird, R., Armstrong, R., Hassaguer, O. (1987). *Dynamics of Polymer Liquids*, volume 1. John Wiley & Sons, New York.

Boger, D.V. (1977). A highly elastic constant-viscosity fluid. *J. Non-Newtonian Flui. Mech.*, 3(1), 87.

Boulafentis, T., Lacassagne, T., Cagney, N., Balabani, S. (2023). Experimental insights into elasto-inertial transitions in Taylor-Couette flows. *Phil. Trans. R. Soc.*, A381, 20220131.

Brochard, F. and de Gennes, P.G. (1977). Dynamical scaling for polymers in theta solvents. *Macromolecules*, 10, 1157.

Chossat, P. and Iooss, G. (1994). *The Couette-Taylor Problem*. Springer Verlag, New York.

Crumeyrolle, O. (2002). Étude expérimentale des liquides viscoélastiques dans le système de Couette-Taylor. PhD Thesis, Université du Havre, Le Havre.

Crumeyrolle, O., Mutabazi, I., Grisel, M. (2002). Experimental study of inertioelastic Couette-Taylor instability modes in dilute and semidilute polymer solutions. *Phys. Fluids*, 14, 1681–1688.

Crumeyrolle, O., Latrache, N., Ezersky, A., Mutabazi, I. (2003). Instability modes observed in a viscoelastic Couette–Taylor flow. *Mech. Ind.*, 4, 397–409.

Crumeyrolle, O., Latrache, N., Ezersky, A., Mutabazi, I. (2005). Instabilities with shear-thinning polymer solutions in the Couette-Taylor system. *J. Phys. Conf. Ser.*, 14, 78–93.

Demay, Y. and Iooss, G. (1985). Calcul des solutions bifurquées pour le problème de Couette – Taylor avec les deux cylindres en rotation. *J. Méca. Théo. App.*, 193–216.

Denn, M. and Roisman, J. (1969). Rotational stability and measurement of normal stress functions in dilute polymer solutions. *AIChE J.*, 15, 454.

Dutcher, C.S. and Muller, S.J. (2011). Effects of weak elasticity on the stability of high Reynolds number co- and counter-rotating Taylor-Couette flows. *J. Rheo.*, 55, 1271.

de Gennes, P.G. (1984). *Scaling Concepts in Polymer Physics*. Cornell University Press, New York.

Groisman, A. and Steinberg, V. (1996a). Couette-Taylor flow in a dilute polymer solution. *Phys. Rev. Lett.*, 77, 1480.

Groisman, A. and Steinberg, V. (1996b). Mechanism of elastic instability in Couette flow of polymer solutions: Experiment. *Phys. Fluids*, 10, 2451–2463.

Gyr, A. and Bewersdorff, H.W. (1995). *Drag Reduction of Turbulent Flows by Additives*. Kluwer Academic Publishers, Dordrecht.

Kelai, F. (2010). Étude expérimentale des instabilités viscoélastiques dans le système de Couette-Taylor. PhD Thesis, Université du Havre, Le Havre.

Kuramoto, Y. (1984). *Chemical Oscillations, Waves and Turbulence*. Springer-Verlag, Berlin.

Lacassagne, T., Cagney, N., Gillissen, J.J.J., Balabani, S. (2020). A route to elasto-inertial turbulence in Taylor-Couette flow. *Phys. Rev. Fluids*, 5, 113303.

Larson, R.G. (1992). Instabilities in viscoelastic flows. *Rheol. Acta*, 31, 213.

Larson, R.G., Shaqfeh, E.S.G., Muller, S.J. (1990). A purely elastic instability in Taylor–Couette flow. *J. Fluid Mech.*, 218, 573.

Latrache, N. and Mutabazi, I. (2021). Transition to turbulence via flame patterns in viscoelastic Taylor-Couette flow. *Eur. Phys. J. E.*, 44, 63.

Latrache, N., Crumeyrolle, O., Abcha, N., Mutabazi, I. (2008). Destabilization of inertio-elastic mode via spatiotemporal intermittency in a Couette-Taylor viscoelastic flow. *J. Phys. Conf. Ser.*, 137, 012022.

Latrache, N., Crumeyrolle, O., Mutabazi, I. (2012). Transition to turbulence in a flow of a shear-thinning viscoelastic solution in a Taylor-Couette cell. *Phys. Rev. E*, 86, 056305.

Latrache, N., Abcha, N., Crumeyrolle, O., Mutabazi, I. (2016). Defect-mediated turbulence in ribbons of viscoelastic Taylor-Couette flow. *Phys. Rev. E*, 93, 043126.

Lepiller, V., Goharzadeh, A., Prigent, A., Mutabazi, I. (2008). Weak temperature gradient effect on the stability of the circular Couette flow. *Eur. Phys. J. B.*, 61, 445–455.

Moazzen, M., Lacassagne, T., Thomy, V., Amir Bahrani, S. (2023). Friction dynamics of elasto-inertial turbulence in Taylor-Couette flow of viscoelastic fluids. *Phil. Trans. R. Soc.*, A381, 20220300.

Tagg, R., Edwards, W.S., Swinney, H.L., Marcus, P.S. (1989). Nonlinear standing waves in Couette-Taylor flow. *Phys. Rev. A*, 39, 3734.

Tagg, R., Edwards, W.S., Swinney, H.L. (1990). Convective versus absolute instability in flow between conterrrotating cylinders. *Phys. Rev. A*, 42, 831.

Tobias, S.M., Proctor, M.R.E., Knobloch, E. (1998). Convective and absolute instabilities of fluid flows in finite geometry. *Physica D*, 113, 43.

Van Hecke, M., Storm, C., van Saarloos, W. (1999). Sources, sinks and wavenumber selection in coupled CGL equations and experimental implications for counter-propagating wave systems. *Physica D*, 134, 1.

Zaleski, S., Tabeling, P., Lallemand, P. (1985). Flow structures and wave-number selection in spiraling vortex flows. *Phys. Rev. A*, 32(1), 655–658.

6

Application of the WKB Method to the Study of Flow Stability Generated by Low-frequency Modulation of Boundary Conditions

This chapter focuses on the study of two configurations of low-frequency phase modulation of the boundary conditions. The first one concerns a Newtonian fluid layer confined between two horizontal walls to which sinusoidal temperatures of the same amplitude and frequency are imposed. The second one concerns a fluid layer confined in a Taylor–Couette system where the outer cylinder and the inner cylinder are subjected to sinusoidal rotations.

For each configuration, we conduct a linear stability analysis of the oscillating basic solution, and we determine the asymptotic law relative to the critical control parameter for low frequencies. These asymptotic laws are compared to the numerical results obtained by complete numerical resolution of the conservation equations using the Chebyshev spectral collocation methods and the Floquet theory. For low frequencies within ranges where numerical results can be obtained, this comparison shows strong agreement.

We finally conclude that the asymptotic Wentzel–Kramers–Brillouin (WKB) method can overcome the difficulty of the numerical calculation to provide solutions for the problems of instabilities in the presence of low-frequency modulation.

Chapter written by Saïd ANISS and Mouh ASSOUL.

6.1. Introduction

6.1.1. *Instabilities of unsteady base states*

The study of the stability of time-dependent equilibrium positions leads to specific mechanisms of instability. The best known among these mechanisms is that of parametric resonance. The example of the simple pendulum, whose point of suspension is subject to periodical and vertical motion, is a good illustration of this type of instability (Stoker 1950). The equation of motion of such a system, linearized around the equilibrium positions, is reduced to the linear parametric oscillator corresponding to Mathieu's equation with a damping term. The Floquet theory applied to this oscillator indicates that, on the one hand, the pendulum's oscillation frequency can be similar to that of the suspension point (harmonic or synchronous solutions) or half of this frequency (sub-harmonic solutions). Pendulum oscillations generally depend on the amplitude and frequency of the motion of the point of suspension. On the other hand, the Floquet theory can also be used to prove that the pendulum position of stable equilibrium can be destabilized, while the unstable one can be stabilized. Based on this example, an original idea emerged, consisting of the use of this type of parametric instability in hydrodynamics to control flow instabilities through modulation frequency and amplitude. This requires the introduction of a time-periodic parameter into the system of equations governing the motion of the system under study. This parameter modulates in amplitude and frequency the weak oscillations of the basic solution. The study of the instabilities of time-periodic flows was the object of several works including the Stokes layer (Kerczek and Davis 1972, 1974), the plane Poiseuille flow with a time-periodic pressure gradient (Grosch and Salwen 1968) and the Couette flow with oscillating boundaries (Kelly and Cheers 1970). For further details on this subject, the reader can refer to the synthesis article by Davis (1976) or to the book by Drazin and Reid (1981). The following section presents several essential results on other instabilities of periodic flows, which are also covered by a rich literature.

In hydrodynamics, parametric instabilities or resonances occur when the entire system oscillates, thus generating driving forces that are superimposed to gravitational forces (gravitational modulation) or, on the contrary, when oscillations are imposed on the boundary conditions (modulation of boundary conditions).

6.1.2. *Gravitational modulation*

6.1.2.1. *Gravitational modulation of Rayleigh–Bénard convection*

This first type of modulation is often used in thermal convection. In Rayleigh–Bénard convection, gravitational modulation is achieved by putting into vertical oscillation a horizontal fluid layer submitted to a vertical temperature gradient corresponding to top or bottom heating. This modulation was initiated by Gresho and Sani (1970) for periodic oscillations. A summary on gravitational modulation of thermal convection was written by Gershuni and Zhukhovitskii (1976) and Gershuni and Lyubimov (1998). The configuration studied by Gresho and Sani (1970) was examined in Hele-Shaw geometry by Aniss et al. (2000a) who showed that parametric resonance can only occur for small values of the Prandtl number. The case of non-periodic oscillation was examined by Boulal et al. (2007, 2008) considering that the oscillation involves two immeasurable frequencies. As shown by all of these works, modulation may have a stabilizing or destabilizing effect, and the convection threshold may correspond to harmonic, sub-harmonic or quasi-periodic solutions. It is important to note that volume forces can be modulated in a magnetic liquid in the presence of an oscillating magnetic field (Aniss et al. 2000b, 2001).

6.1.2.2. *Gravitational modulation of a flow on inclined plane*

Gravitational modulation was also used for free surface fluid flows on inclined walls. Oscillations are considered to have normal and lateral components (Woods and Lin 1995; Lin et al. 1996; Garih et al. 2013; El Jaouahiry 2020). In this case, the amplitude of oscillation normal to the inclined plane stabilizes the gravitational instability, which occurs at low wave numbers. This stabilizing effect is reinforced by the introduction of lateral oscillation, which contributes to the complete suppression of gravitational instability. The harmonic and sub-harmonic zones, initially stable in the absence of modulation, are destabilized by lateral oscillation.

6.1.3. *Modulation of boundary conditions*

The second type of modulation is used in several flow configurations: Taylor–Couette flow with oscillation of inner and outer cylinders, thermal convection by imposing time-periodic boundary conditions on temperature and also problems of flow stability of a fluid layer adjacent to oscillating wall.

6.1.4. *Modulation of boundary conditions in Taylor–Couette geometry*

Modulation in centrifugal flows was introduced in the case of incompressible viscous fluid confined between two coaxial cylinders (Taylor–Couette flow). The outer cylinder is considered to be fixed, while the inner one has an angular rotation velocity with a steady component and another one varying sinusoidally in time. This modulation produces a basic flow that is time and space dependent. This problem was experimentally initiated by Donnelly (1964) who concluded that the modulation of the rotation velocity of the inner cylinder stabilizes the basic flow, which means it delays the instability threshold. The theoretical analysis of linear and nonlinear stability relative to the experimental problem of Donnelly (1964) was conducted by

Hall (1975) for the case when the gap, that is, the distance between the two cylinders, is very small. Hall (1975) showed that the instability threshold is slightly below that of the non-modulated case, when modulation amplitude and frequency tend to zero. This destabilization effect, introduced by modulation, was confirmed by a linear stability analysis, based on Floquet theory, conducted by Riley and Laurence (1976). It is worth mentioning that recently, a satisfactory agreement between theory and experiment was discovered by Gassa et al. (2014) when they reexamined the Donnelly configuration. Based on a spectral analysis, these authors proved the existence of synchronization between pattern and excitation at high modulation frequencies, contrary to low modulation frequencies where this synchronization is more significant. The case in which the gap between the two coaxial cylinders is not small was treated by Carmi and Tustaniwskyj (1981). The latter showed there is a very significant alteration of the instability threshold of the basic flow for low modulation frequencies. This result is in strong agreement with the experimental results obtained by Walsh et al. (1988). The importance of the destabilization effect for low frequencies was often discussed in these works. An answer to this question was given by Kuhlmann et al. (1989) who showed, through a numerical simulation using finite differences, that the destabilization effect is very strong for low frequencies.

The particular case, in which the average rotation velocity imposed to the two cylinders is zero, was treated by Riley and Laurence (1976) and by Carmi and Tustaniwskyj (1981). Later, Aouidef et al. (1994) conducted a theoretical and experimental study of the pulsed flow stability when the two inner and outer cylinders oscillate at the same amplitude and angular velocity (phase modulation). On the one hand, they showed that the flow is described by two Stokes layers along the walls and a central core at rest for high frequencies. On the other hand, for low frequencies, the entire fluid undergoes solid body rotation. A criterion of instability was determined for these two limit cases. It was also shown that the destabilization effect for low and high frequencies is less significant than for a certain intermediate frequency. The configuration (Aouidef et al. 1994) was also studied for a fluid following a linear Maxwell model (Riahi et al. 2014). This work highlighted that elasticity has no effect for low frequencies, but it has a destabilizing effect for high frequencies. The same conclusion is reached when the inner cylinder is modulated, while the outer cylinder is fixed (Riahi et al. 2015) and also for out of phase modulation of the two cylinders (Riahi et al. 2016). Recently, Riahi et al. (2019) took an interest in reverse and non-reverse flows in the case of in-phase modulation, but with different amplitudes. The authors showed that an increase in modulation amplitude leads to the emergence of families of reverse and non-reverse flows and of other interesting bifurcation phenomena, including co-dimension two bifurcation points. For out-of-phase modulation (Riahi et al. 2022), the authors showed that there is a suppression of reverse flows in certain frequency ranges, and this type of Taylor vortex flow can only occur within the high frequency range.

6.1.5. *Modulation of boundary conditions relative to temperature*

The critical threshold corresponding to the onset of Rayleigh–Bénard convection motion can be controlled by a modulation of temperatures imposed on the horizontal surfaces of the fluid layer. This is achieved by adding time-dependent components to the usual temperature boundary conditions. Venezian (1969) examined the case of the temperature gradient imposed to the fluid layer that has a steady component and a sine-fluctuating one with low amplitude. Using a perturbation method and conducting a linear stability analysis limited to two-dimensional perturbations and to the free-free case, Venezian (1969) showed that temperature modulation at the boundary can have a stabilizing or destabilizing effect. A similar problem had been previously studied by Gershuni and Zhukhovitski (1963) for temperature fluctuations following a periodical rectangular law. Then, the same problem was approached by Rosenblat and Herbert (1970) by considering a temperature modulation at low frequencies and at arbitrary amplitudes. In the same context, Rosenblat and Tanaka (1971) showed that modulation has a stabilizing effect, particularly for low frequencies.

After having used the Galerkin method, they obtained an asymptotic solution using the WKB perturbation method. Yih and Li (1972) considered that modulation amplitude and frequency are arbitrary and they used the Galerkin method to conduct a linear stability analysis. They showed that the critical convection threshold may correspond to harmonic or sub-harmonic solutions. Consequently, Finucane and Kelly (1976) conducted an experimental study and also a theoretical analysis taking into account nonlinear terms, and they showed that low-frequency modulation is destabilizing. Dowden (1981) considered also the case of low frequencies, but with a base temperature depending only on time. With a linear stability theory, he showed that modulation at high amplitudes has a very significant destabilizing effect. A synthesis of the study of base solutions, depending only on time, is provided by Gershuni and Zhukhovitskii (1976). Perturbations that are three-dimensional with respect to the base solution were subjected to a weakly nonlinear analysis by Roppo et al. (1984). Using an expansion of velocity, pressure and temperature in a series of powers of a parameter related to the amplitude of a convective motion, they established the stability curves related to the frequency of temperature modulation, for various values of the Prandtl number. Without being exhaustive, other studies focused on the modulation of boundary conditions related to temperature, for example by considering a fluid following the linear Maxwell model (Oukada et al. 2006), a binary viscoelastic fluid (Bhadauria and Kiran 2014), an electrically conductive fluid (Bhadauria 2008), also by considering specific geometries such as the Hele-Shaw cell (Aniss et al. 2005) or a porous medium (Bhadauria 2008).

6.1.6. *Modulation of boundary conditions related to the heat flux imposed in Bénard–Marangoni convection*

Another type of modulation of the temperature boundary conditions was approached, in Bénard–Marangoni convection, by Gershuni et al. (1992, 1994, 1996). In these studies, the free surface of the fluid layer is considered to be excited by a time-periodic heat flux. As surface tension depends on the temperature, thermocapillary instabilities of parametric type are developed at the free surface. Gershuni et al. (1992, 1994, 1996) showed that, for very high Prandtl numbers, only harmonic modes exist at the threshold of instability, while for very low Prandtl numbers, there is competition between harmonic and sub-harmonic modes. In their analysis, the authors took into account the thickness of the fluid layer, the thermal conditions at the free surface (Newton type) and also the superposition of gravitational forces with those resulting from surface tension gradients. Moreover, it is important to note that Gershuni et al. (1996) took into consideration the surface deformation and showed that there are two types of perturbation that can develop an instability: perturbations related to convection and to the free surface. For certain values of the Galileo number, surface instability is more influential than the convective one.

6.1.7. *Objectives of the chapter – modulation of boundary conditions at low frequency and WKB method*

In most modulation problems presented in the previous sections, when numerical methods are used to study the effect of oscillation frequency when it tends to zero, we expect a singular problem with boundary layer. Indeed, most direct numerical methods yield poorly conditioned solutions and marginal stability results are obtained up to a certain limit value of the frequency, not sufficiently low, where numerical procedures stop.

A method that is very often used and that allowed approaching the instability threshold in the low-frequency regime is the WKB method (Landau and Lifshitz 1981; Nayfeh 1981). Indeed, the WKB theory can be used to determine the overall behavior of the solution of a linear differential equation for which the highest derivative is associated with a small parameter (William 2014). In this chapter, this small parameter is the modulation frequency. This asymptotic approach was used in many fields of research. We can mention, for example, the work of Or (1997) for modulated Rayleigh–Bénard convection, and also the work of Smorodin and Velarde (2000, 2001, 2008).

Two case studies will be presented in this chapter. The first case concerns parametric thermal instability in a fluid layer where phase-modulated temperatures are imposed on the top and bottom horizontal walls. In the second case, we will examine the flow stability resulting from a phase modulation of the inner and outer cylinders of a Taylor–Couette system. While reexamining these two problems, we will show that the use of WKB method leads us to asymptotic laws that are entirely different from those obtained in the literature.

6.2. Application of WKB method for studying the effect of low-frequency phase modulation of temperature on the convection threshold of a fluid layer

6.2.1. *Thermal configuration*

Let us consider a horizontal Newtonian layer of an incompressible fluid confined between two horizontal walls of infinite extent in the x^* and y^*directions (see Figure 6.1). The horizontal boundaries of the fluid layer are at $z^* = 0, \mathrm{d}$. The layer is isothermal at a temperature T_m in the absence of modulation. The modulated temperature imposed on horizontal walls is given in the following dimensional form: $T_m + T_s \cos(\omega^* t^*)$.

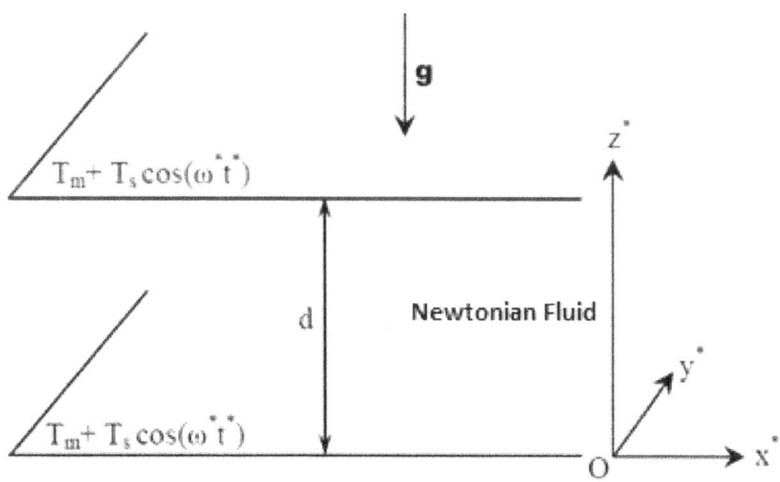

Figure 6.1. *Fluid layer confined between two horizontal walls subjected to in-phase modulation of temperature*

This in-phase modulation configuration was studied by Souhar and Aniss (2016) who first dealt with solving the system of linear equations corresponding to the onset of convection using the Floquet theory and a technique allowing the transformation of an eigenvalue problem into a problem with initial conditions. Then, Souhar and Aniss (2016) determined an asymptotic law of the convection threshold in terms of the critical Rayleigh number and the critical wave number for low frequencies, using considerations based on scale analysis in quasi-steady state. This method was used previously by Aouidef et al. (1994) and later on by Souhar and Aniss (2016). Our objective is to review the asymptotic analysis for low frequencies, presented in Souhar and Aniss (2016), using this time, for the time resolution, the WKB method combined with the numerical one, known as Chebyshev spectral collocation method, for space resolution.

6.2.2. *Formulation of the problem of base solution stability*

The conservation equations are those of continuity, conservation of momentum and conservation of energy:

$$\nabla \cdot \mathbf{V}^* = 0 \qquad [6.1]$$

$$\rho_e \frac{\partial \mathbf{V}^*}{\partial t^*} = -\nabla p^* + \mu \Delta \mathbf{V}^* - \rho g \mathbf{z}^* \qquad [6.2]$$

$$\frac{\partial T^*}{\partial t} + \mathbf{V}^* \cdot \nabla T^* = \kappa \Delta T^* \qquad [6.3]$$

where μ and κ denote, respectively, the dynamic viscosity and the thermal diffusivity of the fluid. The variables $\{z^*, t^*, T^*\}$ are rendered dimensionless by $\{d, d^2/\kappa, T_s\}$. The base state corresponds to a rest state in which the dimensionless temperature, $T_b = \frac{T^*-T_m}{T_s}$, satisfies the diffusion equation:

$$\frac{\partial T_b}{\partial t} = \frac{\partial^2 T_b}{\partial z^2} \text{ and } T_b = \cos(\omega t) \text{ in } z = 0,1 \qquad [6.4]$$

where $\omega = \omega^* d^2/\kappa$ is a dimensionless frequency. The solution to equation [6.4], on longer time scales, is the following:

$$T_b = T_1(z)\cos(\omega t) + T_2(z)\sin(\omega t) \qquad [6.5]$$

with

$$T_1(z) = \frac{\cos(\sigma(1-z))\cosh(\sigma z) + \cosh(\sigma(1-z))\cos(\sigma z)}{\cosh(\sigma) + \cos(\sigma)}$$

$$T_2(z) = \frac{\sin(\sigma(1-z))\sinh(\sigma z) + \sinh(\sigma(1-z))\sin(\sigma z)}{\cosh(\sigma) + \cos(\sigma)}$$

where $\sigma = \sqrt{\omega/2}$ is a frequency parameter. It should be noted here that the basic temperature solution is similar to the basic pulsed flow in a Newtonian fluid confined between two coaxial cylinders oscillating in phase with the same frequency about their axis (Aouidef et al. 1994). Moreover, the basic temperature gradient has no steady component that generally triggers the onset of Rayleigh–Bénard convection.

During the formulation of the problem of linear stability of the base solution, the following dimensionless quantities are introduced:

$$(x, y, z) = (x^*/d,\ y^*/d, z^*/d), t = t^*\kappa/d^2, T = T'/T_s,$$

$$\mathbf{V}(u, v, w) = \mathbf{V}^*(du^*/\kappa, \mathrm{d}v^*/\kappa, \mathrm{d}w^*/\kappa).$$

Moreover, in this formulation we use a normal mode analysis: $(w, T) = (\widetilde{w}(z,t), \widetilde{T}(z,t))\,exp(ikx)$, where k denotes the wave number. We thus obtain the following eigenvalue problem (Souhar and Aniss 2016):

$$\mathbf{B}\frac{\partial \mathbf{u}}{\partial t} = \mathbf{A}\mathbf{u} \qquad [6.6]$$

where the components of vector $\mathbf{u} = \begin{pmatrix} \widetilde{w} \\ \widetilde{T} \end{pmatrix}$ are the vertical velocity and the temperature. Operators **A**, **B** and M are defined by:

$$\mathbf{A} = \begin{pmatrix} M^2 & -q^2 Ra \\ -\dfrac{\partial T_b}{\partial z} & M \end{pmatrix}, \quad \mathbf{B} = \begin{pmatrix} \mathrm{Pr}^{-1} M & 0 \\ 0 & 1 \end{pmatrix}, \quad M = \frac{\partial^2}{\partial z^2} - k^2 \qquad [6.7]$$

where $\mathrm{Pr} = \nu/\kappa$ is the Prandtl number, Ra is the Rayleigh number defined by $Ra = \frac{\beta g T_s d^3}{\nu\kappa}$, ν being the kinematic viscosity. It is important to note that this number is not defined from a temperature difference, but from a modulation amplitude as indicated in Yih and Li (1972).

The first derivative with respect to z of the basic state temperature $T_b(z,t)$ is given by:

$$\frac{\partial T_b}{\partial z} = \frac{dT_1}{\partial z}\cos(\omega t) + \frac{dT_2}{\partial z}\sin(\omega t) \qquad [6.8]$$

At z = 0,1, we consider the two types of boundary conditions:

$$\text{Free-free: } \tilde{w} = \frac{\partial^2 \tilde{w}}{\partial z^2} = \tilde{T} = 0 \qquad [6.9]$$

$$\text{Rigid-rigid: } \tilde{w} = \frac{\partial \tilde{w}}{\partial z} = \tilde{T} = 0 \qquad [6.10]$$

6.2.3. *Low-frequency asymptotic analysis WKB*

Let us consider that the dimensional frequency of the modulation of boundary temperature is quite low, $\omega^* << 1$. Hence, the modulation period of the temperature, T_p, is much longer than the time scales of thermal diffusivity and kinematic viscosity:

$$T_p >> \max\left(\frac{d^2}{a}, \frac{d^2}{\nu}\right) \qquad [6.11]$$

The dimensionless frequency satisfies the following condition:

$$\omega << \min(1, Pr^{-1}) \qquad [6.12]$$

After the temporal change of variable $\tau = \omega t$, the system [6.6] of period $2\pi/\omega$ with respect to time t, has a period 2π with respect to τ. Therefore, it takes the following form:

$$\omega \mathbf{B} \frac{\partial \mathbf{u}}{\partial \tau} = \mathbf{Au} \qquad 0 < \omega << 1 \qquad [6.13]$$

For sufficiently small values of ω, the WKB method can be used to approximate the solution to the system of equations [6.13] as follows (Gershuni and Lyubimov 1998):

$$\tilde{w}(z,\tau) = W(z,\tau) e^{\frac{1}{\omega}\int_0^{\tau} \lambda(\tau')d\tau'} \qquad \tilde{T}(z,\tau) = \theta(z,\tau) e^{\frac{1}{\omega}\int_0^{\tau} \lambda(\tau')d\tau'} \qquad [6.14]$$

Here, λ is a 2π-periodic function of τ. At small frequency ω, the amplitudes W, θ, the Rayleigh number Ra and the function λ are written as series expansion in ω:

$$\begin{aligned} W &= W_0 + \omega W_1 + \omega^2 W_2 + \cdots \\ \theta &= \theta_0 + \omega \theta_1 + \omega^2 \theta_2 + \cdots \\ \lambda &= \lambda_0 + \omega \lambda_1 + \omega^2 \lambda_2 + \cdots \\ Ra &= Ra_0 + \omega Ra_1 + \omega^2 Ra_2 + \ldots \end{aligned} \qquad [6.15]$$

The series expansion of the derivative of the basic temperature [6.8] in ω is written as:

$$\frac{\partial T_b}{\partial z} = (1-2z)\frac{\omega}{2}\sin(\tau) + o(\omega^2) \quad [6.16]$$

After substitution of [6.14], [6.15] and [6.16] in the system of equations [6.13] and in the boundary conditions [6.9] and [6.10], we obtain a hierarchy of successive approximations. Then, if we conserve the lowest order terms with respect to ω and we proceed to the following changes of variables:

$$\bar{W}_0 = \omega W_0 \qquad \bar{R}a_0 = \omega Ra_0 \quad [6.17]$$

the system of equations [6.13] becomes:

$$\lambda_0 \mathbf{B}\mathbf{u}_0 = \mathbf{A}_0\mathbf{u}_0 \quad [6.18]$$

where $\mathbf{u}_0 = \begin{pmatrix} \bar{W}_0 \\ \theta_0 \end{pmatrix}$ is a function vector. The operator $\mathbf{A}_0$ is given by:

$$\mathbf{A}_0 = \begin{pmatrix} M^2 & -q^2\bar{R}a_0 \\ -\frac{1-2z}{2}\sin(\tau) & M \end{pmatrix} \quad [6.19]$$

As for the boundary conditions [6.9] and [6.10], they become:

$$\text{Free-free case: } \bar{W}_0 = \frac{\partial^2 \bar{W}_0}{\partial z^2} = \theta_0 = 0 \quad \text{in} \quad z = 0{,}1 \quad [6.20]$$

The boundary conditions on $\bar{W}_0$ and θ_0, in [6.20], are also expressed from the system [6.18] and written as follows:

$$\frac{\partial^2 \theta_0}{\partial z^2} = \frac{\partial^4 \theta_0}{\partial z^4} = \frac{\partial^4 w_0}{\partial z^4} = 0 \quad \text{in} \quad z = 0{,}1 \quad [6.21]$$

$$\text{Rigid-rigid case: } \bar{W}_0 = \frac{\partial \bar{W}_0}{\partial z} = \theta_0 = 0 \quad \text{in} \quad z = 0, 1 \quad [6.22]$$

The boundary conditions on the velocity $\bar{W}_0$, in [6.22], are expressed on the temperature and write in z = 0, 1 as follows:

$$\begin{gathered} \frac{\partial^2 \theta_0}{\partial z^2} = 0, \lambda_0 Pr^{-1}\frac{\partial^2 \bar{W}_0}{\partial z^2} = \left(\frac{\partial^4}{\partial z^4} - 2k^2\frac{\partial^2}{\partial z^2}\right)\bar{w}_0 \\ \lambda_0\frac{\partial \theta_0}{\partial z} = \left(4\frac{\partial^3}{\partial z^3} - k^2\frac{\partial}{\partial z}\right)\theta_0 \end{gathered} \quad [6.23]$$

It is important to note that, for numerical reasons, we chose to have 12 boundary conditions for each free-free and rigid-rigid case. Indeed, the additional boundary conditions [6.21] and [6.23] are deduced from the system of equations [6.18] using [6.20] and [6.22], respectively. The system of equations [6.18] with the free-free and rigid-rigid boundary conditions given, respectively, by [6.20]–[6.21] and [6.22]–[6.23] is an ordinary eigenvalue problem with λ_0 as eigenvalue and the amplitudes $\bar{W}_0$ and θ_0 as eigenfunctions. The parameters of this problem are $\bar{R}a_0 = \omega Ra_0$, with the wave number k, the time τ and the function λ_0. According to the WKB method, the threshold of instability representing the convection onset is determined using the following integral condition:

$$\int_0^{2\pi} \lambda_{0r}\,(\tau')d\tau' = 0 \qquad [6.24]$$

where λ_{0r} is the real part of λ_0. The resolution of this ordinary eigenvalue problem considering the integral condition [6.24] allows us to determine first of all the marginal stability curve $(\bar{R}a_0, k)$ that yields the critical parameters, $\bar{R}a_{0c}$ and k_c, for the low frequencies ω. Then, using the change of variable on the Rayleigh number in [6.17], we determine the evolution of Ra_{0c}, corresponding to the instability threshold, as a function of low frequencies of modulation ω.

6.2.4. *Numerical procedure*

The system of equations [6.18] associated with free-free and rigid-rigid boundary conditions [6.20]–[6.21] and [6.22]–[6.23] can be solved using the Chebyshev spectral collocation method (Trefethen 1996; Weideman and Reddy 2000), which spatially discretizes the problem between −1 and 1. Hence, the problem that was initially defined between z = 0 and z = 1, will be defined between −1 and 1 with respect to ξ after the change of variable $\xi = 2z - 1$. Here, the considered collocation points are Chebyshev–Gauss–Lobatto points bounded between −1 and 1 and defined as follows:

$$\xi_i = \cos\left(\frac{\pi(i-1)}{N-1}\right), \qquad i = 1 \ldots N \qquad [6.25]$$

These are exactly N real extrema of the Chebyshev polynomial:

$$T_N(\xi) = \cos(N \arccos(\xi)) \qquad [6.26]$$

According to this numerical approach, the solutions $\bar{W}_0$ and θ_0 of the system of equations [6.18] are built considering them a superposition of Lagrangian spatial functions with constant coefficients:

$$\bar{W}_0(\xi) = \sum_{j=1}^{N} \bar{W}_j L_j(\xi) \qquad \theta_0(\xi) = \sum_{j=1}^{N} \theta_j L_j(\xi) \qquad [6.27]$$

where $\bar{W}_j$ and θ_j denote constant amplitudes and $L_j(\xi) = \prod_{i=0,i\neq j}^{N} \left(\frac{\xi-\xi_i}{\xi_j-\xi_i}\right)$ are Lagrange interpolation polynomials that are defined for Chebyshev–Gauss–Lobatto points and expressed as follows:

$$L_j(\xi) = \frac{(-1)^j(\xi^2-1)T'_{N-1}(\xi)}{c_j N^2(\xi-\xi_j)} \qquad [6.28]$$

where $c_1 = c_N = 2$ and $c_2 = \ldots = c_{N-1} = 1$. The amplitudes $\bar{W}_j$ and θ_j correspond, respectively, to solutions $\bar{W}_0(\xi)$ and $\theta_0(\xi)$ in each of the N points of collocation ξ_i so that $L_j(\xi_i) = \delta_{ij}$, where δ_{ij} is the Kronecker symbol. Hence,

$$\bar{W}_0(\xi_i) = \begin{cases} \bar{W}_j & i = j \\ 0 & i \neq j \end{cases} \qquad \theta_0(x_i) = \begin{cases} \theta_j & i = j \\ 0 & i \neq j \end{cases} \qquad [6.29]$$

At the same points, the first derivative of $L_j(\xi)$ represents a square differentiated matrix D_{ij} of size $N \times N$ so that $L_j^{(1)}(\xi_i) = D_{ij}$. This matrix is defined by:

$$D_{ij} = \begin{cases} \dfrac{c_i(-1)^{i+j}}{c_j(\xi_i-\xi_j)} & i \neq j \\ \dfrac{\xi_i}{2(\xi_i^2-1)} & 1 \leq i = j \leq N-1 \\ \dfrac{2N^2+1}{6} & i = j = 0 \\ -\dfrac{1+2N^2}{6} & i = j = N \end{cases} \qquad [6.30]$$

The derivative matrices of orders $(n > 1)$, of $L_j(\xi_i)$, are expressed in terms of D_{ij}, so that $L_j^{(n)}(\xi_i) = (D_{ij})^n$.

Then, the nth order derivatives in each of the N points of collocation ξ_i of solutions $\bar{W}_0(\xi)$ and $\theta_0(\xi)$, expressed in [6.27], are written as:

$$\frac{d^n}{d\xi^n}\bar{W}_0(\xi_i) = \sum_{j=1}^{N} \bar{W}_j\, D_{ij}^n, \qquad \frac{d^n}{d\xi^n}\theta_0(\xi_i) = \sum_{j=1}^{N} \theta_j\, D_{ij}^n \qquad [6.31]$$

We now apply the spatial discretization to our problem in order to transform it into a matrix system that is composed of a solution vector of size $2N$ components and of square matrices of size $2N \times 2N$. For this, we substitute the solutions [6.27] and their derivatives [6.31] in the system of equations [6.18] by considering the set of points of collocation ξ_i of index $i = 4\ldots N-3$. We then obtain $2N-12$ equations, corresponding to the number of lines of matrices of the system, which have $2N$ columns. The resulting matrix system is written as:

$$\left[\lambda_0 Pr^{-1}(4D_{ij}^2 - k^2), 0_{ij}\right]\begin{bmatrix}\bar{W}_j\\ \theta_j\end{bmatrix} = \left[16D_{ij}^4 - 8k^2D_{ij}^2 + k^4, -\bar{R}a_0k^2\right]\begin{bmatrix}\bar{W}_j\\ \theta_j\end{bmatrix} \qquad [6.32]$$

$$\left[0_{ij}, \lambda_0 I_{ij}\right]\begin{bmatrix}\bar{W}_j\\ \theta_j\end{bmatrix} = \left[\frac{\xi_i}{2}\sin(\tau), 4D_{ij}^2 - k^2\right]\begin{bmatrix}\bar{W}_j\\ \theta_j\end{bmatrix} \qquad [6.33]$$

The last 12 lines remaining to the matrices of the system [6.32] and [6.33], to be square of size $2N \times 2N$, will be reserved to 12 boundary conditions for each of the two free-free ([6.20] and [6.21]) and rigid-rigid ([6.22] and [6.23]) cases. We also substitute the solutions [6.27] and their derivatives [6.31] in these boundary conditions by considering the point of collocation ξ_i of index $i = 1$ for $\xi = 1$ and of index $i = N$ for $\xi = -1$. On the other hand, the index j preserves the same values $j = 1\ldots N$. These boundary conditions are expressed as follows:

Free-free case:

$$\begin{aligned}
&\left[I_{1j}, 0_{1j}\right]\begin{bmatrix}\bar{W}_j\\ \theta_j\end{bmatrix} = 0; \quad \left[I_{Nj}, 0_{Nj}\right]\begin{bmatrix}\bar{W}_j\\ \theta_j\end{bmatrix} = 0; \quad \left[D_{1j}^2, 0_{1j}\right]\begin{bmatrix}\bar{W}_j\\ \theta_j\end{bmatrix} = 0\\
&\left[D_{Nj}^2, 0_{Nj}\right]\begin{bmatrix}\bar{W}_j\\ \theta_j\end{bmatrix} = 0; \quad \left[0_{1j}, I_{1j}\right]\begin{bmatrix}\bar{W}_j\\ \theta_j\end{bmatrix} = 0; \quad \left[0_{Nj}, I_{Nj}\right]\begin{bmatrix}\bar{W}_j\\ \theta_j\end{bmatrix} = 0\\
&\left[0_{1j}, D_{1j}^2\right]\begin{bmatrix}\bar{W}_j\\ \theta_j\end{bmatrix} = 0; \quad \left[0_{Nj}, D_{Nj}^2\right]\begin{bmatrix}\bar{W}_j\\ \theta_j\end{bmatrix} = 0; \quad \left[D_{1j}^4, 0_{1j}\right]\begin{bmatrix}\bar{W}_j\\ \theta_j\end{bmatrix} = 0\\
&\left[D_{Nj}^4, 0_{Nj}\right]\begin{bmatrix}\bar{W}_j\\ \theta_j\end{bmatrix} = 0; \quad \left[0_{1j}, D_{1j}^4\right]\begin{bmatrix}\bar{W}_j\\ \theta_j\end{bmatrix} = 0; \quad \left[0_{Nj}, D_{Nj}^4\right]\begin{bmatrix}\bar{W}_j\\ \theta_j\end{bmatrix} = 0
\end{aligned} \qquad [6.34]$$

Rigid-rigid case:

$$[I_{1j}, 0_{1j}]\begin{bmatrix}\bar{W}_j\\ \theta_j\end{bmatrix} = 0; \quad [I_{Nj}, 0_{Nj}]\begin{bmatrix}\bar{W}_j\\ \theta_j\end{bmatrix} = 0; \quad [D_{1j}, 0_{1j}]\begin{bmatrix}\bar{W}_j\\ \theta_j\end{bmatrix} = 0$$

$$[D_{Nj}, 0_{Nj}]\begin{bmatrix}\bar{W}_j\\ \theta_j\end{bmatrix} = 0; \quad [0_{1j}, I_{1j}]\begin{bmatrix}\bar{W}_j\\ \theta_j\end{bmatrix} = 0; \quad [0_{Nj}, I_{Nj}]\begin{bmatrix}\bar{W}_j\\ \theta_j\end{bmatrix} = 0$$

$$[0_{1j}, D_{1j}^2]\begin{bmatrix}\bar{W}_j\\ \theta_j\end{bmatrix} = 0; \quad [0_{Nj}, D_{Nj}^2]\begin{bmatrix}\bar{W}_j\\ \theta_j\end{bmatrix} = 0$$

$$[\lambda_0 Pr^{-1} 4D_{1j}^2, 0_{1j}]\begin{bmatrix}\bar{W}_j\\ \theta_j\end{bmatrix} = [16D_{1j}^4 - 8k^2 D_{1j}^2, 0_{1j}]\begin{bmatrix}\bar{W}_j\\ \theta_j\end{bmatrix} \qquad [6.35]$$

$$[\lambda_0 Pr^{-1} 4D_{Nj}^2, 0_{Nj}]\begin{bmatrix}\bar{W}_j\\ \theta_j\end{bmatrix} = [16D_{1j}^4 - 8k^2 D_{Nj}^2, 0_{Nj}]\begin{bmatrix}\bar{W}_j\\ \theta_j\end{bmatrix}$$

$$[0_{1j}, \lambda_0 D_{1j}]\begin{bmatrix}\bar{W}_j\\ \theta_j\end{bmatrix} = [0_{1j}, 4D_{1j}^3 - k^2 D_{1j}]\begin{bmatrix}\bar{W}_j\\ \theta_j\end{bmatrix}$$

$$[0_{Nj}, \lambda_0 D_{Nj}]\begin{bmatrix}\bar{W}_j\\ \theta_j\end{bmatrix} = [0_{Nj}, 4D_{Nj}^3 - k^2 D_{Nj}]\begin{bmatrix}\bar{W}_j\\ \theta_j\end{bmatrix}$$

The matrix system ([6.32] and [6.33]) associated with the free-free and rigid-rigid boundary conditions ([6.34] and [6.35]) is a matrix problem with eigenvalues whose parameters are $\bar{R}a_0 = \omega Ra_0$, the wave number k, the time τ and the function λ_0. This system is written as follows:

$$\lambda_0 \widetilde{\mathbf{B}}\widetilde{\mathbf{u}} = \widetilde{\mathbf{A}}\widetilde{\mathbf{u}} \qquad [6.36]$$

where λ_0 is the eigenvalue and $\widetilde{\mathbf{u}}$ is the solution of the eigenvector of size $2N$ given by:

$$\widetilde{\mathbf{u}} = (\bar{W}_1, \bar{W}_2, \ldots, \bar{W}_{N-1}, \bar{W}_N, \theta_1, \theta_2, \ldots, \theta_{N-1}, \theta_N)^T \qquad [6.37]$$

It is important to note that square matrices $\widetilde{\mathbf{A}}$ and $\widetilde{\mathbf{B}}$ have constant coefficients. The numerical resolution of the matrix system [6.36] consists of obtaining $2N$ eigenvalues λ_0 corresponding to the nontrivial solution $\widetilde{\mathbf{u}}$. This allows us to obtain, by fixing $\bar{R}a_0$ and k and by varying the time τ from 0 to 2π, n values of λ_0 with $n = 2\pi/h$, where h is the time step. Then, by fixing k and varying $\bar{R}a_0$, the integral [6.24] is numerically calculated. The values of $\bar{R}a_0$ that satisfy the integral condition [6.24] are retained. Hence, we can determine the marginal stability curve $(k, \bar{R}a_0)$ whose minimum corresponds to critical parameters k_c and $\bar{R}a_{0c}$. Finally, by considering the second change of variable of [6.17] for the critical case, namely $Ra_{0c} = \omega^{-1}\bar{R}a_{0c}$, we define the evolution of Ra_{0c} as a function of the inverse of small modulation frequencies, ω.

It is important to note that, on the one hand, we have validated the numerical procedure (spatial discretization) in recent works (Echchadli and Aniss 2021, 2022), and, on the other hand, we have validated the WKB method by reproducing the results of Gershuni and Lyubimov (1998) presented in Chapter V, section V on the parametric excitation of convective instability at finite frequencies.

6.2.5. *Asymptotic laws of the critical Rayleigh number and wave number at low frequencies*

Let us now examine the results of the asymptotic WKB method at low frequencies and the numerical results obtained by direct resolution of the system [6.6]. As an illustration, all of the calculations were made for $Pr = 7$. First of all, we show in Figure 6.2 the marginal stability curves corresponding to the evolution of the Rayleigh number $\bar{R}a_0$ as a function of the wave number k for free-free (Figure 6.2a) and rigid-rigid (Figure 6.2b) boundary conditions. These marginal stability curves were determined by solving the system [6.36] deriving from the asymptotic method WKB considered for a low-frequency temperature modulation, $\omega << 1$.

The critical parameters corresponding to the minimum of each of the two curves in Figure 6.2 are:

$$\text{Free-free case: } \bar{R}a_{0c} = 26{,}419.7 \; k_c = 3 \qquad [6.38]$$

$$\text{Rigid-rigid case: } \bar{R}a_{0c} = 57{,}986.45 \; k_c = 3.9 \qquad [6.39]$$

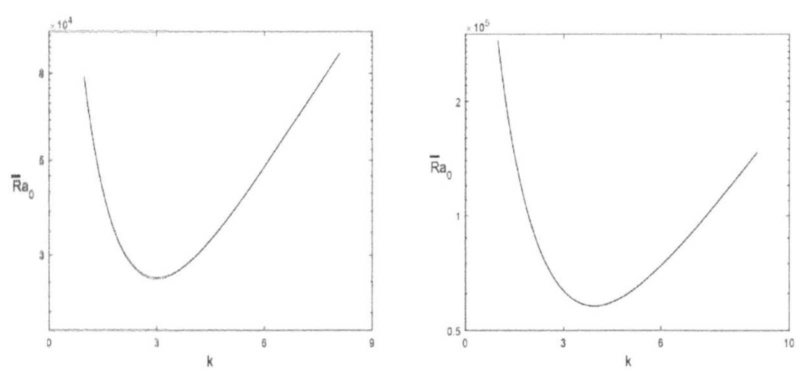

a) b)

Figure 6.2. *Marginal stability curves determined by the asymptotic WKB method for a low-frequency temperature modulation $\omega << 1$ for: (a) free-free case; (b) rigid-rigid case*

The asymptotic behavior of the critical Rayleigh number Ra_{0c} and the wave number k_c is obtained by using the changes of variables [6.17]:

$$\text{Free-free case: } Ra_{0c} = 26{,}419.7\omega^{-1}, \qquad k_c = 3 \qquad [6.40]$$

$$\text{Rigid-rigid case: } Ra_{0c} = 57{,}986.45\ \omega^{-1} \quad k_c = 3.9 \qquad [6.41]$$

By means of a direct numerical resolution of the system [6.6] using Chebyshev spectral collocation methods (Trefethen 1996; Weideman and Reddy 2000) and the Floquet theory that we implemented based on Or (1997), we show in Figure 6.3, for free-free (Figure 6.3a) and rigid-rigid (Figure 6.3b) boundary conditions, the numerical evolution of the critical Rayleigh number Ra_c, corresponding to the convection threshold, as a function of the inverse of the modulation frequency, $1/\omega$. This evolution is compared to the results obtained by the asymptotic WKB method for low frequencies and given by [6.40] and [6.41]. It is important to note that there is a potential instability for an intermediate frequency and less instability for low and high frequencies.

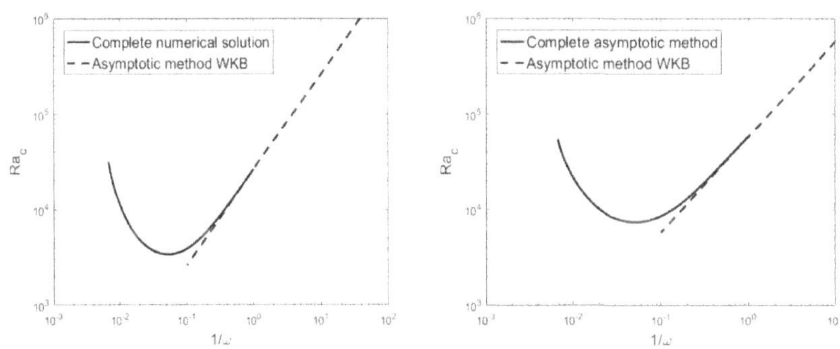

a) b)

Figure 6.3. *Numerical evolution of the critical Rayleigh number Ra_c as a function of the inverse of the modulation frequency $1/\omega$. Comparison between the results of direct numerical calculations and asymptotic numerical results using the WKB method: (a) free-free case; (b) rigid-rigid case*

Similarly, we show in Figure 6.4, for free-free (Figure 6.4a) and rigid-rigid (Figure 6.4b) boundary conditions, the numerical evolution of the critical wave number k_c, corresponding to the onset of thermal convection, as a function of the inverse of the frequency modulation, $1/\omega$. We also compare the direct numerical calculations and the results of the low-frequency asymptotic WKB method.

The comparisons in Figures 6.3 and 6.4 show that the more the dimensionless frequency decreases, $\omega \to 0$, the more the critical Rayleigh number Ra_{0c} and the wave number k_c, obtained with the WKB method, coincide with those obtained from direct numerical calculations. Hence, the behavior of the threshold of convective instability for low-frequency modulation is described by laws [6.40] and [6.41], obtained by WKB method. These laws allow us to bridge the gap of low frequencies, which is not the case of direct numerical resolution.

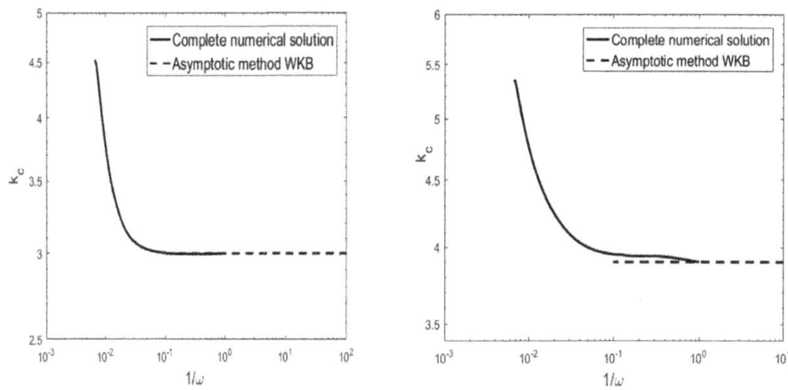

a) b)

Figure 6.4. *Numerical evolution of the critical wave number k_c as a function of the inverse of the modulation frequency, $1/\omega$. Comparison between the results of direct numerical resolution and the asymptotic numerical results obtained with the WKB method: (a) free-free case; (b) rigid-rigid case*

6.3. Application of the WKB method to the study of pulsed flow stability generated by in-phase low-frequency oscillation, in Taylor–Couette geometry

6.3.1. *Configuration of the studied flow*

Let us consider a Newtonian fluid of density ρ and dynamic viscosity μ confined between two coaxial cylinders of radii R_1 and $R_2 = R_1 + d$ having, respectively, the angular velocities $\Omega_1 = \Omega_2 = \Omega_o \cos(\omega^* t^*)$, where Ω_o and ω^* are, respectively, the dimensional modulation amplitude and frequency (Figure 6.5). We examine the case of the two cylinders oscillating in phase. This configuration studied by Aouidef et al. (1994) is reviewed here for low frequencies, knowing that the authors of this study have conducted, in this case, a scale analysis in quasi-steady state to establish the threshold of instability in terms of critical Reynolds number and wave number, as a function of dimensionless frequency.

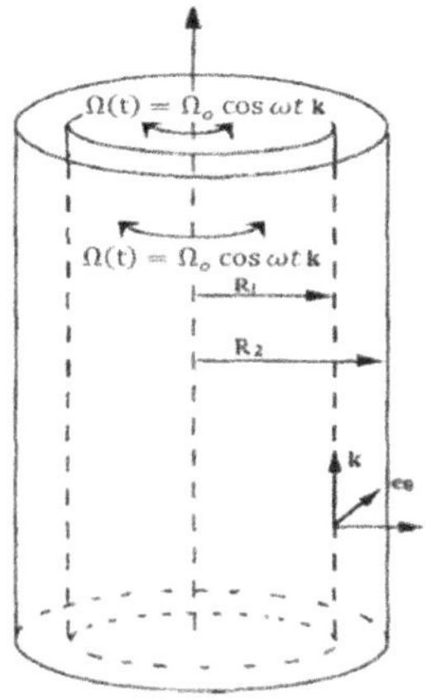

Figure 6.5. *Geometry of modulated Taylor–Couette flow*

6.3.2. *Formulation of the linear stability problem of the basic solution*

The equations governing the basic flow are the equation of momentum conservation and the continuity equation:

$$\rho\left(\frac{\partial \mathbf{V}^*}{\partial t^*} + \mathbf{V}^* \cdot \nabla \mathbf{V}^*\right) = -\nabla P^* + \mu \Delta \mathbf{V}^* \qquad [6.42]$$

$$\nabla \cdot \mathbf{V}^* = 0 \qquad [6.43]$$

where ρ is the density of the fluid and $\mathbf{V}^*$ is the velocity vector. In a system of cylindrical coordinates$(r^*,\ \theta,\ z^*)$, the components of the basic flow velocity in the radial, azimuthal and axial direction are denoted, respectively, by (U^*, V^*, W^*). Assuming that this flow is azimuthal and axisymmetric, the velocity field is written as: $\mathbf{V}^* = (0, V^*, 0)$. Under these hypotheses, the equations of conservation [6.42] and [6.43] allow us to obtain the following system:

$$-\rho\frac{V^{*2}}{r^*} = -\frac{\partial P^*}{\partial r^*} \qquad [6.44]$$

$$\rho\frac{\partial V^*}{\partial t^*} = \mu\left(\frac{\partial^2 V^*}{\partial r^{*2}} + \frac{\partial}{\partial r^*}\left(\frac{V^*}{r^*}\right)\right) \qquad [6.45]$$

$$0 = -\frac{\partial P^*}{\partial z^*} \qquad [6.46]$$

where the pressure P^* and V^* do not depend on θ. We consequently introduce the following dimensionless variables:

$$r^* = R_1 + dx, t = \frac{t^*}{\frac{d^2}{\nu}}, V_B = \frac{V^*}{R_1 \Omega_o} \; P_B = \frac{P^*}{\rho \, R_1 \, d \, \Omega_o^2}$$

Equation [6.46] shows that pressure does not depend on z^*. Considering that the gap, d, is small compared to the radius of the inner cylinder, R_1, the terms of order d/R_1 are suppressed. Therefore, equation [6.45] is written as:

$$\frac{\partial V_B}{\partial t} = \frac{\partial^2 V_B}{\partial x^2} \quad [6.47]$$

with the following boundary conditions:

$$V_B(x = 0, t) = \cos(\omega t) \text{ at } x = 0, V_B(x = 1, t) = \cos(\omega t) \text{ at } x = 1 \quad [6.48]$$

The parameter $\omega = \omega^* d^2/\nu$ is the frequency number defined as the ratio of the characteristic time of viscous diffusivity to the modulation frequency. The pressure of the basic flow, $P_B(x, t)$, is obtained from equation [6.44], and can be rewritten as follows:

$$V_B^2 = -\frac{\partial P_B}{\partial x} \quad [6.49]$$

The solution to equation [6.47] associated with the boundary conditions [6.48] is:

$$V_B(x, t) = V_1(x) \cos(\omega t) + V_2(x) \sin(\omega t) \quad [6.50]$$

Functions V_1 and V_2 are given by:

$$V_1(x) = \frac{\cos(\sigma x) \cosh(\sigma(1 - x)) + \varepsilon \cos(\sigma(1 - x)) \cosh(\gamma r)}{\cos(\sigma) + \cosh(\sigma)} \quad [6.51]$$

$$V_2(x) = \frac{\sin(\sigma x) \sinh(\sigma(1 - x)) + \varepsilon \sin(\sigma(1 - x)) \cosh(\gamma r)}{\cos(\sigma) + \cosh(\sigma)} \quad [6.52]$$

where $\sigma = \sqrt{\omega/2}$. This dimensionless parameter represents the ratio of two characteristic lengths $\sigma = d/\delta_N$, where $\delta_N = \sqrt{2\nu/\omega}$ is the thickness of a Stokes layer of a Newtonian fluid.

The flow stability described by equations [6.49] and [6.50] with regard to axisymmetric perturbations is achieved by considering that the field of velocities can be decomposed into the sum of a basic flow and an infinitesimal perturbation:

$$\mathbf{V} = (0, V_B, 0) + \big(u(x,z,t), v(x,z,t), w(x,z,t)\big), P = P_B + p(x,z,t) \qquad [6.53]$$

The above expressions are then inserted in equations [6.42] and [6.43], leading to a system of equations reflecting the evolution of perturbations, in which we linearize and use the normal modes, $(u, v) = (\tilde{u}(x,t), \tilde{v}(x,t))\, exp(ikz)$, k being the wave number along the z axis, and the following system is obtained:

$$\mathbf{B}\frac{\partial \mathbf{u}}{\partial t} = \mathbf{A}\mathbf{u} \qquad [6.54]$$

with the vector $\mathbf{u} = \begin{pmatrix} \tilde{u} \\ \tilde{v} \end{pmatrix}$.

Operators **A**, **B** and M are defined by:

$$\mathbf{A} = \begin{pmatrix} M^2 & -2k^2T_a^2V_B \\ -\dfrac{\partial V_B}{\partial x} & M \end{pmatrix}, \quad \mathbf{B} = \begin{pmatrix} M & 0 \\ 0 & 1 \end{pmatrix}, \quad M = \frac{\partial^2}{\partial x^2} - q^2 \qquad [6.55]$$

where $T_a = (R_1\Omega_0 d/\nu)(d/R_1)^{1/2}$ is the Taylor number, which represents a measure of the ratio between the destabilizing centrifugal forces and the stabilizing viscous forces. The boundary conditions of system [6.54] are given as:

$$u = v = \frac{\partial u}{\partial x} = 0 \; x = 0.1 \qquad [6.56]$$

6.3.3. *Asymptotic analysis for low frequencies – WKB method*

After making the temporal change of variable $\tau = \omega t$, the system [6.54] of period $2\pi/\omega$ with respect to time t has the period 2π with respect to τ. Therefore, it has the following form:

$$\omega\mathbf{B}\frac{\partial \mathbf{u}}{\partial \tau} = \mathbf{A}\mathbf{u} \qquad [6.57]$$

For $0 < \omega << 1$, the WKB method suggests that the solution to system [6.57] can be approximated as follows:

$$\tilde{u}(x,\tau) = U(x,\tau)e^{\frac{1}{\omega}\int_0^\tau \lambda(\tau')d\tau'},\ \tilde{v}(x,\tau) = V(x,\tau)e^{\frac{1}{\omega}\int_0^\tau \lambda(\tau')d\tau'} \qquad [6.58]$$

Here, λ is a 2π periodic function. At low frequency ω, the amplitudes U, V, the Taylor number Ta and the function λ are expanded as a series of powers of ω:

$$\begin{aligned} U &= U_0 + \omega U_1 + \omega^2 U_2 + \cdots \\ V &= V_0 + \omega V_1 + \omega^2 V_2 + \cdots \\ Ta &= Ta_0 + \omega Ta_1 + \omega^2 Ta_2 + \cdots \\ \lambda &= \lambda_0 + \omega\lambda_1 + \omega^2\lambda_2 + \cdots \end{aligned} \qquad [6.59]$$

The asymptotic expansions of the basic velocity V_B and of its derivative are given by:

$$V_B = \cos(\omega t) + \omega\frac{x}{2}(1-x)\sin(\omega t) + h.o.t \qquad [6.60]$$

$$\frac{\partial V_B}{\partial x} = \frac{\omega}{2}(1-2x)\sin(\omega t) + h.o.t \qquad [6.61]$$

After substitution of [6.59], [6.60] and [6.61] in the system [6.57], we keep the terms of lowest order with respect to ω and by proceeding to the following changes of variables:

$$\bar{U} = \omega U_0,\ \bar{T}a = Ta_0\sqrt{\omega} \qquad [6.62]$$

we obtain:

$$\lambda_0 \mathbf{B}\mathbf{u}_0 = \mathbf{A}_0\mathbf{u}_0 \qquad [6.63]$$

where $\mathbf{u}_0 = \begin{pmatrix} \bar{U} \\ V_0 \end{pmatrix}$ is a vector of functions. The operator $\mathbf{A}_0$ is given by:

$$\mathbf{A}_0 = \begin{pmatrix} M^2 & -2k^2\bar{T}a\cos(\tau) \\ -\frac{1}{2}(1-2x)\sin(\tau) & M \end{pmatrix} \qquad [6.64]$$

The boundary conditions associated with the system [6.63] are:

$$\overline{U} = V_0 = \frac{\partial \overline{U}}{\partial x} = 0 \; x = 0{,}1 \qquad [6.65]$$

We use the same numerical procedure as the one presented in section 6.2.4 to determine the marginal stability curve (Ta_0, k), which makes it possible to obtain the critical values (Ta_{0c}, k_c). Then by using the change of variable in [6.62], we determine the evolution of Ta_c as a function of frequency, ω.

6.3.4. *Asymptotic laws of critical Taylor and wave numbers at low frequencies*

The numerical resolution of system [6.63], obtained for low frequencies and independent of ω, through the numerical procedure presented in section 6.2.4, leads to the marginal stability curve, presented in Figure 6.6, having a critical Taylor number, and a critical wave number: $\bar{T}a_c = 248.357$ and $k_c = 3.92$.

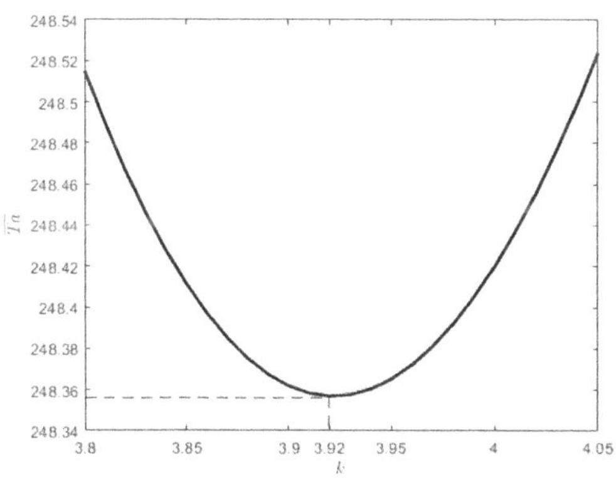

Figure 6.6. *Marginal stability curve, obtained by the resolution of system [6.63], yielding the critical threshold: critical Taylor and wave numbers*

Using the change of variable [6.62], we obtain the asymptotic law yielding the critical Taylor and wave numbers:

$$Ta_{0c} = 248.357\omega^{-\frac{1}{2}}, \qquad k_c = 3.92 \qquad [6.66]$$

Figure 6.7 shows the curve representing the critical Taylor and wave numbers as a function of $\omega^{-1/2}$. This curve is the result of a complete resolution of the system of equations [6.54] using Chebyshev spectral collocation methods for spatial resolution (Trefethen 1996; Weideman and Reddy 2000; Choujaa et al. 2021; Riahi et al. 2022) and the iterative method established by Or (1997) for temporal resolution. For low frequencies, there is perfect agreement with the results obtained by WKB method and given by the asymptotic law [6.66]. Furthermore, Table 6.1 provides a comparison, for low frequencies, between the numerical results of WKB method and those of direct resolution. Recall that the results obtained by Aouidef et al. (1994) with a scale analysis in quasi-steady state are: $Ta = 193.23\ (\omega/2)^{-\frac{1}{2}}$ $\sim 273.26\ \omega^{-\frac{1}{2}}, k_c = 3.92$. Let us note that the method used by Aouidef et al. (1994) does not take into account the temporal variation of the basic solution in the system obtained for low frequencies and which was rendered quasi-steady while it requires a resolution under unsteady conditions.

	Numerical method		**Asymptotic method: WKB**	
ω	k_c	T_{ac}	k_c	$T_{ac} = \mathbf{248.3570}/\sqrt{\omega}$
0.1	–	–	3.92	785.3740
0.2	–	–	3.92	555.3433
0.3	3.92	435.9170	3.92	453.4359
0.4	3.92	393.4236	3.92	392.6870
0.5	3.92	352.2511	3.92	351.2299
0.6	3.91	321.9591	3.92	320.6276
0.7	3.91	289.5089	3.92	296.8435
0.8	3.91	279.6929	3.92	277.6716
0.9	3.90	267.1888	3.92	261.7913

Table 6.1. *Comparison between the numerical results obtained by direct resolution and the asymptotic results of the WKB method for various low values of* ω

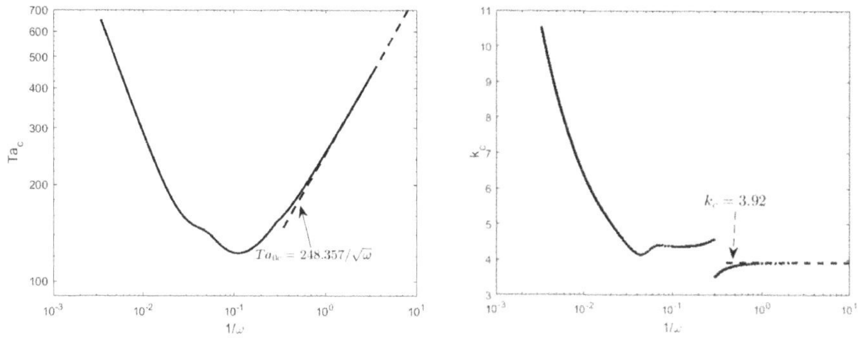

Figure 6.7. *Solid line: critical threshold, critical Taylor and wave numbers as a function of the dimensionless frequency ω, obtained by direct resolution of the system [6.54]; dotted line: asymptotic solution at low frequencies using WKB method*

6.4. Conclusion

This study examines in-phase modulation, where in the first case study the applied temperature gradient has no stationary component, but rather a periodic component. The same is true for the second case study, the base velocity has no stationary component and corresponds to a pulsed flow. In both cases, the asymptotic results obtained with the asymptotic WKB method serve to correct the models presented in other research works that used methods based on scale analysis while in a quasi-steady state.

Finally, we consider WKB method to be an effective and potentially useful means to bridge the gap left uncalculated by many numerical studies in problems of hydrodynamic stability in the presence of gravitational modulation and periodic boundary conditions at low frequencies. The asymptotic WKB method seems promising and is potentially useful to examine other types of low-frequency modulated instabilities.

6.5. References

Aniss, S., Souhar, M., Belhaq, M. (2000a). Asymptotic study of the convective parametric instability in Hele-Shaw cell. *Phys. Fluids A*, 12, 262–268.

Aniss, S., Souhar, M., Belhaq, M. (2000b). Effect of a magnetic modulation on the onset of convection in a magnetic liquid layer heated from above. *Comptes Rendus de l'Académie de Sciences – Serie IIb – Mécanique*, 328(5), 399–406.

Aniss, S., Belhaq, M., Souhar, M. (2001). Effects of a magnetic modulation on the stability of a magnetic liquid layer heated from above. *ASME J. Heat Transfer*, 123(3), 428–433.

Aniss, S., Belhaq, M., Souhar, M., Velarde, M.G. (2005). Asymptotic study of Rayleigh-Benard convection under time periodic heating in Hele-Shaw cell. *Phys. Scr.*, 71(4), 395–401.

Aouidef, A., Normand, C., Stegner, A., Wesfreid, J.E. (1994). Centrifugal instability of pulsed flow. *Phys. Fluids.*, 11, 3665.

Bhadauria, B.S. (2008). Combined effect of temperature modulation and magnetic field on the onset of convection in an electrically conducting-fluid-saturated porous medium. *ASME J. Heat Transfer*, 130(5), 052601.

Bhadauria, B.S. and Kiran, P. (2014). Heat and mass transfer for oscillatory convection in a binary viscoelastic fluid layer subjected to temperature modulation at the boundaries. *Heat Mass Transfer*, 58, 166–175.

Boulal, T., Aniss, S., Belhaq, M., Rand, R. (2007). Effect of quasiperiodic gravitational modulation on the stability of a heated fluid layer. *Phys. Rev. E*, 76(52), 56320.

Boulal, T., Aniss, S., Belhaq, M., Azouani, A. (2008). Effect of quasi-periodic gravitational modulation on the convective instability in Hele-Shaw cell. *Int. Journal of Non-Linear Mechanics*, 76(52), 56320.

Carmi, S. and Tustaniwskyi, J.I. (1981). Stability of modulated finite-gap cylindrical Couette flow: Linear theory. *J. Fluid Mech.*, 108, 19.

Choujaa, M.H. Aniss, S., Ouazzani, M.T., Naciri, J.K., Riahi, M. (2021). Stability of an oscillatory Taylor–Couette flow in an upper convected Maxwell fluid. *Phys. Fluids*, 33, 074105. doi: 10.1063/5.0057277.

Davis, S.H. (1976). The stability of time-periodic flows. *Annu. Rev. Flui. Mech.*, 8, 57–74.

Donnelly, R.J. (1964). Experiments on the stability of viscous flow between rotating cylinders. *Proc. R. Soc. A*, 281(1384), 130.

Dowden, J. (1981). The stability of a periodically heated layer of fluid. *J. Fluid Mech.*, 110, 149–159.

Drazin, P.G. and Reid, W.H. (1981). *Hydrodynamic Stability*. Cambridge University Press, Cambridge.

Echchadli, M. and Aniss, S. (2021). Thermal convection instability of two miscible viscous fluids in a Hele-Shaw cell. *J. Porous Media*, 24, 31–47.

Echchadli, M. and Aniss, S. (2022). Thermal convection instability of two miscible viscous fluids in a rotating annular Hele–Shaw cell. *Phys Fluids*, 34(8), 082111.

El Jaouahiry, A. and Aniss, S. (2020). Linear stability analysis of a liquid film down on an inclined plane under oscillation with normal and lateral components in the presence and absence of surfactant. *Phys. Fluids*, 32, 034105.

Finucane, R.G. and Kelly, R.E. (1976). Onset of instability in a fluid layer heated sinusoidally from below. *Int. J. Heat Mass Transfer*, 19(1), 71–85.

Garih, H., Strzelecki, A., Casalis, G., Estivalezes, J.L. (2013). Detailed analysis of the vibration induced instability of a liquid film flow. *Phys. Fluids*, 25, 014101.

Gassa Feugaing, C.M., Crumeyrolle, O., Yang, K.S., Mutabazi, I. (2014). Destabilization of the Taylor Couette flow of the inner cylinder rotation. *European Journal of Mechanics B/Fluids*, 44, 82–87.

Gershuni, G.Z. and Lyubimov, D.V. (1998). *Thermal Vibrational Convection*. John Wiley & Sons, New York.

Gershuni, G.Z. and Zhukhovitskii, E.M. (1963). On parametric excitation of convective instability. *J. Appl. Math. Mech. (PMM)*, 27, 1197–1204.

Gershuni, G.Z. and Zhukhovitskii, E.M. (1976). *Convective Instability of Incompressible Fluid*. Keter Publisher, Jerusalem.

Gershuni, G.Z., Nepomnyashchy, A.A., Velarde, M.G. (1992). On dynamic excitation of Marangoni instability. *Phys. Fluids, A*, 4, 2394–2398.

Gershuni, G.Z., Nepomnyashchy, A.A., Smorodin, B.L., Velarde, M.G. (1994). On parametric excitation of thermocapillary and thermogravitational convective instability. *Microgravity Quart*, 4, 215–220.

Gershuni, G.Z., Nepomnyashchy, A.A., Smorodin, B.L., Velarde, M.G. (1996). On parametric excitation of Marangoni instability in a liquid layer with free deformable surface. *Microgravity Quart*, 6, 203–209.

Gresho, P.M. and Sani, R.L. (1970). The effects of gravity modulation on the stability of a heated fluid layer. *J. Fluid Mech.*, 40, 783–806.

Grosch, C.E. and Salwen, H. (1978). The stability of steady and time dependent plane Poiseuille flow. *J. Fluid Mech.*, 34, 177–205.

Hall, P. (1975). The stability of unsteady cylinder flows. *J. Fluid Mech.*, 67, 29–63.

Kelly, R.E. and Cheers, A.M. (1970). On the stability of oscillating, plane Couette flow. *Q. J. Mech. Appl. Math.*, 23, 127–136.

Kerczek, Von, C. and Davis, S.H. (1972). The stability of oscillatory Stokes layers. *Stud. Appl. Math.*, 51, 239–252.

Kerczek, Von, C. and Davis, S.H. (1974). Linear stability theory of oscillatory Stokes layers. *J. Fluid Mech.*, 62, 753–573.

Landau, L.D. and Lifshitz, E.M. (1981). *Theoretical Physics, Quantum Mechanics*. Pergamon Press, Oxford.

Lin, S.P., Chen, J.N., Woods, D.R. (1996). Suppression of instability in a liquid film flow. *Phys. Fluids*, 8, 3247.

Nayfeh, A.H. (1981). *Introduction to Perturbation Techniques*. John Wiley & Sons, New York.

Or, A.C. (1997). Finite wavelength instability in a horizontal liquid layer on an oscillating plane. *J. Fluid Mech*, 335, 213–232.

Oukada, B., Ouazzani, M.T., Aniss, S. (2006). Effects of a temperature modulation in phase at the frontier on the convective instability of a viscoelastic layer. *C. R. Mec.*, 334(3), 205–211.

Riahi, M., Aniss, S., Ouazzani Touhami, M., Skali Lami, S. (2014). Stability of a pulsed Taylor Couette flow in a viscoelastic fluid. *Journal of the Society of Rheology, Japan*, 42(5), 321–327.

Riahi, M., Aniss, S., Ouazzani Touhami, M., Skali Lami, S. (2015). Pulsed Taylor–Couette flow in a viscoelastic fluid under inner cylinder modulation. *European Physical Journal Plus*, 130, 253.

Riahi, M., Aniss, S., Ouazzani Touhami, M., Skali Lami, S. (2016). Centrifugal instability of pulsed Taylor–Couette flow in a Maxwell fluid. *European Physical Journal E*, 39, 82.

Riahi, M., Aniss, S., Ouazzani, M.T. (2019). Families of reversing and non-reversing Taylor vortex flows between two cooscillating cylinders with different amplitudes. *Physics of Fluids*, 31, 014101. doi: 10.1063/1.5064656.

Riahi, M., Choujaa, M.H., Aniss, S. (2022). Reversing and non-reversing Taylor vortex flows in modulated Taylor–Couette flow with counter-oscillating cylinders. *European Journal of Mechanics/B Fluids*, 95, 148–159.

Riley, P.J. and Laurence, R. (1976). Linear stability of modulated circular flow. *J. Fluid Mech.*, 75, 625.

Roppo, M.N., Davis, S.H., Rosenblat, S. (1984). Bénard convection with time-periodic heating. *Phys. Fluids*, 27(4), 796–803.

Rosenblat, S. and Herbert, D.M. (1970). Low frequency modulation of thermal instability. *J. Fluid Mech.*, 43(part 2), 385–398.

Rosenblat, S. and Tanaka, D.A. (1971). Modulation of thermal convection instability. *Phys. Fluids*, 14(7), 1319–1322.

Smorodin, B.L. and Velarde, M.G. (2000). Electrothermoconvective instability of an ohmic liquid layer in an unsteady electric field. *Journal of Electrostatics*, 48(3–4), 261–277.

Smorodin, B.L. and Velarde, M.G. (2001). On the parametric excitation of electrothermal instability in a dielectric liquid layer using an alternating electric field. *Journal of Electrostatics*, 50(3), 205–226.

Smorodin, B.L. and Velarde, M.G. (2008). Convective instability of an Ohmic liquid layer in an unsteady thermal field. *Phys. Fluids*, 20, 044101.

Souhar, K. and Aniss, S. (2011). Effect of temperature modulation on the onset of convection in a Hele-Shaw cell. *Journal of Porous Media*, 14(6), 533–539.

Souhar, K. and Aniss, S. (2016). Effect of phase thermal modulation without stationary temperature gradient on the threshold of convection. *J. Heat Transfer*, 138(10), 102502.

Stoker, J.J. (1950). *Nonlinear Vibrations in Mechanical and Electrical Systems*. Intersciences, New York.

Trefethen, L.N. (1996). *Finite Difference and Spectral Methods for ODE and PDE*. Cornell University Press, New York.

Venezian, G. (1969). Effect of modulation on the onset of thermal convection. *J. Fluid Mech.*, 35(2), 243–254.

Weideman, J.A.C. and Reddy, S.C. (2000). A MATLAB differentiation matrix suite. *ACM Transactions of Mathematical Software*, 26, 465–519.

William, P. (2014). *Asymptotic Analysis and Perturbation Theory*. CRC Press, Boca Raton.

Woods, D.R. and Lin, S.P. (1995), Instability of a liquid film flow over a vibrating inclined plane. *J. Fluid Mech.*, 294, 391–407.

Yih, C.S. and Li, C.H. (1972). Instability of unsteady flows or configurations. Part 2. Convective instability. *J. Fluid Mech.*, 54, 143–152.

7

Multiplicity of Convective Flows in a Horizontal Annular Layer

The studies of natural convection in a horizontal annular layer, fluid or porous medium were first the subject of many experimental works, which were followed by two-dimensional (2D) and then three-dimensional (3D) numerical works and by various analytical approaches, and lastly by theoretical works concerning the stability of various flows observed in this geometric configuration. These studies are motivated by the very wide range of technological applications in the industrial and academic domain. This chapter reviews the main experimental, numerical and theoretical works conducted over the past four decades, including not yet published illustrations of these flows. For the case of two horizontal cylinders of inner radius *Ri* and outer radius *Re*, verifying $R_e - R_i = e << R_i$, we have quite recently obtained an analytical 2D solution of the convective flow. This solution is based on the hypothesis of parallel flow that completes all of the previously developed analytical approaches. We will use this analytical solution, which was numerically validated, for the analytical study of the stability of the basic solution in the limit case of small gap annular spaces. The previous stability studies of this convective flow have mainly been numerical.

7.1. Introduction

Natural convection flows in horizontal annular geometry in fluid medium or in porous medium saturated by a single-component fluid have been extensively studied. The following section focuses on natural convection in horizontal, annular, either in fluid or porous cells saturated by a Newtonian single-component fluid. There is significant similarity between natural convection flows emerging in fluid

---

Chapter written by Abdelkader MOJTABI and Marie-Catherine CHARRIER-MOJTABI.

and porous annular cells. This consists not only of the comparable streamlines in the two configurations, but also in similar transitions between single-cell flow, emerging at very small temperature differences between the two coaxial and horizontal cylinders and multicell flows that emerge for higher values of the Rayleigh number.

7.1.1. *In fluid medium*

The first experimental works, having led to a good description of natural convection in the horizontal annular column filled with air at atmospheric pressure, are due to Grigull and Hauf (1966). These authors managed to get remarkable photos of the thermal field for various values of the temperature difference between the two cylinders using a Mach–Zehnder interferometer. Then they deduced the values of global Nusselt number and local Nusselt number at the inner cylinder as a function of the temperature differences imposed to the two coaxial cylinders. Streamlines have been visualized using cigarette smoke. These photos of thermal fields and streamlines can be viewed in the remarkable book by Van Dyke (1982).

The second significant contribution is due to Kuehn and Goldstein (1976). These authors conducted experiments in which the horizontal annular geometry is filled with air or with water at atmospheric pressure. The study was made for a single radius ratio of the two cylinders by means of a Mach–Zehnder interferometer used for the measurement of a temperature field and for calculating the heat transfer in the annular space. Then, they conducted 2D numerical simulations using the finite difference method to solve the equations of natural convection in this geometry. The numerical simulation results and the experimental results are in good agreement for moderate values of the Rayleigh number.

Mojtabi and Caltagirone (1979a) studied the linear stability of the basic solution that emerges in the horizontal annular cylinder for moderate Rayleigh numbers. For annular cylinders with a high aspect ratio, and for cylinder length L, compared to their inner and outer radii, the flow is invariant, in any vertical plane orthogonal to the cylinder horizontal axis, except for the vicinity of endplates $z = 0$ or $z = L$, the axis **z** being along the horizontal axis of the coaxial cylinders. The basic solution for this steady 2D flow was obtained with the perturbation method, as a function of the Rayleigh number, expanded to the third order. The critical Rayleigh numbers and wave numbers associated with the loss of stability of the 2D solution were obtained for three radii ratios: $R = \frac{R_e}{R_i} = 1.2, \sqrt{2}$ and 2. This linear stability study was also conducted numerically with the Galerkin method. The comparison of the results obtained with these two methods showed they were in very good agreement. This work was completed by the same authors by analyzing the energy stability of this convective flow (Mojtabi and Caltagirone 1979b). To study the nonlinear stability of this basic flow in relation with non-infinitesimal perturbations, the energy method was used, and the resulting Euler Lagrange equations were solved using the Galerkin method.

Charrier-Mojtabi et al. (1979) conducted numerical studies of the various convective 2D flow structures, in the horizontal annular geometry for Rayleigh numbers varying between 100 and 5×10^4 and radii ratios of 1.2 and 2 for the air (Pr = 0.7) and for liquid sodium (Pr = 0.002).

Cheddadi et al. (1992) used laser-Doppler velocimetry to measure the velocity field when the annular geometry was filled with air under atmospheric pressure. The numerical simulations conducted by finite differences, an implicit alternating-direction implicit (A.D.I.) method, are in good agreement with the experimental results.

In 1999, Dyko et al. (1999) resumed the theoretical and experimental works of Mojtabi (1979) where various convective flows were highlighted. They developed 3D numerical simulations that confirmed the experimental results of Mojtabi (1979).

Three different cells were used in the work of Mojtabi (1979). In addition to air and water, several silicone oils with Prandtl numbers ranging between 10 and 10^3 were used in these experiments. The first cell with a radii ratio R = 2.18 and length 250 mm has been constructed for the visualization of isotherms using a Mach–Zehnder interferometer. The dimensions of this cell were chosen according to the dimensions of the available interferometer. Using air, several photos, representing interference fringes, hence isotherms, were obtained for several values of the Rayleigh number: photos 1 and 2 (Figure 7.1).

a) $Pr = 0.7 \quad Ra = 1.01 \times 10^4$ b) $Pr = 0.7 \quad Ra = 1.01 \times 10^4$

Figure 7.1. *(a) Isotherms visualized by Mach–Zehnder interferometry, in a cross-section plane perpendicular to the cell axis; (b) on the left, isotherms visualized by Mach–Zehnder interferometry; on the right, numerically determined isotherms The second cell was intended to visualize streamlines. The inner and outer cylinders have diameters of 80 and 160 mm, respectively, and a length of 400 mm*

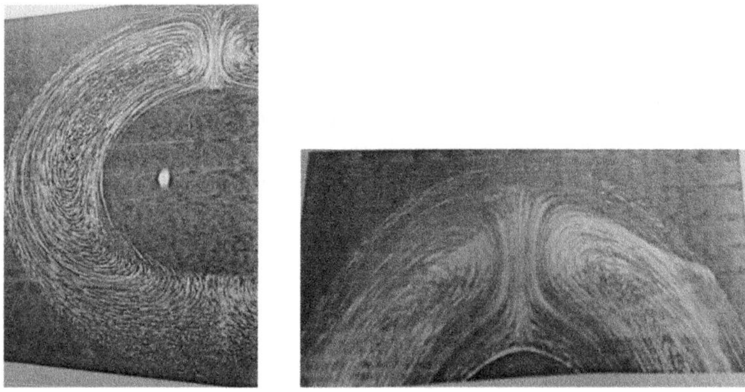

c) $Pr = 100$, $Ra = 4.36 \times 10^4$ d) $Pr = 1000$, $Ra = 9.14 \times 10^3$

Figure 7.2. *Visualization of streamlines in a horizontal annular layer filled with silicone oil (Pr = 100 and 1,000): photos 3 and 4*

Streamlines were visualized for air with cigarette smoke, for water with commercial detergent and for silicone oils with alumina powder. Figure 7.2 presents photos 3 and 4 for $Pr = 100, 1{,}000$, R = 2.1 and for two values of the Rayleigh number $Ra = 4.36 \times 10^4$ and $Ra = 9.14 \times 10^3$.

The third cell, with a radius ratio $R = 2$, of inner radius $Ri = 80$ mm and greater length $L = 800$ mm, is instrumented with 20 thermocouples, out of which 16 are located in various points of the fluid domain to detect any modification of the fluid temperature that may induce a change in the convective flow structure. The analysis of the thermal field for various values of the Rayleigh number shows that the flow is steady for Ra less than 11×10^4. These local temperature measurements cannot be used to conclude on the 2D or 3D nature of the studied convective flow. This shows the importance of 3D numerical simulations conducted by Dyko et al. (1999), which provided for the first time a 3D quantitative analysis of various 2D and 3D natural convection flows emerging in the horizontal annular cavity.

7.1.2. *In porous medium*

Similar works in the porous annular layer began later on. A first, important contribution belongs to Caltagirone (1976), including not only a theoretical and numerical part, but also an experimental part, with temperature measurements conducted in the median zone of a cell of length $L = 80$ cm and inner radius of 4 cm and outer radius of 8 cm, which leads to a radius ratio $R = 2$ and an aspect ratio of the cell $A = L/Ri = 20$. Several numerical methods have been used, for the low values of the filtration Rayleigh number Ra, the perturbation method expanded to the third order as a function of *Ra* was used to describe the pseudo-conduction state also obtained in the annular cell in fluid medium. The thermal field is then quasi-conductive, but the velocity field is not zero. For higher values of the Rayleigh number, the regime is purely convective, and the field of isotherms is no longer quasi-circular. This is a steady 2D convection state invariant in any cross-section perpendicular to the axis of the annular cell. The governing equations of this 2D flow were solved using an A.D.I. finite difference method, and a good agreement was obtained between numerical and experimental results. These numerical studies do not allow us to take into account the instabilities observed in experiments for higher values of the Rayleigh number. For this, a linear stability study of this 2D steady flow was conducted using the Galerkin method. The results obtained show the importance of 3D effects for the onset of the fluctuating convection. The critical Rayleigh number Ra_c is thus determined as a function of radii ratio R by solving an eigenvalue problem. A 3D numerical model based on the finite element method was thus developed in order to highlight the various types of time evolutions of convective structures. A steady 2D convection state and a fluctuating 3D state were obtained numerically. The theoretical results of linear stability and 2D and 3D numerical results are in good agreement with the experimental results obtained for this configuration.

This work was then resumed and continued further by Charrier-Mojtabi et al. (1991) and by Charrier-Mojtabi (1993). The authors presented a 2D numerical study conducted by two different numerical methods, namely, Fourier–Galerkin and Chebychev collocation methods with high orders of approximation. These works also presented an experimental study using the Christiansen effect for the visualization of thermal fields.

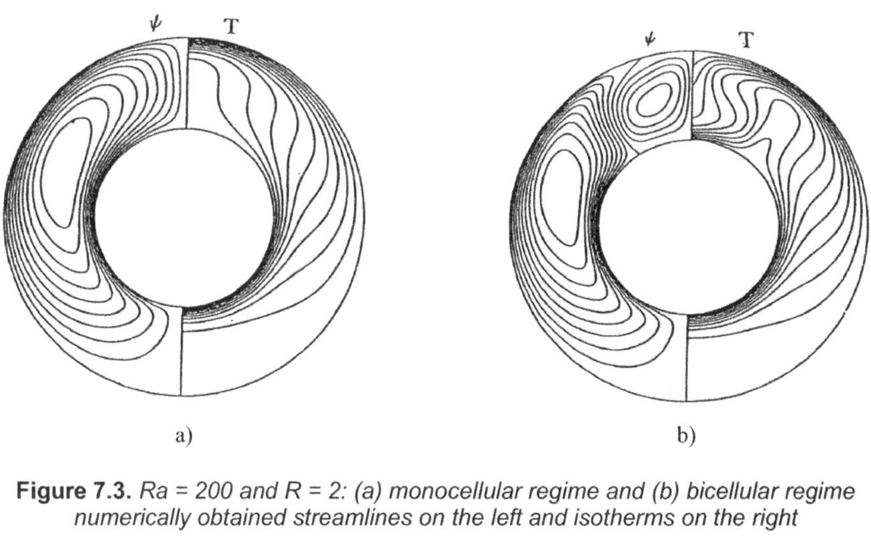

a) b)

Figure 7.3. *Ra = 200 and R = 2: (a) monocellular regime and (b) bicellular regime numerically obtained streamlines on the left and isotherms on the right*

The isotherms (Figures 7.3(a) and(b)) represented on each right half and the streamlines on each left half are perfect reproductions of photos shown in Figures 7.4(e) and (f) of experiments obtained for very close values to filtration Rayleigh numbers.

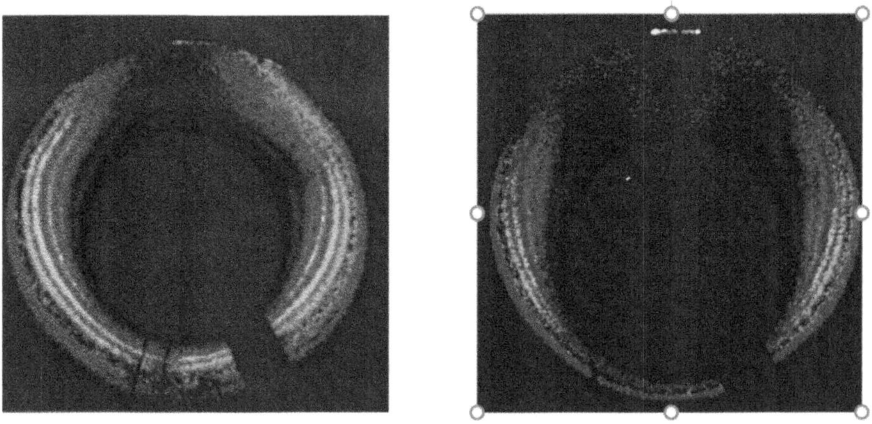

a) b)

Figure 7.4. *Isotherms obtained by Christiansen effect: (a) monocellular regime Ra = 232, R = 2; (b) bicellular regime Ra = 264, R = 2. For a color version of this figure, see www.iste.co.uk/saouab/mechanics.zip*

The photos taken show, for the first time, the existence of bicellular convective structures, and this for the same radius ratio R = 2 and for close filtration Rayleigh numbers: Ra = 232 and Ra = 264. This multicellular regime was obtained only numerically by Rao et al. (1987) and Himasekhar and Bau Haim (1988). For low values of the radii ratios R, R = $2^{1/4}$, it was numerically established that, depending on the initial conditions used in the numerical simulations for Ra = 500, the convective flow obtained at steady state could be either unicellular bicellular, tricellular or multicellular.

In 1997, Charrier-Mojtabi published a detailed numerical study based on various mathematical formulations and studied the 2D and 3D convective flows emerging in a horizontal annular porous layer. In 2D, this natural convection problem is formulated in terms of stream function–temperature or velocity–temperature. In 3D, in addition to the velocity–temperature formulation, we also find the potential vector–temperature formulation. Moreover, this article introduced the pressure–temperature formulation.

Based on this later formulation, we obtain the equation that must be satisfied by the pressure field by taking the divergence of the Darcy equation that no longer features the velocity field, as the continuity equation is verified. To simplify the resolution of this problem, the annular domain in cylindrical coordinates (r, θ, z) is transformed into a parallelepipedic domain (X, Y, Z) using a conformal transformation that associates point $M(r, \theta, z)$ with point M(X = ln(r), Y = θ, Z = z). The resolution of the two coupled equations in pressure and temperature was obtained with a Fourier–Galerkin approximation for the two azimuthal and axial periodic directions and a Chebychev collocation approximation in the confined radial direction. For a radius ratio of the two cylinders, R = 2, a good agreement was found with the experimental results and the stability analysis results of Caltagirone (1976). Figures 7.5(a) and (b) present not yet published 3D results.

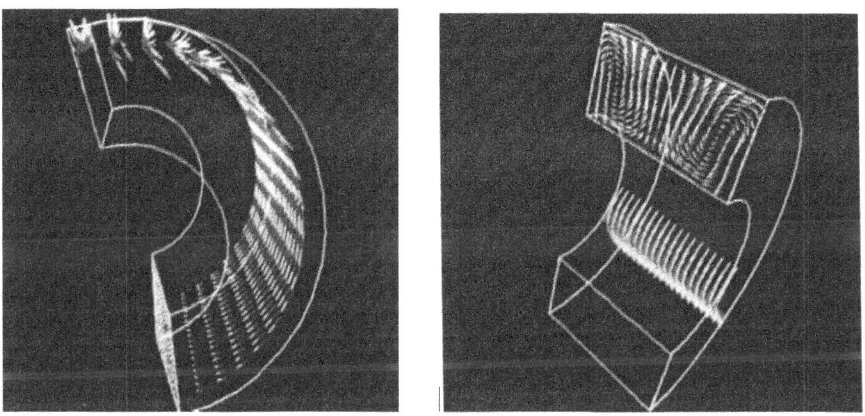

a) b)

Figure 7.5. *3D velocity field according to r = cst for (a) R = 2, Ra = 100, A = L/Ri = 2. The 3D aspect is observed at the top of the annular layer. (b) R = 2, Ra = 100, A = L/Ri = 2. At the top of the annular layer in a vertical plane containing the cell axis, two Rayleigh–Bénard cells can be observed. For a color version of this figure, see www.iste.co.uk/saouab/mechanics.zip*

Many approximate solutions were obtained from asymptotic expansions as a function of the Rayleigh filtration number (Ra) and for low values of the radii ratios depending on the parameter $\varepsilon = \frac{R_e - R_i}{R_i} = R - 1$. These works will be examined in the next section.

7.2. Mathematical formulation

The length of the horizontal annular cell is L. The inner cylinder of radius R_i is maintained at the temperature T_h and the outer cylinder of radius R_e is maintained at the temperature T_c, with $T_h > T_c$. This geometry is characterized by two dimensionless numbers, namely, the radius ratio $R = \frac{R_e}{R_i}$ and the longitudinal aspect ratio $A = \frac{L}{R_i}$.

The porous matrix is defined by its porosity ε and by its constant permeability K, the porous layer being assumed isotropic and homogeneous. The fluid saturating the porous medium is characterized by its density ρ, its thermal expansion coefficient β_T, its volumetric heat $(\rho c)_f$ and its dynamic viscosity μ. The porous matrix saturated by the fluid is assimilated to an isotropic fictitious fluid of thermal conductivity k* and volumetric heat: $(\rho c)^* = \varepsilon(\rho c)_f + (1-\varepsilon)(\rho c)_s$, where $(\rho c)_s$ is the volumetric heat of the porous solid structure. The filtration velocity is sufficiently low to use the Darcy law. Moreover, the solid and the fluid are assumed in thermal equilibrium, which allows the use of a single energy equation. A further assumption is that the fluid is incompressible and viscous dissipations are negligible. Using the Boussinesq approximation, the fluid density is written as:

$$\rho = \rho_0[1 - \beta_T(T' - T_0)] \qquad [7.1]$$

According to these assumptions, the continuity equation, the momentum and energy conservation equations can be written in their dimensional form as follows:

$$\begin{cases} \boldsymbol{\nabla}.\boldsymbol{V'} = 0 \\ \boldsymbol{V'} = -\dfrac{K}{\mu}\left(\boldsymbol{\nabla}P' - \rho_0[1-\beta_T(T'-T_0)]\boldsymbol{g}\right) \\ (\rho c)^*\dfrac{\partial T'}{\partial t'} + (\rho c)_f\boldsymbol{V'}.\boldsymbol{\nabla}T' = k^*\,\boldsymbol{\nabla}^2 T' \end{cases} \qquad [7.2]$$

In equations [7.1] and [7.2], $\boldsymbol{V'}$ is the filtration velocity, P' is the pressure, T' is the local temperature, and ρ_0 is the density at temperature T_0. The boundary conditions of the problem are given by:

$$\begin{cases} r = R_i, \ \ T' = T_h, \ \ \boldsymbol{V'}.\boldsymbol{n} = 0 \\ r = R_o, \ \ T' = T_c, \ \ \boldsymbol{V'}.\boldsymbol{n} = 0 \\ \theta = 0, \pi\,, \ V'_\theta = \dfrac{\partial T'}{\partial \theta} = 0 \end{cases} \qquad [7.3]$$

The system of equations is written in a dimensionless form, the scale of length, time, velocity, temperature and pressure being, respectively,

R_i, $\frac{(\rho c)^* R_i^2}{k^*}$, $\frac{k^*}{R_i(\rho c)_f}$, $(T_h - T_c)$ and $\frac{k^* \mu}{K(\rho c)_f}$. The resulting dimensionless governing equations are:

$$\begin{cases} \boldsymbol{\nabla}.\boldsymbol{V} = 0 \\ \boldsymbol{V} + \boldsymbol{\nabla} P = Ra\, T \boldsymbol{x} \\ \frac{\partial T}{\partial t} + \boldsymbol{V}.\boldsymbol{\nabla} T = \nabla^2 T \end{cases} \qquad [7.4]$$

With the associated dimensionless boundary conditions, we have:

$$\begin{cases} r = 1, \quad T = 1, \; V_r = 0 \\ r = R, \quad T = 0, \quad V_r = 0 \\ \theta = 0, \pi, \quad V_\theta = \frac{\partial T}{\partial \theta} = 0 \end{cases} \qquad [7.5]$$

where $\boldsymbol{x} = -\cos(\theta)\boldsymbol{e_r} + \sin(\theta)\boldsymbol{e_\theta}$; Ra denotes the filtration Rayleigh number defined by: $Ra = g\beta_T K R_i (T_h - T_c)(\rho c)_f / \nu k^*$ where $\nu = \mu/\rho$ denotes the kinematic viscosity of the fluid.

7.3. Various analytical solutions obtained in natural convection

Bau (1984) studied the global heat transfer between the two cylinders using the perturbation method with the Rayleigh number as expansion parameter. The expansion terms are calculated up to the order 20. Himasekhar and Bau Haim (1988) extended this expansion to the order 32. This solution, characterizing the monocellular flow, was then used to study its linear stability. Then, they demonstrated the presence of multicellular solutions. Mojtabi and Charrier-Mojtabi (1992) used a formal expansion of temperature and stream function depending on the geometric parameter $\varepsilon = \frac{R_e - R_i}{R_i} = R - 1$ in the following form:

$$\Psi = \sum_{n=1}^{N} \sum_{i=0}^{I} \varepsilon^i f_n^i(x) \sin(n\theta) \text{ and } T = \sum_{n=0}^{N} \sum_{i=0}^{I} \varepsilon^i g_n^i(x) \cos(n\theta)$$

where $r = \varepsilon x + 1$. The functions $f_n^i(x)$ and $g_n^i(x)$ are then polynomial solutions of a linear system of second-order differential equations. The degree of polynomial functions increases with the order of approximation I. For orders of approximation I > 2, the use of Maple software is required. In this work, the results obtained at the order 6 for T and Ψ are in very good agreement with the numerical results obtained by finite differences method or by the spectral method. Mojtabi et al. (2021) extended this result to order 15 in a recent paper.

The global Nusselt numbers at the inner and outer cylinders given by:

$$Nu_g = -\frac{\ln(1+\varepsilon)}{\pi\varepsilon}\int_0^{\pi}\frac{\partial T}{\partial x}_{[x=0}\,\mathrm{d}\theta = -\frac{(1+\varepsilon)\ln(1+\varepsilon)}{\pi\varepsilon}\int_0^{\pi}\frac{\partial T}{\partial x}_{[x=1}\,\mathrm{d}\theta$$

were calculated for several values of radii ratios: $R = 2^{\frac{1}{32}},\ 2^{\frac{1}{16}},\ 2^{\frac{1}{6}},\ 2^{\frac{1}{4}},\ \varepsilon \approx 0.0219,\ 0.0443, 0.090, 0.189$. The values of Nu_g remain in the vicinity of 1, which shows that the convective state is quasi-conductive.

The global Nusselt number obtained by this expansion is a function of Ra^2, result that is in agreement with the results of Himasekhar and Bau Haim (1988). It should be noted that the fourth-order approximation (equation [7.6]) obtained by Mojtabi and Charrier-Mojtabi (1992) already provides a good approximation of the global Nusselt number:

$$Nu_g = 1 + 17Ra_{th}^2\,(\varepsilon^2 - \varepsilon^3)/40{,}320 + O(\varepsilon^4) \qquad [7.6]$$

where the Rayleigh number (Ra_{th}) is based on the thickness of the annular cylinder. The relationship between Ra based on the inner radius and Ra_{th} is: $Ra_{th}=Ra(R-1) = \varepsilon Ra$. This relation equation [7.6] is valid for a wide range of variations in the Rayleigh number.

7.4. Analytical solution of the convective flow for small annular spaces

Based on equation [7.6], we deduce that the thermal transfer remains conductive even for radii ratios equal to 1.2 and for moderate values of the Rayleigh number. Therefore, the temperature field depends only on the radial coordinate r as the Nusselt number (Nu_g) remains very close to 1. The isotherms are therefore cylindrical surfaces of the same axis as the one of the porous annular cavities and radii $r \in [R_i, R_e]$. Outside the vicinity of the two surfaces $\theta = 0$ and $\theta = \pi$, the streamlines are nearly arcs of circle in any plane orthogonal to the horizontal axis of the annular cavity (parallel flow approximation). Under these conditions, the components of the velocity field of the convective flow verify the following:

$$|V_r| << |V_\theta| \ \ \forall\, \theta \in \text{in the vicinity of the two surfaces } \theta = 0 \text{ and } \theta = \pi.$$

As $\frac{\partial T}{\partial \theta} = 0$ and Vr << 1, the energy equation is written as: $\nabla^2 T = 0$. Considering the thermal boundary conditions, the solution to this equation is:

$$T = 1 - \ln(r)/\ln(R) \qquad [7.7]$$

Introducing the stream function ϕ defined by $V_r = -\partial\phi/r\partial\theta$ and $V_\theta = \partial\phi/\partial r$, the continuity equation is automatically verified. The dimensionless Darcy equation of system [7.4] is written as:

$$\nabla^2\phi = Ra\sin(\theta)\frac{\partial T}{\partial r} \qquad [7.8]$$

Setting $\phi = Ra\sin(\theta)f(r)$, equation [7.8] leads to:

$$\frac{d}{dr}\left(r\frac{df}{dr}\right) - \frac{f}{r} - r\frac{dT}{dr} = 0 \qquad [7.9]$$

The inner and outer circles of radii $r = 1$ and $r = R$, in dimensionless form, connected by vertical segments associated with $\theta = 0$ and $\theta = \pi$ form a streamline associated with $\phi = 0$, in which $f(r = 1) = 0$ and $f(r = R) = 0$.

Taking into account the boundary conditions associated with equation [7.9], we obtain the analytical expression of this equation:

$$f(r) = \frac{R^2(r^2-1)\ln(R) - r^2(R^2-1)\ln(r)}{2r(R^2-1)\ln(R)} \qquad [7.10]$$

The stream function is therefore given by:

$$\phi = Ra\sin(\theta)\frac{R^2(r^2-1)\ln(R) - r^2(R^2-1)\ln(r)}{2r(R^2-1)\ln(R)} \qquad [7.11]$$

Based on the expression of the stream function, we deduce the expression of two radial and azimuthal components of convective velocity:

$$\begin{cases} V_r = -Ra\, f(r)\, cos(\theta)\,/r \\ V_\theta = \dfrac{F}{2r^2(R^2-1)\, ln(R)} \end{cases} \qquad [7.12]$$

The function F is given by:

$$F = Ra\sin(\theta)(R^2(r^2+1)\ln(R) - r^2(R^2-1)(\ln(r)+1)) \qquad [7.13]$$

The analytical solution obtained is in good agreement with the results of direct simulations, obtained under Comsol multiphysics, for low radii ratios and moderate Rayleigh numbers and without any other restrictive hypothesis (Mojtabi et al. 2021).

7.5. Linear stability analysis of the analytically obtained single-cell flow

Linear stability studies conducted until now are related to the transition between this unicellular 2D convective flow, obtained either numerically or from an expansion as a function of the Rayleigh number at order 3, and 3D flows. The stability criteria obtained from the basic solution led to stability curves in which critical Rayleigh numbers Ra_c and wave numbers k_c were determined as a function of the radii ratio R; k is the wave number along the axial direction z orthogonal to the plane of the basic flow.

We now have available an analytical expression of the basic flow (see section 7.4), which is valid whatever the axial coordinate **z** not belonging to the vicinity $z = 0$ and $z = L$ and for low radii ratios of coaxial cylinders. This chapter resumes the linear stability study using the basic solution determined in section 7.4.

Until now, the only studies highlighting the existence of unicellular and multicellular 2D solutions have been purely numerical (Charrier-Mojtabi et al. 1991; Charrier-Mojtabi 1997). In the latter numerical study, the author showed the presence of unicellular and bicellular 2D solutions and 3D solution for the same values of the Rayleigh number (Ra_e), based on the layer thickness and for values of Ra_e much greater than the values of the critical Rayleigh number of transitions between the 3D unicellular flow and the 3D flow. Figures 7.6(a) and (b) highlight the presence of two unicellular and bicellular 2D flows for the same Rayleigh number and two radius ratios $R = \sqrt{2}$, and $R = 2$. This refers to the evolution of the global Nusselt number as a function of the Rayleigh number Ra_e. Numerical simulations were obtained using 30 collocation points according to r and 95 collocation points according to θ. For the radius ratio $R = \sqrt{2}$, the onset of the 2D bicellular flow is obtained for Ra_e between 111.5 and 112 (see Figure 7.6(a)), which result in perfect agreement with that of Himasekhar and Bau Haim (1988) using the Galerkin method, namely, $Ra_e = 112$.

For the radius ratio $R = 2$, the linear stability study of the unicellular solution related to the 2D perturbations, a component of which is orthogonal to the flow plane, by Mojtabi (1979) had led to $Ra_{emono2D->3D} = 40.83$. The onset of the bicellular 2D flow is obtained for Ra^* between 60.5 and 61.5 (see Figure 7.6(b)). Rao et al. (1987) found that $Ra_e = 65.5 \pm 0.5$ using the low-approximation Galerkin method. These two figures show that the bicellular flow leads to a more significant heat transfer.

Until present, no theoretical or analytical study was conducted to determine the possible transitions between basic 2D unicellular convective flow and 2D multicellular flows.

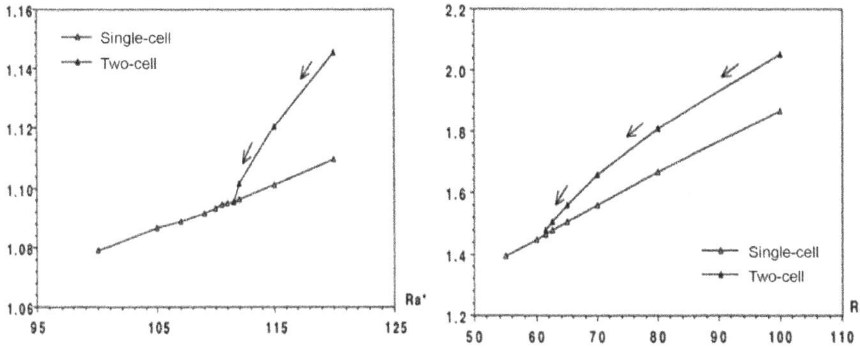

a) b)

Figure 7.6. *Transition between the unicellular 2D convective flow and the bicellular 2D flow for* $R = \sqrt{2}$ *(a) and* $R = 2$ *(b)*

7.5.1. *Stability of this solution with respect to (r, z) axisymmetric perturbation*

This section focuses on (r, z) axisymmetric 2D perturbations. The study is restricted to the top part of the annular layer, that is, $\theta = \pi$. Let us assume that velocity and temperature perturbations are written in the following form:

$$\begin{cases} \boldsymbol{v} = u(r,z,t)\boldsymbol{e}_r + w(r,z,t)\boldsymbol{e}_z \\ \vartheta = \vartheta(r,z,t) \end{cases} \qquad [7.14]$$

with $(u, v, \vartheta) = (u(r), v(r), \vartheta(r))e^{(isz-\sigma t)}$, where s denotes the wave number characterizing the periodicity of the perturbation in the direction of the z axis of the two coaxial cylinders, the latter being orthogonal to the flow plane defined by the basic solution studied. We assume the principle of exchange of stability, $\sigma \in R$ (Guershuni and Zhukhovitskii 1972). It is therefore admitted that infinitesimal perturbations are monotonically increasing or decreasing functions of time. The marginal state is obtained for $\sigma = 0$. Setting $Ra_m = Ra^{1/2}$ and replacing ϑ by $\frac{\vartheta}{Ra_m}$, we obtain the second-order linear differential system linking functions $u(r)$ and $\vartheta(r)$:

$$\begin{cases} Ra_m\left(\frac{D}{r} - s^2\right)\vartheta - \left(D^2 + \frac{D}{r} - s^2 - \frac{1}{r^2}\right)u = 0 \\ \left(D^2 + \frac{D}{r} - s^2 - V_r\right)\vartheta - Ra_m\frac{\partial T}{\partial r}u = 0 \end{cases} \qquad [7.15]$$

$V_r = Ra_m{}^2 \frac{f(r)}{r}$ and T=1 – $\ln(r)/\ln(R)$ denote, respectively, the velocity component along $\boldsymbol{e}_r$ and the temperature of the base flow. D represents the partial derivative with respect to r.

This study focuses on the stability of the top part of the annular layer, $\theta = \pi$, since, as shown by the above mentioned stability studies, it is in the vicinity of $\theta = \pi$, in which instability is generated. The Galerkin method is used to transform this differential system into an algebraic system. For this, the unknown functions $u(r)$ and $\vartheta(r)$ are written as follows:

$u = \sum_{n=1}^{N} a_n\, u_n(r)$, $\vartheta = \sum_{n=1}^{N} b_n \vartheta_n\,(r)$, the functions u_n and ϑ_n must verify the following boundary conditions: $u_n(r = 1, R) = 0$ and $\vartheta_n(r = 1, R) = 0$. Then the functions u_n and ϑ_n are defined by the same linearly independent test functions forming a base:

$$\frac{(r-1)(R-r)}{(R-1)^2}, \frac{r(R-(R+1)r+r^2)}{(R-1)^3}, \frac{\left((r-1)(R-r)\right)^2}{(R-1)^4}, \frac{(r-1)^i((r-1)(R-r))^2}{(R-1)^{4+i}},$$

with $i = 1 \ldots I$.

The results of this stability study and the comparison with the previous works are summarized by Table 7.1.

R	N = 1		N = 2		N = 3	
	Ra_{ec}	Sc	Ra_{ec}	Sc	Ra_{ec}	Sc
	Linear theory (numerical results)					
1.2	40.323	3.169	39.819	3.148	39.819	3.148
$\sqrt{2}$	41.283	3.185	40.833	3.165	40.834	3.165
	Linear theory (perturbation methods)					
1.2	40.403	3.172	39.895	3.151	39.895	3.151
$\sqrt{2}$	41.385	3.195	40.912	3.175	40.834	3.175
	Linear theory (analytical solution, parallel flow)					
1.2	40.060	3.165	40.345	3.171	39.841	3.150
$\sqrt{2}$	40.197	3.170	41.436	3.197	40.984	3.160

Table 7.1. *Critical Rayleigh number and critical wave number of the transition between the unicellular 2D convective flow obtained through various methods and 3D flow*

7.5.2. Linear stability study of this basic solution in the limit case $\varepsilon \to 0$ (axisymmetric 2D perturbation (x, z))

The initial dimensionless formalism (section 7.4) shows two dimensionless numbers of the problem, which are the Rayleigh number Ra, based on the radius of the inner cylinder R_i and the radii ratio R. The following section introduces the Rayleigh number based on the thickness $Ra_e = \varepsilon Ra$ and the parameter $\varepsilon = R - 1$. Moreover, we make the following change of variable, $x = (r-1)/\varepsilon$, to replace the variable r that varies between 1 and R with the variable x that varies between 0 and 1 in the differential system [7.15]:

$$\begin{cases} Ra_m\left(\dfrac{\varepsilon D}{1+\varepsilon x} - s^2\right)\vartheta - \left(D^2 + \dfrac{\varepsilon D}{1+\varepsilon x} - s^2 - \dfrac{\varepsilon^2}{(1+\varepsilon x)^2}\right)u = 0 \\ \left(D^2 + \dfrac{\varepsilon D}{1+\varepsilon x} - s^2 - V_r D\right)\vartheta - Ra_m \dfrac{\partial T}{\partial x} u = 0 \end{cases} \quad [7.16]$$

Here, the operator $D = \frac{d}{dx}$, $s = \pi m(R_e - R_i)/L$, m the number of convective cells developed in this direction and

$$\frac{\partial T}{\partial x} = -1 + o(\varepsilon),\ V_r = Ra\frac{f(r)}{r} = o(\varepsilon),\ \vartheta = \vartheta_0 + o(\varepsilon) \text{ and } u = u_0 + o(\varepsilon).$$

The differential system [7.16] leads to the following system:

$$\begin{cases} Ra_m s^2 \vartheta_0 + (D^2 - s^2)u_0 = 0 \\ (D^2 - s^2)\vartheta_0 + Ra_m u_0 = 0 \end{cases} \quad [7.17]$$

This system [7.17] is identical to the one obtained for the study of the linear stability of the bottom heated plane horizontal porous layer (Roger Lapwood problem). By writing the perturbations ϑ_0 and u_0 in the form of an expansion in $\sin(n\pi x)$ verifying the set of boundary conditions in $x = 0$ and $x = 1$, we obtain the following using the Galerkin method:

$$Ra_m^2 = Ra_e = \frac{(n^2\pi^2 + s^2)^2}{s^2} \quad [7.18]$$

By minimizing Ra_e with respect to s (wave number of the perturbation) and to n, we obtain the linear critical Rayleigh number and the associated critical wave number:

$$Ra_{ec} = 4\pi^2 \text{ and } s_c = \pi \quad [7.19]$$

Hence, when $\varepsilon \to 0$, the natural convection flow in porous annular geometry for 2D axisymmetric perturbation is unstable for $Ra_e > 4\pi^2$.

This result is only valid for a very large extension column along the horizontal **z** axis. For a finite extension column, which is the case of all laboratory experiments, the critical wave number $s_c = \pi$ corresponding to $Ra_{ec} = 4\pi^2$ can only be obtained for a particular cell ratio, since $s = \pi m(R_e - R_i)/L$. The critical wave number $s_c = \pi$ is only obtained for particular columns of length equal to a multiple of annular column thickness (values reached for $\frac{L}{R_e - R_i} = 1, 2, 3 \ldots$).

Similarly, if we study the variation of $Ra_{ec} = (\pi^2/s + s)^2$ as a function of $\frac{L}{R_e - R_i}$ for $m = 1, m = 2, m = 3 \ldots$, we obtain the curve shown in Figure 7.7. It can be noted that when $\frac{L}{R_e - R_i}$ is increasingly below 1, Ra_{ec} strongly increases. As an example for a cell used in our experiments with $\frac{L}{R_e - R_i} = 0.237$, we find $Ra_{ec} = 187$.

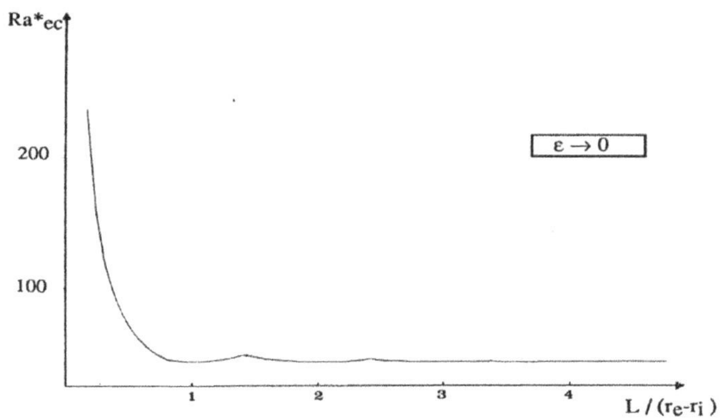

Figure 7.7. *Variation of Ra_{ec} as a function of $\frac{L}{R_e - R_i}$*

7.5.3. Linear stability equations in relation with 2D perturbations in the (r, θ) plane

To study the stability of this convective flow (section 7.4), we transform the formulation [7.2] in terms of velocity, pressure and temperature in a formulation in terms of stream function and temperature. Under these conditions, the continuity equation is automatically verified, and the problem can be reduced to only two scalar differential equations having as unknowns only the stream function Ψ and the temperature field T:

$$\begin{cases} \nabla^2\Psi = Ra\,(\sin(\theta)\dfrac{\partial T}{\partial r} + \cos(\theta)\dfrac{\partial T}{r\partial\theta}) \\ \nabla^2 T = \dfrac{\partial\Psi}{r\partial\theta}\dfrac{\partial T}{\partial r} - \dfrac{\partial\Psi}{\partial r}\dfrac{\partial T}{r\partial\theta} \end{cases} \quad [7.20]$$

with $V_r = -\partial\Psi/r\partial\theta$ and $V_\theta = \partial\Psi/\partial r$. This problem admits as solution, the basic solution previously obtained for equations [7.7] and [7.11], which is: $T_0 = 1 - \ln(r)/\ln(R)$ and $\Psi_0 = Ra\sin(\theta)\frac{R^2(r^2-1)\ln(R) - r^2(R^2-1)\ln(r)}{2r(R^2-1)\ln(R)}$. The Rayleigh number Ra here is based on the inner radius of the annular cylinder.

Let us introduce the infinitesimal perturbations φ and ϑ of the stream function and the temperature: the perturbed fields $(\tilde{\phi}, \tilde{T})$ are defined by: $\tilde{\phi} = \Psi_0 + \varphi$, $\tilde{T} = T_0 + \vartheta$. In these conditions, after having neglected the second-order terms of perturbations, the linear stability equations are written as:

$$\begin{cases} \nabla^2\varphi = Ra\,(\sin(\theta)\dfrac{\partial\vartheta}{\partial r} + \cos(\theta)\dfrac{\partial\vartheta}{r\partial\theta}) \\ \dfrac{\partial\vartheta}{\partial t} - \dfrac{\partial\varphi}{r\partial\theta}\dfrac{\partial T_0}{\partial r} - \dfrac{\partial\Psi_0}{r\partial\theta}\dfrac{\partial\vartheta}{\partial r} + \dfrac{\partial\Psi_0}{r\partial r}\dfrac{\partial\vartheta}{\partial\theta} + \dfrac{\partial\varphi}{\partial r}\dfrac{\partial T_0}{r\partial\theta} = \nabla^2\vartheta \end{cases} \quad [7.21]$$

with $\boldsymbol{v} = -\frac{\partial\varphi}{r\partial\theta}\boldsymbol{e}_r + \frac{\partial\varphi}{\partial r}\boldsymbol{e}_\theta$.

We make the following change of variable, $x = (r-1)/\varepsilon$, to replace the variable r that varies between 1 and R with the variable x that varies between 0 and 1 as in section 7.5.2. The system [7.20] yields the following system:

$$\begin{cases} r^2\dfrac{\partial^2\Psi}{\partial x^2} + +\varepsilon r\dfrac{\partial\Psi}{\partial x} + \varepsilon^2\dfrac{\partial^2\Psi}{\partial\theta^2} = Ra_e(r^2\sin(\theta)\dfrac{\partial T}{\partial x} + \varepsilon r\cos(\theta)\dfrac{\partial T}{\partial\theta}) \\ r^2\dfrac{\partial^2 T}{\partial x^2} + +\varepsilon r\dfrac{\partial T}{\partial x} + \varepsilon^2\dfrac{\partial^2 T}{\partial\theta^2} = r\varepsilon^2(\dfrac{\partial\Psi}{\partial\theta}\dfrac{\partial T}{\partial x} - \dfrac{\partial\Psi}{\partial x}\dfrac{\partial T}{\partial\theta}) \end{cases} \quad [7.22]$$

where $r = 1 + \varepsilon x$.

Without being restricted to the summit part as in section 7.5, perturbations are written as follows:

$$\varphi = \varphi(r,\theta)e^{-\sigma t} \text{ and } \vartheta = \vartheta(r,\theta)e^{-\sigma t} \quad [7.23]$$

The boundary conditions on the perturbations of temperature and stream function are, respectively, given by:

$$\begin{cases} \vartheta(x=0,\theta)=\vartheta(x=1,\theta)=0\ \ \forall\,\theta\in[0,\pi] \\ \dfrac{\partial\vartheta}{\partial\theta}(x,\theta=0)=\dfrac{\partial\vartheta}{\partial\theta}(\mathrm{x},\theta=\pi)=0\ \ \forall\,x\in[0,1] \end{cases} \quad [7.24]$$

$$\begin{cases} \varphi(x=0,\theta)=\varphi(x=1,\theta)=0\ \ \forall\,\theta\in[0,\pi] \\ \varphi(x,\theta=0)=\varphi(x,\theta=\pi)=0\ \ \forall\,x\in[0,1] \end{cases} \quad [7.25]$$

Under these conditions, after having neglected the second-order terms of perturbations, the linear stability equations are written as follows:

$$\begin{cases} r^2\dfrac{\partial^2\varphi}{\partial x^2}++\varepsilon r\dfrac{\partial\varphi}{\partial x}+\varepsilon^2\dfrac{\partial^2\varphi}{\partial\theta^2}=Ra_e(r^2\sin(\theta)\dfrac{\partial\,\vartheta}{\partial x}+\varepsilon r\cos(\theta)\dfrac{\partial\,\vartheta}{\partial\theta}) \\ r^2\dfrac{\partial^2\,\vartheta}{\partial x^2}++\varepsilon r\dfrac{\partial\,\vartheta}{\partial x}+\varepsilon^2\dfrac{\partial^2\,\vartheta}{\partial\theta^2}=r\varepsilon^2(\dfrac{\partial\Psi_0}{\partial\theta}\dfrac{\partial T}{\partial x}+\dfrac{\partial\varphi}{\partial\theta}\dfrac{\partial T_0}{\partial x}-\dfrac{\partial\varphi}{\partial x}\dfrac{\partial T_0}{\partial\theta}-\dfrac{\partial\Psi_0}{\partial x}\dfrac{\partial\vartheta}{\partial\theta}) \end{cases} \quad [7.26]$$

This system can be solved using the Galerkin method where the test functions are polynomials verifying that the set of boundary conditions [7.24] and [7.25] has the following general form:

$$\begin{aligned} \varphi_1 &= \textstyle\sum_{n=1}^{N}\sum_{m=1}^{M} a_{n,m}\, x^n(1-x)\theta^m(\pi-\theta) \\ \vartheta &= \textstyle\sum_{n=1}^{N}\sum_{m=0}^{M-1} b_{n,m}\, x^n(1-x)g_m(\theta) \end{aligned} \quad [7.27]$$

The functions g_m are given by:

$$g_0=cst, g_1=\theta^2\left(\theta-\frac{3\pi}{2}\right), g_2=\theta^2(\pi-\theta)^2\,, g_{2+i}=\theta^{2+i}(\pi-\theta)^2$$

It is also possible to use for this study polynomial functions in x and Fourier functions in θ:

$$\begin{aligned} \varphi_1 &= \textstyle\sum_{n=1}^{N}\sum_{m=1}^{M} a_{n,m}\, x^n(1-x)\sin(m\theta) \\ \vartheta &= \textstyle\sum_{n=1}^{N}\sum_{m=0}^{M-1} b_{n,m}\, x^n(1-x)\cos(m\theta) \end{aligned} \quad [7.28]$$

7.6. Conclusion

This study examined a large part of the experimental, numerical and theoretical works published during these four past decades on natural convection in horizontal cylindrical annular in fluid and porous medium. This geometry can be found in many industrial applications.

In the first part, we presented the first experimental results having led to a good description of the natural convection flow in horizontal annular configuration in fluid medium. To illustrate this part, we inserted several photos obtained by our team using Mach–Zehnder interferometry to characterize the thermal field. Streamlines were visualized for the air using cigarette smoke, for water using a detergent and for various silicone oils using alumina powder. We obtained a good agreement between our numerical simulation results and our experimental results.

The second part dealt with similar works in porous annular geometry. Various numerical studies were obtained using the finite differences method or the Galerkin method, and then spectral collocation methods and the finite element method. Analytical approaches were also developed, first the perturbation method, which was extensively used. The temperature field and the stream function are written as integer series of the Rayleigh number. This method of expansion to the second or third order yields a correct approximation of the convective solution only for low values of the Rayleigh number. For a cylindrical annular whose thickness is small compared to its inner radius, the temperature field and the stream function are obtained by an asymptotic expansion as a function of the geometric parameter $\varepsilon = R - 1$. Using this expansion, we have formally obtained an analytical solution to this convection problem. Formal calculation (Maple software) allowed us to effectively obtain the solution at the 15th order. The results obtained using this expansion are in very good agreement with those obtained by the Galerkin method or the finite differences method. The use of the parallel flow hypothesis to the two circles limiting the annular space, numerically observed in the case of a small annular space, allowed us to obtain an analytical solution to this problem.

In the last part of this work, we studied the stability of the analytical solution obtained under the parallel flow hypothesis. We recently obtained (Mojtabi et al. 2021) an analytical solution of the 2D unicellular flow in the case of small annular spaces and we studied, in the present work, the stability of this unicellular solution in relation with axisymmetric 2D perturbations in the direction of the **z** axis of the cylinders and of the radial direction, r (section 7.5.1). We also assumed that the flow was destabilized in the top part of the annular layer ($\theta = \pi$), considering the experimental observations and the numerical simulations conducted. We analyzed the loss of stability of this 2D unicellular solution to the benefit of the 3D flow, whose structure was described in the first part of this study. The comparison of the stability results of this analytical solution valid for small annular spaces and the stability results of the solution obtained by the perturbation method valid for low values of the Rayleigh number numerically shows a good agreement between these different approaches (Table 4.1) for $R = 1.2$ and $R = \sqrt{2}$.

We then proposed a linear stability study in the case of $\varepsilon = R - 1$ tends toward zero and by limiting the study to the vicinity of $\theta = \pi$. We used 2D perturbations in the $\boldsymbol{e}_z$ and $\boldsymbol{e}_r$ directions. By expanding the basic flow and the perturbations to order o(ε), we found the stability equations of the problem of porous horizontal plane layer heated from below and having the solution $Ra_{ec} = 4\pi^2$ and $s_c = \pi$. An examination of the analytical results obtained shows that if $\frac{L}{R_e - R_i}$ <<1, the Rayleigh number becomes very high compared to $Ra_{ec} = 4\pi^2$. The lateral confinement of the cell stabilizes 2D unicellular flow with respect to the perturbations in the two directions $\boldsymbol{e}_z$ and $\boldsymbol{e}_r$. In view of this work, it would seem interesting to conduct an experimental study in order to analyze the multicellular flows that are likely to develop, as the critical Rayleigh number leading to the 3D flow becomes very high compared to $4\pi^2$.

In the last part of this work, we presented the stability equations of 2D unicellular flow in relation to the 2D perturbations in the flow plane, whatever θ. To facilitate this stability study, by the Galerkin method, we presented various bases of test functions, polynomial and Fourier functions, for the perturbations.

7.7. References

Bau, H.H. (1984). Thermal convection in a horizontal, eccentric annulus containing a saturated porous medium: An extended perturbation expansion. *Intl J. Heat Mass Transfer*, 27, 2277–2787.

Caltagirone, J.P. (1976). Thermoconvective instabilities in a porous medium bounded by two concentric horizontal cylinders. *Journal of Fluid Mechanics*, 76(2), 337–362.

Charrier-Mojtabi, M.C. (1993). Etude numérique, théorique et expérimentale des écoulements thermoconvectifs bidimensionnels et tridimensionnels en couche annulaire poreuse horizontale. PhD Thesis, Université Bordeaux 1, Talence.

Charrier-Mojtabi, M.C. (1997). Numerical simulation of two and three-dimensional free convection flows in a horizontal porous annulus using a pressure and temperature formulation. *Int. J. Heat and Mass Transfer*, 40(7), 1521–1533.

Charrier-Mojtabi, M.C. and Mojtabi, A. (1990). Natural convection in a horizontal porous annulus. In *Transport Phenomena in Porous Media*, Ingham, D.B. and Pop, I. (eds). Pergamon Press, Oxford.

Charrier-Mojtabi, M.C. and Mojtabi, A. (1998). Natural convection in a horizontal porous annulus. In *Transport Phenomena in Porous Media*, Ingham, D.B. and Pop, I. (eds). Pergamon Press, Oxford.

Charrier-Mojtabi, M.C., Mojtabi, A., Caltagirone, J.P. (1979). Numerical solution of a flow due to natural convection in horizontal cylindrical annulus. *Journal of Heat Transfer, ASME*, 101(1), 171–173.

Charrier-Mojtabi, M.C., Mojtabi, A., Azaiez, M., Labrosse, G. (1991). Numerical and experimental study of multicellular free convection flows in an annular porous layer. *Int. J. Heat and Mass Transfer*, 34(12), 3061–3074.

Cheddadi, A., Caltagirone, J.P., Mojtabi, A., Vafai, K. (1992). Free two-dimensional convective bifurcation in a horizontal annulus. *Journal of Heat Transfer, ASME*, 114(1), 99–106.

Dyko, M.P., Vafai, K., Mojtabi, A. (1999). A numerical and experimental investigation of stability of natural convective flows within a horizontal annulus. *Journal of Fluid Mechanics*, 381, 27–61.

Grigull, U. and Hauf, W. (1966). Natural convection in horizontal cylindrical annuli. *International Heat Transfer Conference 3*, August 7–12, Chicago.

Guershuni, G.L. and Zhukhovitskii, E.M. (1972). *Convective Stability of Incompressible Fluids*. Israel Program for Scientific Translation, Jerusalem.

Himasekhar, K. and Bau Haim, H. (1988). Two dimensional bifurcation phenomena in thermal convection in horizontal concentric annuli containing saturated porous media. *Journal of Fluid Mechanics*, 187, 267–300.

Kuehn, T.H. and Goldstein, R.J. (1976). An experimental and theoretical study of natural convection in the annulus between horizontal concentric cylinders. *Journal of Fluid Mechanics*, 74(4), 695–719.

Mojtabi, A. (1979). Instabilités thermoconvectives dans des espaces confinés et semi-confinés. PhD Thesis, Université Pierre et Marie Curie, Paris.

Mojtabi, A. and Caltagirone, J.P. (1979a). Study of the stability of a free-convection flow in a horizontal annular space. *Journal de Mécanique*, 18(2), 225–241.

Mojtabi, A. and Caltagirone, J.P. (1979b). Energy stability of natural convective flow in a horizontal annular space. *Phys. Fluids*, 22, 1208–1209.

Mojtabi, A. and Charrier-Mojtabi, M.C. (1992). Analytical solution of steady natural convection in an annular porous medium evaluated with a symbolic algebra code. *Journal of Heat Transfer*, 114(4), 1065–1067.

Mojtabi, A., Sioud, K., Bergeon, A., Charrier-Mojtabi, M.C. (2021). Numerical and analytical studies of Soret-driven convection flow inside an annular horizontal porous cavity. *Fluids*, 6, 357.

Rao, Y.F., Fukuda, K., Hasegawa, S. (1987). Steady and transient analysis of natural convection in a horizontal porous annulus with Galerkin method. *Journal of Heat Transfer*, 109, 919–927.

Van Dyke, M. (1982). *An Album of Fluid Motion*. The Parabolic Press, Stanford, CA.

8

Instability Modeling in Structural Mechanics: Buckling and Wrinkling

This chapter presents a multiple scale method, based on the Fourier series with slowly varying coefficients, to describe the formation of instability patterns. Shell finite element codes can predict the details of membrane wrinkles, but with difficulties due to the large number of unknowns and to the existence of many solutions. Macroscopic models are also available, but they only take into account the effect of wrinkles on membrane behavior. The method based on Fourier series with slowly varying coefficients was introduced in order to model the main characteristics of wrinkles, while using equations with partial derivatives only at the macroscopic level. In this method, the sought-after solution has the form of several terms of the Fourier series whose coefficients vary more slowly than the oscillations. Several examples of buckling and wrinkling are presented and discussed.

8.1. Introduction

This chapter focuses on the macroscopic description of cellular instabilities, such as the membrane wrinkling (Rossi et al. 2005; Lecieux and Bouzidi 2010; Trouffard et al. 2010), the buckling of long structures (Potier-Ferry 1983; Damil and Potier-Ferry 1986; Boucif et al. 1991; Abdelmoula et al. 1992; Cochelin et al. 1994), Rayleigh–Bénard convection (Newell and Whitehead 1969; Segel 1969), the buckling of thin elastic films on a flexible substrate (Chen and Hutchinson 2004; Audoly and Boudaoud 2008; Wang et al. 2008), the micro-buckling of fibers, which is the mechanism that initiates the compression fracture of long-fiber composites

---

Chapter written by Noureddine DAMIL and Michel POTIER-FERRY.

For a color version of all of the figures in this chapter, see www.iste.co.uk/saouab/mechanics.zip.

(Kyriakides et al. 1995; Waas and Schultheisz 1996; Drapier et al. 2001), and many other instabilities in various scientific fields (Pomeau and Zaleski 1981; Cross et al. 1983; Wesfreid and Zaleski 1984; Hunt et al. 1988; Cross and Hohenberg 1993; Hunt and Wadee 1998; Hunt et al. 2000; Leotoing et al. 2002; Hoyle 2006; Dawes 2010; Wadee and Gardner 2012). In all of these cases, the system response is similar to a weakly modulated spatial oscillation. This is why the considered macroscopic models will be similar to slowly varying functions relative to this oscillation.

The evolution of space-time instabilities can be macroscopically described by generic Landau–Ginzburg amplitude equations (Iooss et al. 1989; Damil and Potier-Ferry 1992). The classical multiscale asymptotic approach leads to a Landau–Ginzburg amplitude equation.

In this chapter, we establish a variant of this approach that takes into account the coupling between local and global instabilities and deals similarly with mean field and fluctuation. The final model is a generalized continuous medium, where generalized stresses are Fourier coefficients of the microscopic stress.

A new approach was presented in Damil and Potier-Ferry (2006, 2008, 2010), which relies on the concept of Fourier series with slowly varying coefficients. According to this technique, Fourier coefficients are the unknowns of the macroscopic problem. It was established that the models obtained in this way are consistent with the Landau–Ginzburg technique, but they can also remain valid beyond the bifurcation point and the coupling between global and local instabilities can be taken into consideration (Damil and Potier-Ferry 2006). Budd et al. (2001) and Budd and Kuske (2005) include several sophisticated extensions of the asymptotic Landau–Ginzburg approach with destabilizing–restabilizing nonlinearities that are thus valid relatively far from the bifurcation point. It is however not possible to obtain a clear and reliable consideration of the boundary conditions, which are the third drawback, as it is intrinsically related to the use of any model reduction. Furthermore, this chapter deals with the treatment of boundary conditions in reduced macroscopic models as the asymptotic Landau–Ginzburg approach or those deduced from the reference method (Damil and Potier-Ferry 2006, 2008, 2010).

The simplest model involves two real macroscopic variables: the mean field and the envelope of oscillations. It was evaluated by Damil and Potier-Ferry (2010) and Hu et al. (2011) and its capacity to provide a good estimation of the pattern distribution was demonstrated. Unfortunately, the hypothesis of a real oscillation envelope locks the oscillation phase, which leads to certain inaccuracies near the boundary. Moreover, phase locking also involves a locking of the period that is fixed in this model with an actual envelope. A multiple scale modeling was proposed in order to bypass the problem of boundary conditions (Hu et al. 2011): the complete model is considered near the edge, the envelope model in the central part and they are linked by the Arlequin method (Ben Dhia and Rateau 2005). This clarifies the problem of boundary conditions, but the numerical implementation is not simple, as it involves three levels: the complete model, the reduced model and the connection model. A question to be considered is whether a two-scale approach is actually needed and if most instability phenomena could be described only at a macroscopic level. In Mhada et al. (2011), we discussed a third model. The principle is very simple, it associates an actual variable for the mean field with a complex variable for the oscillation amplitude. The introduction of the complex amplitude allows the prediction of the phase shift and the shift in the oscillation period.

Several models can be built according to this principle, particularly as a function of the considered number of harmonics. These models will be compared and evaluated based on several numerical tests in the case of a beam on a nonlinear elastic foundation that is quite similar to the Swift–Hohenberg equation (Swift and Hohenberg 1977; Dawes 2010). The objective is not necessarily to define a model that should be optimal in all cases. Indeed, the best model depends on the requirements of the user, who is sometimes interested only in a rough estimate of the space distribution of instability patterns and may sometimes want a very accurate solution everywhere and especially near the edge.

These reduced models rely on the restrictive hypothesis of a uniform wavelength and a uniform orientation of the wrinkles. In Khalil et al. (2020), we released these hypotheses and presented a new general framework of the macroscopic modeling for a membrane wrinkling occurring in various and variable directions. Model reduction relies on the same multiple scale approach using the method of Fourier series with slowly varying coefficients. In other terms, the reduction procedure introduced in Damil and Potier-Ferry (2006) is extended to the case when the assumed wave vector is not uniform, particularly when the assumed orientation of wrinkles may vary in the domain. This leads to reduced models satisfied by the envelopes that can be solved by standard numerical methods.

The complete modeling of this type of phenomenon can lead to very large-scale numerical models when there are many wrinkles, so that the reduced-order modeling is a very interesting approach. Nowadays, many commercial finite element codes can be used for such calculations, for example, for structural elements of nonlinear thin shells. Numerical analyses can describe the details of membrane response: size, wave length, wrinkle orientation, instability threshold, etc., but this leads to large size numerical models that are additionally very difficult to control when there are many wrinkles and therefore many equilibrium solutions. In Wong and Pellegrino (2006a), the authors presented a general procedure using a shell finite element in the commercial software of finite elements Abaqus to simulate the growth of wrinkles, the results being consistent with the experiments of Wong and Pellegrino (2006b, 2006c). The shell finite elements were adopted in many articles using a large number of degrees of freedom to capture the wrinkle details (e.g. Wang et al. 2009; Flores and Oñate 2011; Lecieux and Bouzidi 2012). In Healey et al. (2013), the authors modified the Föppl von Karman model taking into account the finite strain of the median plane of the highly stretched state. The resulting problem is solved by the Newton–Raphson continuation method (Keller 1987) using a finite element discretization known as compliant. Li and Healey (2016) extended this approach to the neo-Hookean (nHk) model and to the Mooney–Rivlin (MR) model. In Wang et al. (2019), the authors numerically studied the wrinkling and restabilizing behavior using the commercial finite element software Abaqus.

This chapter is organized as follows: section 8.2 reviews the classical multi-scale asymptotic approach of instabilities, Landau–Ginzburg equations, which allow for the deduction of a macroscopic model that controls the spatial evolution of the amplitude of buckling patterns in the case of a beam on an elastic foundation. Sections 8.3 and 8.4 present the multi-scale method for the modeling of instabilities using Fourier series with slowly varying coefficients. Several macroscopic models will be studied in the case of buckling of a beam on an elastic foundation. Section 8.5 presents a 2D Fourier double scale analysis for the global-local instability interaction in sandwich structures. Section 8.6 presents a reduced model based on the multi-scale method in the case of membrane wrinkling with variable orientation. Several numerical simulations will be discussed to evaluate the efficiency and the validity domain of reduced models. This section will numerically study the appearance and development of wrinkles in elastic rectangular membranes subjected to biaxial loads or shearing in radially stretched membranes or in a disk pierced by an offset circular hole under stretching.

8.2. Classical multi-scale asymptotic approach to instabilities

This section reviews the classical multi-scale asymptotic approach (Wesfreid and Zaleski 1984; Damil and Potier-Ferry 1986; Cross and Hohenberg 1993). It can be used to deduce a macroscopic model that governs the spatial evolution of the amplitude of buckling patterns.

Let us consider, as an example, the buckling of an elastic beam of length L, rigidities ES and EI, on a nonlinear elastic Winkler foundation. The potential energy of the system has the following form:

$$P(u,v) = \int_0^L \left(\frac{ES}{2}\gamma^2 + \frac{EI}{2}v''^2 + h(v)\right)dx - \int_0^L fudx \qquad [8.1]$$

$$\gamma = u' + \frac{v'^2}{2} \quad h(v) = c_1\frac{v^2}{2} + c_2\frac{v^3}{3} + c_3\frac{v^4}{4} \qquad [8.2]$$

The unknowns of the problem are axial and transversal components of the displacement vector, $u(x)$ and $v(x)$. The strain γ depends on displacement according to the simplified Von Karman hypotheses. The potential energy of the foundation is given by the function $h(v)$, which is assumed to be a fourth-degree polynomial and where c_1, c_2 and c_3 are characteristics of the foundation.

The extrema of the functional [8.1] are characterized by the following well-known differential system:

$$\frac{dn}{dx} + f = 0 \qquad [8.3]$$

$$\frac{n}{ES} = \frac{du}{dx} + \frac{1}{2}\left(\frac{dv}{dx}\right)^2 \qquad [8.4]$$

$$\frac{d^2}{dx^2}\left(EI\,\frac{d^2v}{dx^2}\right) - \frac{d}{dx}\left(n\frac{dv}{dx}\right) - g = 0 \qquad [8.5]$$

$$-g(v) = c_1 v + c_2\, v^2 + c_3\, v^3 \qquad [8.6]$$

It is important to note that the weak version of the equations of equilibrium [8.3] and [8.5] can be written in the following classical form:

$$\int_0^L (n\delta\gamma + m\delta v'' - g(v)\delta v)\, dx = \int_0^L f\delta u dx \qquad [8.7]$$

where $m = EIv''$.

8.2.1. *Linear stability analysis*

The origin of the instability can be understood by considering the case without horizontal force ($f = 0$), which implies that the normal force is constant, $n(x) = -\lambda$. Let us consider the case of an imposed compression ($\lambda > 0$) and assume that all of the coefficients of materials ES, EI, c_1, c_2, and c_3 are constant.

Classically, the instability mode is a solution of the bending equation linearized over an infinite domain:

$$EI\,\frac{d^4v}{dx^4} + \lambda\,\frac{d^2v}{dx^2} + c_1 v = 0 \qquad [8.8]$$

There is a family of solutions of [8.8] that depend on a wave number q:

$$v(x) = \sin(qx + \phi)\ \lambda(q) = EI\ q^2 + \frac{c_1}{q^2} \qquad [8.9]$$

The critical load λ_c corresponds to a minimum of the curve linking the load and the wave numbers:

$$q = \sqrt[4]{c_1/EI}\ \lambda_c = Min\ \lambda(q) = 2\sqrt{EIc_1} = 2\ EI\ q^2 = \frac{2c_1}{q^2} \qquad [8.10]$$

8.2.2. *Landau–Ginzburg asymptotic approach*

This section reviews the application of the multiple scale asymptotic method to the problem of the beam [8.3]–[8.6] in the case of an imposed and uniform normal force ($f = 0$, $n = -\lambda > 0$). The system [8.3]–[8.6] is reduced to a single equation for the bending:

$$EI\frac{d^4v}{dx^4} + \lambda\frac{d^2v}{dx^2} + c_1 v + c_2 v^2 + c_3 v^3 = 0 \qquad [8.11]$$

In the asymptotic approach, we are seeking solutions close to the bifurcation point, which means that we assume that the load increment $\varepsilon = \lambda - \lambda_c$ and the bending are small, and $v(x)$ is a slowly modulated oscillation. In this case, the multiple scale asymptotic analysis (see, for example, Nayfeh 1973; Sanchez-Palencia 1980) leads to the following expressions:

$$\varepsilon = \lambda - \lambda_c \quad X = x\sqrt{\epsilon} \qquad [8.12]$$

$$v(x) = \sqrt{\varepsilon}\, v^{(1)}(x,X) + \varepsilon\, v^{(2)}(x,X) + \varepsilon^{3/2} v^{(3)}(x,X) + \cdots \qquad [8.13]$$

The two spatial variables x and X are assumed to be independent, which leads to the following rule to take into account rapid variations with x and slow variations with X:

$$\frac{d}{dx} = \frac{\partial}{\partial x} + \sqrt{\varepsilon}\frac{\partial}{\partial X} \qquad [8.14]$$

When the expansion rules [8.12]–[8.14] are applied to [8.11], we obtain the following equations of orders $O(\sqrt{\varepsilon})$, $O(\varepsilon)$, $O(\varepsilon^{3/2})$:

$$L\left(v^{(1)}\right) = EI\frac{\partial^4 v^{(1)}}{\partial x^4} + \lambda_c\frac{\partial^2 v^{(1)}}{\partial x^2} + c_1 v^{(1)} = 0 \qquad [8.15]$$

$$L\left(v^{(2)}\right) = -4EI\frac{\partial}{\partial X}\frac{\partial^3 v^{(1)}}{\partial x^3} - 2\lambda_c\frac{\partial}{\partial X}\frac{\partial v^{(1)}}{\partial x} - c_2\left(v^{(1)}\right)^2 \qquad [8.16]$$

$$\begin{aligned} L\left(v^{(3)}\right) = &-4EI\frac{\partial}{\partial X}\frac{\partial^3 v^{(2)}}{\partial x^3} - 2\lambda_c\frac{\partial}{\partial X}\frac{\partial v^{(2)}}{\partial x} \\ &- 6EI\frac{\partial^2}{\partial X^2}\frac{\partial^2 v^{(1)}}{\partial x^2} - \lambda_c\frac{\partial^2 v^{(1)}}{\partial X^2} - 2c_2 v^{(1)} v^{(2)} \\ &- c_3\left(v^{(1)}\right)^3 \end{aligned} \qquad [8.17]$$

The solution to [8.25] is a modulated oscillation, whose complex amplitude $V_1(X)$ is unknown at this stage. Similarly, the solution to [8.36] depends on the square of this amplitude:

$$v^{(1)}(x,X) = V_1(X)e^{iqx} + \bar{V}_1(X)e^{-iqx} \qquad [8.18]$$

$$v^{(2)}(x,X) = V_0(X) + V_2(X)e^{2iqx} + \bar{V}_2(X)e^{-2iqx} \qquad [8.19]$$

$$V_0(X) = -\tfrac{2c_2}{c_1}|V_1(X)|^2 \; V_2(X) = \tfrac{c_2}{9c_1}V_1(X)^2 \qquad [8.20]$$

It is important to note that the first two terms on the right-hand side of [8.46] are zero because of [8.58] and [8.60] and this implies that the right-hand side of the second equation [8.76] is proportional to the square of $v^{(1)}$. It is not necessary to solve the third equation [8.17]. Considering [8.88]–[8.99], the right-hand side of [8.107] involves three harmonics. The solution $v^{(3)}$ is unbounded with respect to the rapid variable x, if there is a term proportional to the main harmonic of this right-hand side term. To avoid this, we abandon the coefficient of this first harmonics, which leads to a differential equation for the complex amplitude, which is the well-known Landau–Ginzburg equation:

$$4EI\frac{d^2 V_1}{dX^2} + V_1 - V_1|V_1|^2\left(\frac{3c_3}{q^2} - \frac{38c_2^2}{c_1 q^2}\right) = 0 \qquad [8.21]$$

From a multiple scale perspective, the Landau–Ginzburg equation [8.21] is a macroscopic model, as it depends on the slow variable X. In contrast, equations [8.115]–[8.127] describe microscopic evolutions.

This Landau–Ginzburg equation is generic for the bifurcation modes breaking the symmetry $x \longrightarrow -x$, as established in Iooss et al. (1989) and in Damil and Potier-Ferry (1992). The same envelope equation was obtained in many physics problems, as in fluid mechanics. However, it has a disadvantage: it is only valid in the vicinity of a bifurcation point. In the following section, we present another approach, which may remain valid far away from the bifurcation. The objective is to deduce a macroscopic model that is able to couple the evolutions of local and global buckling models. Obviously, this coupling cannot be deduced from the asymptotic approach that involves small perturbations near a specific state.

8.3. Another multiple scale method for modeling instabilities: Fourier series with slowly varying coefficients

8.3.1. *New macroscopic models*

We study phenomena such as the response of the system is the sum of a slowly varying mean field and of a quasi-periodic fluctuation in a spatial direction.

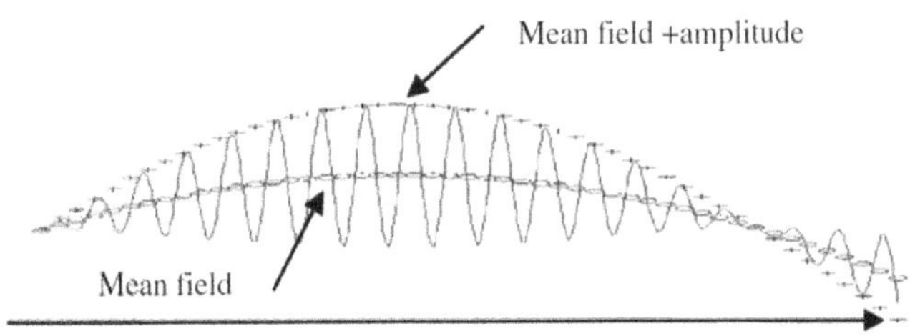

Figure 8.1. *At least two macroscopic fields are required to describe a quasi-periodic response: the mean field and the fluctuation amplitude*

As shown in Figure 8.1, at least two slowly varying functions are required to model these phenomena.

This part presents a general method for deducing the equations satisfied by these slowly varying fields. We assume that the instability wave number q is known. All of the unknowns of the model $U = \{u(x), v(x), n(x), m(x)\}$ are sought in the form of Fourier series, whose coefficients vary more slowly than the harmonics:

$$U(x) = \sum_{j=-\infty}^{+\infty} U_j(x) e^{jiqx} \qquad [8.22]$$

where the macroscopic fields $U_j(x)$ vary slowly over an oscillation period $\left[x, x+\frac{2\pi}{q}\right]$. Naturally, only a finite number of Fourier coefficients will be considered in practice. At least, two functions $U_0(x)$ and $U_1(x)$ are required to describe quasi periodic models: $U_0(x)$ can be identified with the mean value, while $U_1(x)$ represents the envelope or the amplitude of spatial oscillations. The envelope $U_0(x)$ has a real value, while other envelopes are complex. Hence, the envelope of the first harmonics $U_1(x)$ can be written in the form $U_1(x) = r(x)e^{i\varphi(x)}$, where $r(x)$ represents the amplitude modulation and $\varphi(x)$ is the phase modulation. If the phase variation is linear $\varphi(x) = Qx + \varphi_0$, this type of approach can describe quasi-periodic responses whose wave number $q + Q$ is slightly different from the a priori chosen wave number q. The main idea of macroscopic modeling, which is discussed in this chapter, is to deduce certain differential equations satisfied by amplitudes $U_j(x)$.

This method is applied to the classical beam on nonlinear foundation presented in section 8.2.

The macroscopic model can be deduced from the microscopic model [8.3]–[8.6] or [8.7] simply by identifying the Fourier coefficients in each equation. Naturally, the Fourier coefficients $U_j(x)$ are assumed to be constant over a period. Several calculation rules were introduced in Damil and Potier-Ferry (2006) for the management of these Fourier series with slowly varying coefficients.

If $a(x)$ and $b(x)$ are Fourier series with slowly varying coefficients, as in [8.22], the following identities are applied:

$$\int_0^L a(x)b(x)dx = \int_0^L \sum_{j=-\infty}^{+\infty} a_j(x)b_{-j}(x)\,dx \tag{8.23}$$

$$\left(\frac{da}{dx}\right)_j = (a')_j = \left(\frac{d}{dx} + jiq\right)a_j = \left(a_j\right)' + jiqa_j \tag{8.24}$$

$$\begin{aligned}\left(\frac{d^2a}{d\,x^2}\right)_j = (a'')_j &= \left(\frac{d}{dx} + jiq\right)^2 a_j \\ &= \left(a_j\right)'' + 2jiq\left(a_j\right)' + j^2q^2a_j\end{aligned} \tag{8.25}$$

$$(ab)_j = \sum_{j_1=-\infty}^{+\infty} a_{j_1}b_{j-j_1} \tag{8.26}$$

It should be noted that the first identity [8.23] results from Parseval's identity and from the hypothesis of slow variation of coefficients. Applying [8.23] to [8.3]–[8.6] or [8.7], we can deduce the macroscopic version of the principle of virtual work and the behavior equations:

$$\int_0^L \left(\sum_{j=-\infty}^{+\infty} \left(n_j \delta\gamma_{-j} + m_j \delta k_{-j} \right) \right) dx = \int_0^L \left(\sum_{j=-\infty}^{+\infty} \left(nf_j \delta u_{-j} + g_j \delta v_{-j} \right) \right) dx \qquad [8.27]$$

$$\frac{n_j}{ES} = \gamma_j = \left(\frac{d}{dx} + jiq \right) u_j + \frac{1}{2} \sum_{j_1=-\infty}^{+\infty} \left(\frac{d}{dx} + j_1 iq \right) v_{j_1} \left(\frac{d}{dx} + (j - j_1) iq \right) v_{j-j_1} \qquad [8.28]$$

$$\frac{m_j}{EI} = k_j = \left(\frac{d}{dx} + jiq \right)^2 v_j \qquad [8.29]$$

$$-g_j = c_1 v_j + c_2 \sum_{j_1=-\infty}^{+\infty} v_{j_1} v_{j-j_1} + c_3 \sum_{j_1=-\infty}^{+\infty} \sum_{j_2=-\infty}^{+\infty} v_{j_1} v_{j_2} v_{j-j_1-j_2} \qquad [8.30]$$

Similarly, the macroscopic version of the potential energy is written as:

$$\begin{aligned} P(u_j, v_j) &= \int_0^L \left(\frac{ES}{2} \left(\sum_{j=-\infty}^{+\infty} \gamma_j \bar{\gamma}_j \right) + \frac{EI}{2} \left(\sum_{j=-\infty}^{+\infty} k_j \bar{k}_j \right) \right) dx - \int_0^L \sum_{j=-\infty}^{+\infty} f_j \bar{u}_j \, dx \\ &+ \int_0^L \left(\frac{c_1}{2} \left(\sum_{j=-\infty}^{+\infty} v_j \bar{v}_j \right) + \frac{c_2}{3} \left(\sum_{j_1=-\infty}^{+\infty} \sum_{j_2=-\infty}^{+\infty} v_{j_1} v_{j_2} v_{-(j_1+j_2)} \right) \right. \\ &\left. + \frac{c_3}{4} \left(\sum_{j_1=-\infty}^{+\infty} \sum_{j_2=-\infty}^{+\infty} \sum_{j_3=-\infty}^{+\infty} v_{j_1} v_{j_2} v_{j_3} v_{-(j_1+j_2+j_3)} \right) \right) dx \end{aligned} \qquad [8.31]$$

The macroscopic equations of equilibrium result from the application of rules [8.24]–[8.26] to the initial equations of equilibrium [8.3] and [8.5]. The same results can be obtained from the principle of virtual work [8.7] or from the potential energy [8.2]:

$$\left(\frac{d}{dx} + jiq \right) n_j + f_j = 0 \qquad [8.32]$$

$$-\left(\frac{d}{dx}+jiq\right)^2 m_j+\left(\frac{d}{dx}+jiq\right)a_j+g_j=0$$
$$a_j=\sum_{j_1=-\infty}^{+\infty} n_{j_1}\left(\frac{d}{dx}+(j-j_1)iq\right)v_{j-j_1} \quad [8.33]$$

The above part leads to macroscopic models that are generalized continuous media, the macroscopic stresses being defined by the Fourier coefficients n_j, m_j of microscopic stresses n, m. From this perspective, the present technique is similar to a theory of homogenization where a double scale analysis can be used to deduce macroscopic generalized continuous media from microscopic classic continuous media (Forest and Sab 1998; Kouznetsova et al. 2004).

8.3.2. *A model with five harmonics*

As an example, we first express the model coupling global and local instabilities obtained by considering only five terms in [8.27]–[8.30] or in [8.31]: $U_0 \in \mathbb{R}^4$, $U_1 \in \mathbb{C}^4$, $U_2 \in \mathbb{C}^4$, $U_{-1}=\overline{U}_1$, $U_{-2}=\overline{U}_2$. The macroscopic version of the principle of virtual work is as follows:

$$\begin{aligned}\int_0^L \big(&n_0\,\delta\gamma_0+n_1\,\delta\bar{\gamma}_1+\bar{n}_1\,\delta\gamma_1+n_2\,\delta\bar{\gamma}_2+\bar{n}_2\,\delta\gamma_2+m_0\,\delta k_0\\ &+m_1\,\delta\bar{k}_1+\bar{m}_1\,\delta k_1+m_2\,\delta\bar{k}_2+\bar{m}_2\,\delta k_2\big)\,dx\\ &=\int_0^L\big(f_0\,\delta u_0+f_1\,\delta\bar{u}_1+\bar{f}_1\,\delta u_1+f_2\,\delta\bar{u}_2\\ &+\bar{f}_2\,\delta u_2+g_0\,\delta v_0+g_1\,\delta\bar{v}_1+\bar{g}_1\,\delta v_1+g_2\,\delta\bar{v}_2\\ &+\bar{g}_2\,\delta v_2\big)\,dx\end{aligned} \quad [8.34]$$

The macroscopic behavior equations are:

$$\frac{n_0}{ES}=\gamma_0=\frac{du_0}{dx}++\frac{1}{2}\left(\frac{dv_0}{dx}\right)^2+\left|\left(\frac{d}{dx}+iq\right)v_1\right|^2+\left|\left(\frac{d}{dx}+2iq\right)v_2\right|^2 \quad [8.35]$$

$$\begin{aligned}\frac{n_1}{ES}=\gamma_1=\left(\frac{d}{dx}+iq\right)u_1+\left(\frac{d}{dx}-iq\right)v_{-1}\left(\frac{d}{dx}+2iq\right)v_2\\ +\frac{dv_0}{dx}\left(\frac{d}{dx}+iq\right)v_1\end{aligned} \quad [8.36]$$

$$\frac{n_2}{ES}=\gamma_2=\left(\frac{d}{dx}+2iq\right)u_2+\frac{dv_0}{dx}\left(\frac{d}{dx}+2iq\right)v_2+\frac{1}{2}\left(\left(\frac{d}{dx}+iq\right)v_1\right)^2 \quad [8.37]$$

$$\frac{m_0}{EI} = k_0 = \frac{d^2 v_0}{dx^2} \qquad [8.38]$$

$$\frac{m_1}{EI} = k_1 = \left(\frac{d}{dx} + iq\right)^2 v_1 \qquad [8.39]$$

$$\frac{m_2}{EI} = k_2 = \left(\frac{d}{dx} + 2iq\right)^2 v_2 \qquad [8.40]$$

$$\begin{aligned} -g_0 = c_1 v_0 + c_2 (v_0^2 + 2\,|v_1|^2 + 2\,|v_2|^2) \\ + c_3 (v_0^3 + 6v_0\,|v_1|^2 + 6v_0\,|v_2|^2 + 2\,|v_1|^2 v_2 \\ + 2\,|v_1|^2 \bar{v}_2 + v_1^2 \bar{v}_2 + \bar{v}_1^2 v_2) \end{aligned} \qquad [8.41]$$

$$\begin{aligned} -g_1 = c_1 v_1 + c_2 (2v_0 v_1 + 2\bar{v}_1 v_2) \\ + c_3 (3v_0^2 v_1 + 6v_0 \bar{v}_1 v_2 + 3v_1\,|v_1|^2 + 6v_1\,|v_2|^2) \end{aligned} \qquad [8.42]$$

$$\begin{aligned} -g_2 = c_1 v_2 + c_2 (2v_0 v_2 + v_1^2) \\ + c_3 (3v_0 v_1^2 + 3v_0^2 v_2 + 6|v_1|^2 v_2 + 3v_2\,|v_2|^2) \end{aligned} \qquad [8.43]$$

The macroscopic version of the potential energy is the following:

$$\begin{aligned} P(u_0, u_1, u_2, v_0, v_1, v_2) \\ = \int_0^L \left(\frac{ES}{2}(\gamma_0^2 + 2|\gamma_1|^2 + 2|\gamma_2|^2) \right. \\ \left. + \frac{EI}{2}(k_0^2 + 2|k_1|^2 + 2|k_2|^2)\right) dx - \int_0^L f_0 u_0 dx \\ + \int_0^L \frac{c_1}{2}(v_0^2 + 2|v_1|^2 + 2|v_2|^2) dx \\ + \int_0^L \frac{c_2}{3}(v_0^3 + 6v_0|v_1|^2 + 6v_0|v_2|^2 + 3v_1^2 \bar{v}_2 \\ + 3\bar{v}_1^2 v_2) dx + \int_0^L \frac{c_3}{4}(v_0^2 + 2|v_1|^2 + 2|v_2|^2)^2 dx \\ + \int_0^L \frac{c_3}{4}(8|v_1|^2|v_2|^2 + 2|v_2|^4 + 8|v_0 v_1 + \bar{v}_1 v_2|^2 \\ + 2|v_1^2 + 2v_0 v_2|^2) dx \end{aligned} \qquad [8.44]$$

As expected, the model with five envelopes [8.43] couples the mean field $U_0(x)$ and the amplitudes of fluctuations $U_1(x)$, $U_2(x)$. For example, in the equation of the membrane behavior [8.35], the first two terms $\frac{du_0}{dx} + + \frac{1}{2}\left(\frac{dv_0}{dx}\right)^2$ correspond to the macroscopic strain and the other two terms $\left|\left(\frac{d}{dx} + ijq\right) v_j\right|^2$ represent the local instability. They are always positive, which involves an elongation. If the beam is in a state of compression, a local instability can occur and these two terms lead to a reduction of the compression stress. Therefore, this model allows a macroscopic description of the relaxation of stresses due to local buckling. The modified expression of the macroscopic strain [8.35] is similar to the method of Roddeman et al. (1987) for taking into account the membrane wrinkling as additional strains were added to the membrane strain $\frac{du_0}{dx} + + \frac{1}{2}\left(\frac{dv_0}{dx}\right)^2$. In Roddeman et al. (1987), the additional wrinkling strain is characterized by the suppression of compression stresses. According to the present approach, the minimization of the potential energy [8.44] yields equations governing the evolution of wrinkling displacements v_1, v_2, as our approach leads to a generalized continuous medium.

There is a rich literature on membrane wrinkling (see Banerjee et al. (2009) and the references in this chapter), but this chapter uses a novel approach. The classical wrinkling models and the present chapter have in common the introduction of additional kinematic variables, such as inhomogeneous strains used in Banerjee et al. (2009) or the Fourier coefficients of displacements u_1, v_1, u_2, v_2 in the present chapter. A second common point is the introduction of additional stress variables so that the model is a generalized continuous medium: in Banerjee et al. (2009), the intrinsic controlling torque of the Cosserat theory is here the Fourier coefficient of the stress. Nevertheless, the present chapter relies on an entirely different perspective, the macroscopic potential energy [8.44] being completely deduced from the basic model as in the numerical homogenization (Kouznetsova et al. 2004), while the classical wrinkling models are phenomenological and postulate the expression of the additional potential energy.

Even with a limitation of the number of harmonics, the model [8.44] is quite complex, but it can be solved numerically using standard numerical techniques. It is important to note that the same approach was presented in 2D elasticity (Damil and Potier-Ferry 2008).

In Damil and Potier-Ferry (2010), we discussed other simplifications of the model in order to compare it with the Landau–Ginzburg asymptotic analysis.

8.4. Various macroscopic models used to describe the formation of instability patterns

Various macroscopic models allowing the description of the formation of instability patterns are examined in this part. We focus on two questions: the necessity to consider phase changes and boundary conditions to associate with macroscopic models. The analysis is conducted based on numerical simulations for the problem of a compressed beam on a nonlinear foundation, the case being that of coefficient $c_2 = 0$ in [8.11].

The first macroscopic model involves a real envelope, so that the phase is assumed to be constant. The second model is also macroscopic and is some sort of Landau–Ginzburg equation with a complex envelope. The third model results from a multiple scale approach with a numerical connection between the complete model near the boundary and a macroscopic model in the central part.

The objective of this chapter is not to deduce a unique generic model that would be valid in all possible cases. For example, the model [8.44] with $c_2 = 0$ comprises only several harmonics, but it can be unnecessarily complex. In this context, other simplifications are introduced in the potential energy [8.44]. First, we only consider three harmonics $0, \pm q$, with a real envelope $U_0(x) = (u_0, v_0, n_0)$ and a complex envelope $U_1(x) = (u_1, v_1, n_1)$, or in an equivalent manner three real envelopes $U_0(x)$, $U_1^R(x)$, $U_1^I(x)$; $U_1^R(x)$ and $U_1^I(x)$ are, respectively, the real and imaginary parts of $U_1(x)$. The second restriction concerns the axial volume force that is assumed to be constant over each period: in other terms, the volume force contains only the zero harmonics $(f(x) = f_0(x))$. This implies that the normal stress does not fluctuate $(n_1 = 0)$ and this makes it possible to eliminate the unknown $u_1(x)$. In this case, the potential energy depends on the mean field $\big(u_0(x), v_0(x)\big)$ and the bending envelope $v_1(x)$:

$$\begin{aligned} P(u_0, v_0, v_1) = & \int_0^L \left(\frac{ES}{2}\gamma_0^2 + \frac{EI}{2}(k_0^2 + 2|k_1|^2) - f_0 u_0\right) dx \\ & + \int_0^L \frac{c_1}{2}(v_0^2 + 2|v_1|^2) dx \\ & + \int_0^L \frac{c_3}{4}(v_0^4 + 12 v_0^2 |v_1|^2 + 6|v_1|^4) dx \end{aligned} \quad [8.45]$$

Since v_1 is a complex value function, model [8.45] can be used to predict slow phase modulations. In what follows, this model [8.43] will be referred to as "macro-complex". In Damil and Potier-Ferry (2010), additional simplifications were introduced in model [8.45] to define the simplest model coupling the membrane problem and an envelope equation similar to Landau–Ginzburg equation. The first new approximation consists of assuming that the envelope $v_1(x)$ is real, which accounts for the amplitude modulation and sets aside any phase evolution. Second, we assume $v_0(x) = 0$, so that the nonlinear behavior and the instability can only be considered at a local level. More specifically, it will no longer be possible to couple local and global buckling, as with the previous model [8.45]. Finally, the bending term k_1 [8.39] will be simplified to find an envelope equation similar to the Landau–Ginzburg equation. This amounts to neglecting the second-order derivatives of $v_1(x)$ in the bending energy. Finally, the simplified potential energy is given by the formula:

$$P(u_0, v_1) = \int_0^L \frac{ES}{2}\left(u_0' + v_1'^2 + q^2 v_1^2\right)^2 dx + \int_0^L \left(EI\left(6q^2 v_1'^2 + q^4 v_1^2\right) + c_1 v_1^2 + \frac{3c_3}{2} v_1^4 - f_0 u_0\right) dx \quad [8.46]$$

It should be noted that in this approximation, the vertical displacement is given by:

$$v(x) = v_0(x) + 2v_1(x)\sin(qx) \quad [8.47]$$

But it could also be given by $v(x) = v_0(x) + 2v_1(x)\sin(qx + \phi)$. In other terms, the phase ϕ is more or less arbitrary, which explains the main difference between this model and the previous one [8.45]. This very simplified model [8.46] can predict the amplitude modulation, but not the phase modulation. In what follows, the second model [8.46] will be compared to the previous model [8.45] and to the multiple scale model of Hu et al. (2011). It will be referred to as the "macro-real" model.

8.4.1. *Boundary conditions associated with macroscopic models*

It is not easy to associate consistent boundary conditions with macroscopic models such as [8.44], [8.45] or [8.46]. Indeed, the hypothesis of a slowly varying oscillation is not generally valid everywhere, particularly near the edge, given the presence of boundary layers (Schlichting 1979). The initial hypothesis [8.22] must be corrected near the edge of the domain and must be replaced by the following hypothesis:

$$U(x) = \sum_{j=-\infty}^{+\infty} U_j(x) e^{jiqx} + U^{CL}(x) \quad [8.48]$$

where the last term $U^{CL}(x)$ takes into account boundary layers and tends to become zero far from the edge. In principle, this boundary layer term should be included in the consideration of boundary conditions (Daya et al. 2003). A generic procedure was proposed in Hu et al. (2011) to take into account edge effects in cellular instability problems: the initial model [8.3]–[8.6] is used near the edge, a reduced model such as [8.44] or [8.45] or [8.46] in the central part and these two models are connected by the Arlequin method (Ben Dhia and Rateau 2005). In what follows, this numerical technique will be applied when the macroscopic model is the one with real envelope (see [8.46] and will be referred to as "connection model"). Nevertheless, the numerical model will be simpler if we can avoid this connection method and if we are able to associate consistent boundary conditions to the macroscopic model. It is known (Pomeau and Zaleski 1981; Damil and Potier-Ferry 1986) that such macroscopic boundary conditions can be obtained, for example, in the case of a beam or an embedded plate: $v(0) = \frac{dv(0)}{dx} = 0$. For this, it can be assumed that the average value and the oscillating part of the bending:

$$\begin{gathered} v_0(x) \\ Re[v_1(x)e^{iqx}] = v_1^R(x)\cos(qx) - v_1^I(x)\sin(qx) \end{gathered} \tag{8.49}$$

are zero for $x = 0$, as well as their first derivative. This implies the following boundary conditions for the real and imaginary parts of the envelope: $v_1^R(0) = 0$, $\frac{dv_1^R(0)}{dx} - qv_1^I(0) = 0$. If moreover the envelope varies more slowly than the local oscillation $\left(\frac{d}{dx} \ll q\right)$, it can be deduced that at the embedded edge the macroscopic boundary conditions are:

$$\begin{aligned} &v_0(0) = 0, \frac{dv_0(0)}{dx} = 0 \\ &v_1^R(0) = 0, v_1^I(0) = 0 \end{aligned} \tag{8.50}$$

Similarly, let us consider that: $w(x) = Re[v_1(x)e^{iqx}] = v_1^R(x)\cos(qx) - v_1^I(x)\sin(qx)$, satisfies the simply supported boundary conditions $w(0) = \frac{d^2w(0)}{dx^2} = 0$. Using the same approximation $\left(\frac{d}{dx} \ll q\right)$, we readily deduce the following boundary conditions:

$$\begin{aligned} &v_0(0) = 0, \frac{d^2v_0(0)}{dx^2} = 0 \\ &v_1^R(0) = 0, \frac{dv_1^I(0)}{dx} = 0 \end{aligned} \tag{8.51}$$

These Dirichlet–Newman boundary conditions [8.51] can be applied in a similar manner to the end $x = L$ provided that $qL/2\pi$ is an integer number.

It is however difficult to obtain consistent macroscopic boundary conditions in all cases, particularly when an instability or a nonlinear behavior occurs near the edge before the emergence of the global instability. All of these matters will be resumed in the next part by considering several numerical examples.

8.4.2. *Numerical evaluations of macroscopic models*

Here, three models will be compared and evaluated based on two numerical tests in the case of a beam on a nonlinear elastic foundation. The first example is a relatively short beam subjected to uniform compression, whose total length is about five times the instability wavelength. This example can be considered as a validation test. For the second test, we choose the same example as the previous one, except for the length which is this time 20 times the instability wavelength.

8.4.2.1. *Finite element discretization*

The unknowns of the three models, $< r > = \langle u, v \rangle$ for the thin model [8.1], [8.2] with $c_2 = 0$, $< r > = < u_0, v_0, v_1^R, v_1^I >$ for the macroscopic model [8.45] and $< r >= < u_0, v_1 >$ for the macroscopic model [8.46], are discretized in the classic form: $\{r\} = [N]\{q\}$. In the case of the thin model [8.1] and [8.2], referred to as ''reference model'', the element uses a beam element with two nodes and three degrees of freedom per node. The matrix $[N]$ is represented by a Lagrangian shape function for u and a Hermitian shape function for v. In the case of the macroscopic model [8.45], known as ''macro-complex'', the element employed has two nodes and seven degrees of freedom per node, the matrix $[N]$ is represented by a Lagrangian shape function for u_0 and a Hermitian shape function for v_0; v_1^R and v_1^I. In the macroscopic model [8.46], known as ''macro-real'', the employed element has three nodes and two degrees of freedom per node, the matrix $[N]$ is represented by Lagrangian shape functions for u_0 and v_1.

The three models, the "reference model" defined by the potential energy [8.1], [8.2], the "macro real" model defined by the potential energy [8.46] and the "macro complex" model defined by the potential energy [8.45], are solved by the ANM algorithm. In this ANM algorithm (Cochelin et al. 1994, 2007), the solution is determined branch by branch. For each branch, from a given starting point, we search an approximation of the solution branch in the form of a N order truncated Taylor series. A simple formula can be used to calculate the validity domain of this representation and to define the starting point of the next branch. The solution is thus determined branch by branch. Each branch is calculated by inverting a single matrix of tangent rigidity. Point by point calculations were made using the Newton–Raphson method. For the sake of convenience, the "bridging model" was solved only with the Newton–Raphson method.

8.4.2.2. *Buckling of a short embedded beam under uniform compression*

Let us consider an embedded beam of length 10π and characteristics $ES = EI = 1$, $c_1 = 1$, $c_2 = 0$, $c_3 = 1/3$ and the chosen instability wave number is $q = 1$. The beam is subjected to an increasing overall end shortening $u(L) = -\lambda L$ and the volume force is zero: $f_0 = 0$.

For the reference model, the boundary conditions are: $v(0) = v(L) = 0$, $v'(0) = v'(L) = 0$. The macroscopic boundary conditions for the "macro-real" model are: $v_1(0) = v_1(L) = 0$; and for the "macro-complex" model they are: $v_0(0) = v_0(L) = 0$, $v_0'(0) = v_0'(L) = 0$, $v_1{}^R(0) = v_1{}^R(L) = 0$, $v_1{}^I(0) = v_1{}^I(L) = 0$.

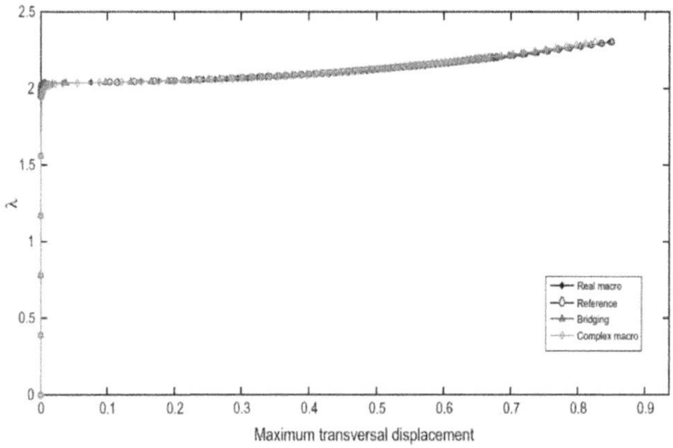

Figure 8.2. *Buckling of a short embedded beam under uniform compression: load–displacement curve*

We choose 100 finite elements for the reference model, five macro-elements for the "macro-real" model and seven macro-elements for the "macro complex" model. Several tests were conducted in order to select the optimal number of elements for each macro model, the chosen grid being the least fine grid that provides a correct solution.

The bifurcation branch (v_{max}, λ) and the instability patterns for $\lambda = 2.055$ and $\lambda = 2.305$ are, respectively, given in Figures 8.2 and 8.3. We have also represented the solution obtained with the bridging method.

The conclusion that can be drawn from this first test is that the three macroscopic models are generally very good, even when the domain is too short for the two spatial scales to be distinct.

The three models provide a correct prediction of the amplitude and spatial distribution of instability patterns, bifurcation point and the shape of the post-buckling solution (see Figures 8.2 and 8.3).

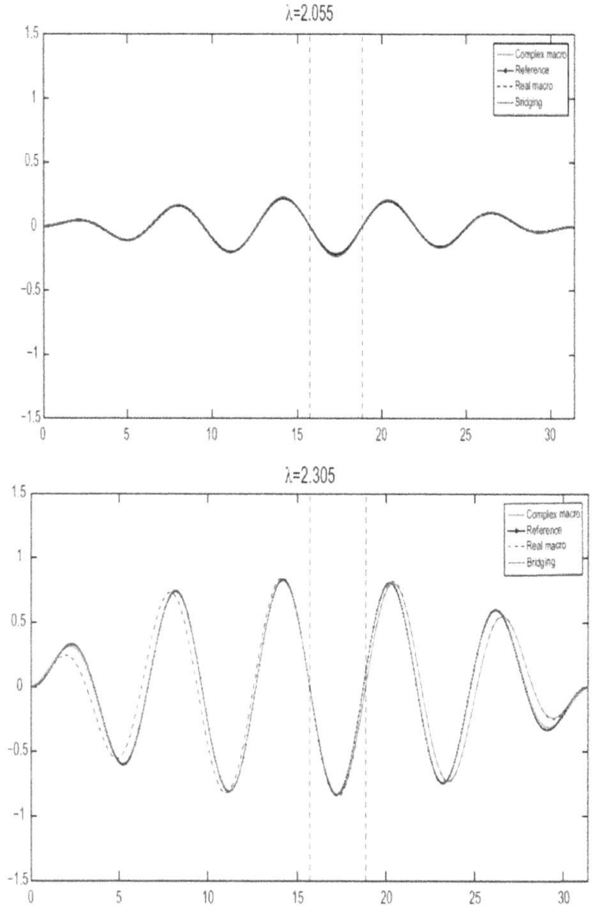

Figure 8.3. *Buckling of a short embedded beam under uniform compression: spatial distribution of instability patterns for* $\lambda = 2.055$ *and* $\lambda = 2.305$

The number of elements required for the three models (reference, macro-real, macro-complex) is indicated in Table 8.1. The most remarkable characteristic is the very small number of elements required with the two macroscopic models: here, only five elements and 22 dof for the macro-real model and 7 elements and 56 dof for the macro-complex model, compared to about 303 dof with the refined model.

Order	Macro real (5 elements 22 dof)	Macro-complex (7 elements 56 dof)	Reference (100 elements, 303 dof)
20	26	24	33
15	36	33	94
10	71	65	182

Table 8.1. *Buckling of a short embedded beam under uniform compression: number of ANM steps required to obtain the solution in Figure 8.2 for the orders* $N = 10;\ 15;\ 20$

The curves represented in Figure 8.2 were obtained with the ANM algorithm. Table 8.1 provides the number of ANM steps required to calculate the solution branch of these curves for various orders and for each model (reference, macro-real, macro-complex). This number of ANM steps can be considered as a measure of the non-linearity of problems: it is clear that macroscopic models are rather easier to solve than the refined model. Finally, the small number of ANM steps is worth noting, about 30 for an order $N = 20$. This means the calculation will be conducted with only 30 tangent matrices, while about 300 matrices are required with Newton–Raphson methods (100 steps and two iterations per step).

8.4.2.3. *Buckling of a short embedded beam under uniform compression*

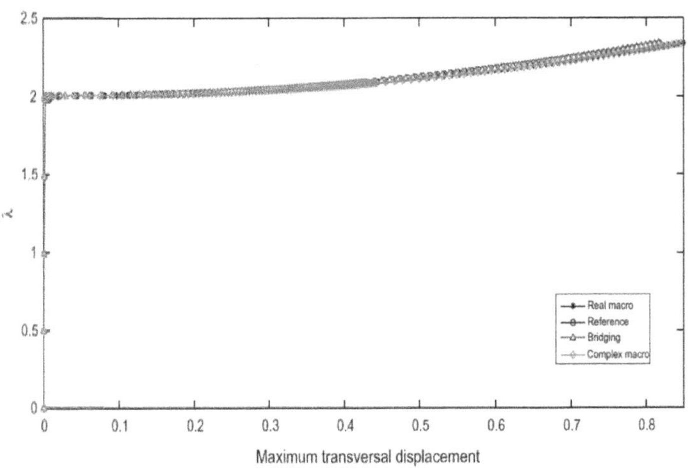

Figure 8.4. *Buckling of a long embedded beam under uniform compression: load–displacement curves*

Let us consider the same example as in section 8.4.2.2, and use the same data except for the length, which is chosen to be equal to 30π. After several trials, we chose to use 120 microscopic elements for the reference model, four macroscopic elements for the two macroscopic models.

The bifurcation branch (v_{max}, λ) and the instability pattern for $\lambda = 2.016$ and $\lambda = 2.211$ are represented in Figures 8.4 and 8.5. We have also represented the solutions obtained with the bridging method.

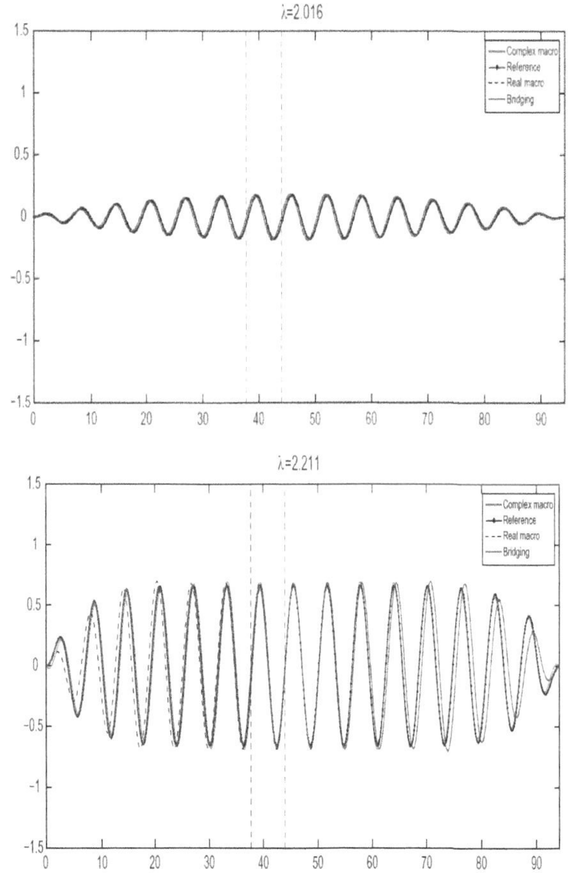

Figure 8.5. *Buckling of a long embedded beam under uniform compression: spatial distribution of instability patterns for* $\lambda = 2.016$ *and* $\lambda = 2.211$

According to Figures 8.4 and 8.5, it can be noted that the three models are in agreement with the reference model for this part of the branch that corresponds to the vicinity of the bifurcation. As in the previous test, the two macroscopic models can be used to describe a complex spatial pattern with a very small number of finite elements (see Table 8.2). Moreover, this number of elements is the same as with the short length, which means that macroscopic discretization does not depend on the number of local oscillations.

Order	Macro real (4 elements 18 dof)	Macro-complex (4 elements 35 dof)	Reference (120 elements, 363 dof)
20	21	23	83
15	29	31	111
10	55	59	219

Table 8.2. *Buckling of a long embedded beam under uniform compression: number of ANM steps required to obtain the solution in Figure 8.4 for the orders* $N = 10;\ 15;\ 20$

8.5. A 2D Fourier double scale analysis of global-local instability interaction in sandwich structures

This section presents a 2D Fourier double scale analysis for studying the interaction between local and global instabilities. A 2D microscopic model of sandwich beam is transformed into a 2D macroscopic model of sandwich beam (Mhada et al. 2013; Potier-Ferry et al. 2016). The 2D macroscopic model is solved by ANM (Cochelin et al. 1994, 2007). A comparison is made with the 1D macroscopic model of the reference (Liu et al. 2012) and with the initial microscopic model.

8.5.1. *The initial 2D sandwich model*

Let us consider a two-dimensional sandwich structure with a soft core and stiff skins. Let us denote by x and z the longitudinal and transversal coordinates in a reference frame of basis vectors $(\boldsymbol{e}_x, \boldsymbol{e}_z)$. The thickness of skins, core and the total thickness of the sandwich beam are, respectively, denoted by h_p , h_c and h_t. The length of the sandwich beam is denoted by L and E_c, G_c and E_p are, respectively, the Young's modulus of the core, the shear modulus of the core and the Young's modulus of the skin.

The equations of equilibrium of the 2D sandwich structure are obtained based on the nonlinear 2D elasticity with a linear stress–strain law. Using the notations in Cochelin et al. (2007) and neglecting the volume forces, the principle of virtual work can be expressed as follows:

$$\int_{\Omega} {}^{t}\{\delta\gamma\}\,\{s\}d\Omega = \lambda \int_{\partial\Omega} {}^{t}\{\delta u\}\,\{F_d\}dS \qquad [8.52]$$

where $\{\delta u\}$ is the virtual displacement, $\{F_d\}$ is the force applied to the boundary and λ is the load parameter. The second Piola Kirchhoff stress tensor $\{s\}$ and the Green–Lagrange strain tensor $\{\gamma\}$ are linked by a linear relation, while the strain is linked to the displacement gradient $\{\theta\}$ by a quadratic relation:

$$\{s\} = [D]\{\gamma\}\;\{\gamma\}\;[H]\{\theta\} + \frac{1}{2}\,[A(\theta)]\{\theta\} \qquad [8.53]$$

Equations [8.52] and [8.53] define the microscopic model. In the next part, the technique of Fourier series with slowly varying coefficients will be applied to equations [8.52] and [8.53] and a 2D macroscopic sandwich model will be deduced.

8.5.2. *A 2D macroscopic sandwich model*

We apply the same procedure as in Damil and Potier-Ferry (2008) based on nonlinear elasticity with a linear stress–strain law (see equations [8.52] and [8.53]). We seek to identify quasi-periodic responses that vary rapidly in one direction. This characteristic direction and the period are described by a wave vector $\boldsymbol{q} = q\boldsymbol{e}_x$ assumed to be given. In practice, this vector results from a linear stability analysis. The vector $\langle W(x)\rangle = <\langle u(x)\rangle, \langle s(x)\rangle, \langle \theta(x)\rangle, \langle \gamma(x)\rangle>$, in [8.22] that comprises here the displacement vector, its gradient, the strain and stress tensors, should have the form of a Fourier series, with slowly varying coefficients $\langle W_m(x)\rangle = <\langle u_m(x)\rangle, \langle s_m(x)\rangle, \langle \theta_m(x)\rangle, \langle \gamma_m(x)\rangle>$. For this application, we keep the harmonics up to level 1:

$$\{W(x)\} = \sum_{m=-1}^{+1} \{W_m(x)\}e^{im\,\boldsymbol{q}.\boldsymbol{x}} \qquad [8.54]$$

We choose a wave vector $\boldsymbol{q}$ parallel to the direction $\boldsymbol{x} = x\boldsymbol{e}_x$ of the sandwich $\boldsymbol{q} = q\boldsymbol{e}_x$, and the wave number q is defined by $q = \frac{\pi}{L}q_0$, where L is the length of the sandwich and q_0 is the chosen wavelength ($q_0 \in \mathbb{N}, q_0 \geq 2$). The macroscopic unknowns $\{W_m(x)\}$, which denote the envelope of the mth order harmonics, varies slowly over a period $[x, x + 2\pi q\,]$.

As in the previous section, the principle of virtual work for the 2D macroscopic model is directly deduced from the principle of virtual work of the microscopic model [8.52] (Damil and Potier-Ferry 2008):

$$\int_{\Omega} \sum_{m=-1}^{+1} {}^t\{\delta\gamma_{-m}\}\{s_m\}\, d\Omega = \lambda \int_{\partial\Omega} \sum_{m=-1}^{+1} {}^t\{\delta u_{-m}\}\{F_{d_m}\} dS \qquad [8.55]$$

This procedure is applied to the behavior law [8.53]. We thus obtain a behavior law for the 0 harmonics (real number) and two behavior laws for the first harmonics (complex number), all of these equations being coupled:

$$\begin{aligned} [D]^{-1}\{s_0\} = \{\gamma_0\} = [H]\{\theta_0\} + \frac{1}{2}[A(\theta_0)]\{\theta_0\} + [A(\theta_{-1})]\{\theta_1\} \\ [D]^{-1}\{s_1\} = \{\gamma_1\} = [H]\{\theta_1\} + [A(\theta_0)]\{\theta_1\} \end{aligned} \qquad [8.56]$$

It can be noted that the principle of the virtual work of the macroscopic model [8.55] and the behavior law of the macroscopic model [8.56] have the same form as in the microscopic models [8.52] and [8.53]:

$$\int_{\Omega} {}^t\{\delta\Gamma\}\{S\} d\Omega = \lambda \int_{\partial\Omega} {}^t\{\delta U\}\{F\} dS \qquad [8.57]$$

$$\{S\} = [D^{gen}]\{\Gamma\} \; \{\Gamma\} = [H^{gen}]\{\Theta\} + \frac{1}{2}[A^{gen}(\Theta)]\{\Theta\} \qquad [8.58]$$

It should be recalled that each vector of the initial model is replaced here by a vector whose size is three times larger, as this latter vector takes into account the 0 harmonics represented by a real vector and the $\pm q$ harmonics, represented by a complex vector or by two real vectors denoted by $\{.\}^R, \{.\}^I$. The definition of these generalized vectors and matrices is as follows: generalized displacement $\{U\}$, generalized stress $\{S\}$, generalized strain $\{\Gamma\}$, gradient of generalized displacement $\{\Theta\}$, generalized force $\{F\}$ and generalized matrices $[D^{gen}]$, $[H^{gen}]$, $[A^{gen}(\Theta)]$:

$$\langle U\rangle = \langle\langle u_0\rangle, \langle u_1^R\rangle, \langle u_1^I\rangle\rangle \qquad \langle S\rangle = \langle\langle s_0\rangle, \langle s_1^R\rangle, \langle s_1^I\rangle\rangle \qquad \langle\Gamma\rangle = \langle\langle\gamma_0\rangle, \langle 2\gamma_1^R\rangle, \langle 2\gamma_1^I\rangle\rangle$$

$$\langle\Theta\rangle = \langle\langle\theta_0\rangle, \langle\theta_1^R\rangle, \langle\theta_1^I\rangle\rangle \quad \langle F\rangle = \langle\langle F_{d,0}\rangle, \langle 2F_{d,1}^R\rangle, \langle 2F_{d,1}^I\rangle\rangle$$

$$[D^{gen}] = \begin{bmatrix} [D] & 0 & 0 \\ 0 & \frac{1}{2}[D] & 0 \\ 0 & 0 & \frac{1}{2}[D] \end{bmatrix}, \quad [H^{gen}] = \begin{bmatrix} [H] & 0 & 0 \\ 0 & 2[H] & 0 \\ 0 & 0 & 2[H] \end{bmatrix}$$

$$[A^{gen}(\Theta)] = 2\begin{bmatrix} \frac{1}{2}[A(\theta_0)] & [A(\theta_1^R)] & [A(\theta_1^I)] \\ [A(\theta_1^R)] & [A(\theta_0)] & 0 \\ [A(\theta_1^I)] & 0 & [A(\theta_0)] \end{bmatrix}$$

It should be noted that the mth component of the displacement gradient is not the gradient of the mth component of displacement. The rule defining the Fourier components of a gradient vector is as follows:

$$\{\nabla u\}_m = \{\nabla(u_m)\} + +mi[Q]\{u_m\} \text{ where } [Q] = \begin{bmatrix} \{q\} & 0 \\ 0 & \{q\} \end{bmatrix}$$

8.5.3. *Interaction between global buckling and local wrinkling in 2D sandwich structures*

We start by the discretization of the $2D$ macroscopic model. As the form of the macroscopic model is similar to that of the microscopic model, it can be discretized with the same set of shape functions. In the microscopic model, the displacement and its gradient can be linked to nodal variables by two standard interpolation matrices $[N]$, $[G]$:

$$\{u(x,z)\}^e = [N(x,z)]\{v\}^e \quad \{\nabla u(x,z)\}^e = [G(x,z)]\{v\}^e$$

where $\{u(x,z)\} = \begin{Bmatrix} u \\ w \end{Bmatrix}$, and u and w are the components of the displacement vector along x and z. For the discretization, the element $Q8$ with 8 nodes and 2 degree of freedom per node is used. Naturally, the generalized displacement will be expressed in terms of generalized nodal displacement via a block diagonal matrix. The matrix defining the generalized gradient can be similarly built:

$$[N^{gen}] = \begin{bmatrix} [N] & 0 & 0 \\ 0 & [N] & 0 \\ 0 & 0 & [N] \end{bmatrix} \quad [G^{gen}] = \begin{bmatrix} [G] & 0 & 0 \\ 0 & [G] & -[Q][N] \\ 0 & [Q][N] & [G] \end{bmatrix}$$

The objective is to evaluate the capacity of the 2D macroscopic model ([8.57] and [8.58]) to capture the wrinkling and to take into account the coupling between local and global instabilities. It also involves comparing the results of the 2D macroscopic model to those of the 2D microscopic model and to the 1D macroscopic model of the reference (Liu et al. 2012) in the case of a sandwich beam subjected to a compression load. The geometric and material parameters of the studied example are as follows: $L = 0.5\ m$, $E_p = 6.9 \times 10^{10}$ Pa, $\frac{E_c}{E_p} = 1 \times 10^{-4}$, $G_c = 2.65 \times 10^6$ Pa, $h_c = 8.4 \times 10^{-3}\ m$ and $\frac{h_t}{h_c} = 0.033$.

The wave number q_0 considered in [8.54] is equal to 35. The 2D microscopic sandwich model is discretized using 1,120 $Q8$ elements with 160 elements over the length and seven elements on the thickness, which leads to 7,390 dof. The 2D macroscopic sandwich model is discretized using 24 $Q8$ elements, out of which six elements on the length and three elements on the thickness, which leads to only 558 dof.

The boundary conditions of the mobile support for the microscopic model are $w = 0$ at both edges and for the macroscopic model, we have $w_0 = 0$, $w_1^I = 0$ at both edges.

The nonlinear equations of the problem are solved using ANM (Cochelin et al. 2007) with the following parameters: truncation order $N = 15$, precision parameter $\varepsilon = 10^{-6}$. The number of ANM steps to obtain the bifurcation branch in Figure 8.6 is equal to 90 for the microscopic model and 80 for the 2D and 1D macroscopic models.

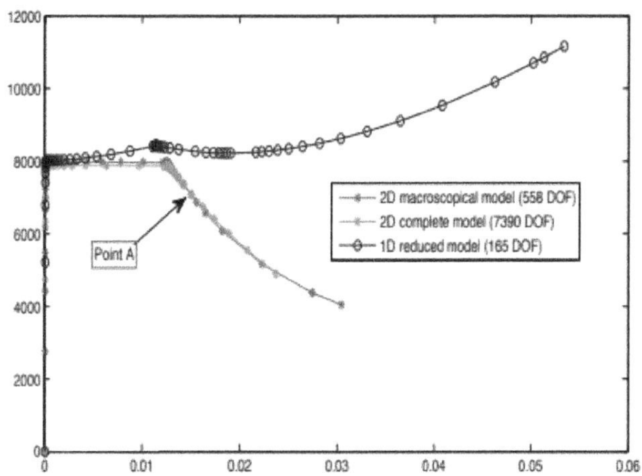

a) Load-displacement curves

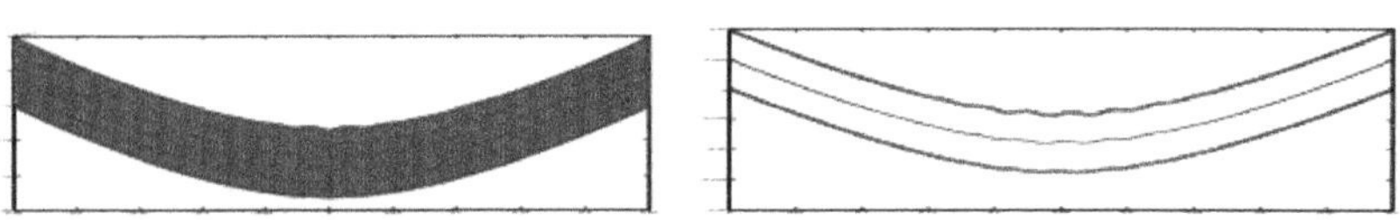

b) 2D microscopic model c) 2D macroscopic model

Figure 8.6. *Load–displacement curves, comparison of the 2D macroscopic model, 2D microscopic model and the 1D macroscopic model of the reference (Liu et al. 2012)*

In Figure 8.6(a), we compare the bifurcation curves for the 2D macroscopic model and the 2D microscopic model in $(x = L/2, z = (h_c + h_p)/2)$. The curve represents the load as a function of the vertical displacement. Figure 8.6(a) shows that the 2D macroscopic model makes a good prediction of both bifurcation points similar to the 2D microscopic model, and this also indicates that the 2D macroscopic model can describe the interaction between local wrinkling and global buckling. This 2D macroscopic model yields better results than the 1D macroscopic model of Liu et al. (2012). It should be noted that the number of degrees of freedom of the 2D microscopic model is 7,390 and the number of degrees of freedom of the 2D macroscopic model is only equal to 558.

8.6. A reduced model based on the multiple scale method in the case of a wrinkling with variable orientation

This section discusses the technique of reduction of the membrane wrinkling model when the wrinkle orientation is not uniform. Indeed, the wrinkle orientation depends on geometry and load, as in the case of a perforated membrane or with nonuniform residual stresses. The initial model is Föppl von Karman (FvK) shell model.

8.6.1. *Context of macroscopic modeling for wrinkling with variable orientations*

The reduced models studied in the previous sections assume that the period $\frac{2\pi}{q}$ and the direction $\boldsymbol{e}_x$ of the wrinkling are constant and given. In this section, we try to relax these conditions, particularly to be able to take into account structures with holes, in which the direction of wrinkles varies and is nearly orthogonal to the boundary of holes. For this, the unknown field $U(x)$, whose components are the membrane displacement, the transversal displacement, the stress and the membrane strains, is assumed in the form of Fourier series with a variable phase:

$$U(x) = \sum_{j=-\infty}^{+\infty} U_j(x) e^{ji\varphi(x)} \qquad [8.59]$$

where the new unknown macroscopic field $U_j(x)$ varies slowly and the phase $\varphi(x)$ is assumed given and variable. Its Taylor series truncated at the first order has the following form:

$$\varphi(x) \approx \varphi_0 + Q(x_0)(x - x_0) \qquad [8.60]$$

The main hypothesis of the procedure concerns the wave vector $Q(x) = \nabla\varphi(x)$ that is not constant but varies more slowly than the phase $\varphi(x)$. In the previous sections, this wave vector Q was given and constant. Hence, the described patterns are quasi periodic and this defines locally a base cell $\left[-\frac{\pi}{q}e + x_0, \frac{\pi}{q}e + x_0\right]$ where $q = |Q(x_0)|$ is the local wave number and $e = \frac{\nabla\varphi(x)}{|\nabla\varphi(x)|}$ is the direction of wrinkles. In the previous sections, we considered that the wrinkles were in the direction $\boldsymbol{e}_x$ of the x axis and had a uniform period $\frac{2\pi}{q}$. In what follows, we assume that the wrinkles are in a fixed and arbitrary given direction $q_x\boldsymbol{e}_x + q_y\boldsymbol{e}_y$, which corresponds to $\varphi(x) = q_x x + q_y y$, or in a given variable direction $\boldsymbol{e}_\theta$ for example in the case of an annular membrane under traction.

The macroscopic model can be deduced from the microscopic models [8.52] and [8.53] simply by identifying the Fourier coefficients in each equation and by using the rule [8.26] and the rule defining the Fourier components of a gradient vector is the following: $\nabla u = \sum_{-\infty}^{+\infty}(\nabla u_j + ij\nabla\varphi \otimes u_j)e^{ij\varphi(x)}$, where $\otimes$ represents the tensor product. For the sake of simplification, in the previous expression and in [8.59], we keep the harmonics up to first level, as in section 8.5. The corresponding macroscopic model is written as [8.57] and [8.58]. For further details on the application of the technique to FvK plates, the reader should refer to Khalil et al. (2000).

8.6.2. *Numerical examples and discussions*

The new technique for model reduction is now evaluated through three numerical examples. The objective of the simulations is to evaluate the new reduction technique based on Fourier series and a variable direction given for the wrinkles. The complete shell problem was solved using the ANM algorithm and the discretization by DKT18 finite elements (Batoz et al. 1976) using the FvK plate model. This microscopic model was compared with that of ABAQUS using the shell element STRI3, the results being quite identical. The macroscopic (or reduced) models were also solved by ANM and discretized by DKT18 finite element. In practice, the approximate estimations of the instability wavelength are often available in the literature, as the Allen formula for the wrinkling of film-substrate systems (Allen 1969).

The first test concerns a rectangular membrane subjected to a biaxial compression-traction load (see Figure 8.7(a)). This is a typical benchmark for Fourier reduction methods, previously discussed in Damil et al. (2013, 2014), Huang et al. (2015) and Attipou et al. (2015), and it shows that reduction is very efficient in the cases with a large number of wrinkles: indeed, the number of finite elements necessary for the macroscopic model results from spatial variations of the envelope and not from the number of wrinkles.

Let us consider a thin rectangular membrane of width $L_x = 200$ mm, length $L_y = 400$ mm and thickness $h = 0.05$ mm, made from an isotropic, homogeneous and elastic material of Young's modulus $E = 70{,}000$ MPa and Poisson coefficient $\nu = 0.3$. The first load corresponds to a traction $F_x = 10$ N/mm along the direction x followed by an increasing compression load λF_y, with $F_y = 1$ N/mm along direction y and λ is a load parameter (see Figure 8.7). During traction, the membrane is subjected to the following boundary conditions: $w = 0$ on all of the edges, $\beta_y = 0$ on the lines $(x = L_x/2, y)$ and $(x = -L_x/2, y)$ and $\beta_x = 0$ on the lines $(x, y = L_y/2)$ and $(x, y = -L_y/2)$. For the compression load, we maintained the same boundary conditions and added $v = 0$ on the lines $(x, y = L_y/2)$ and $(x, y = -L_y/2)$. These boundary conditions are only used for solving the initial complete model. For the boundary conditions of the reduced model, we assumed that $w_1^R = w_1^I = 0$ on all edges, $\beta_{y1}^R = \beta_{y1}^I = 0$ on the lines on $(x = L_x/2, y)$ and $(x = -L_x/2, y)$ and $\beta_{x1}^R = \beta_{x1}^I = 0$ on the lines $(x, y = L_y/2)$ and $(x, y = -L_y/2)$. During the compression load, we maintained the same boundary conditions as during the traction load and we added $v_0 = 0$ on the lines $(x, y = L_y/2)$ and $(x, y = -L_y/2)$. For this example, the wave vector Q is chosen in the direction y and the wave number is given by the formula $q_x = 0$ and $q_y = \sqrt[4]{\frac{F_x \pi^2}{D L_x^2}} = 0{:}2356$ where $D = \frac{Eh^3}{12(1-\nu^2)}$ (Attipou et al. 2015).

The domain occupied by the thin membrane is discretized by 85 nodes in the x direction and by 169 nodes in the y direction, or 86,190 dof for the complete model. For the reduced model, the domain is discretized by seven nodes along the x direction and by 11 nodes along the y direction, or 616 dof.

Figure 8.7(b) shows the load–displacement curve $(\lambda F_y, w/h)$ at the central node of the membrane for the FvK model obtained by complete (micro) and reduced (macro) models, as well as the analytical critical load. It can be noted in this figure that the solutions obtained by complete and reduced models are in very good agreement despite a significant reduction of the size of the problem.

Figure 8.7(c) shows a comparison between the complete (micro) and reduced (macro) models in the context of FvK. It appears that the proposed reduced model can correctly predict the same form and the same number of wrinkles as the complete model.

The second test is a problem of buckling of a rectangular membrane under shear load introduced by Wong and Pellegrino (2006b, 2006c).

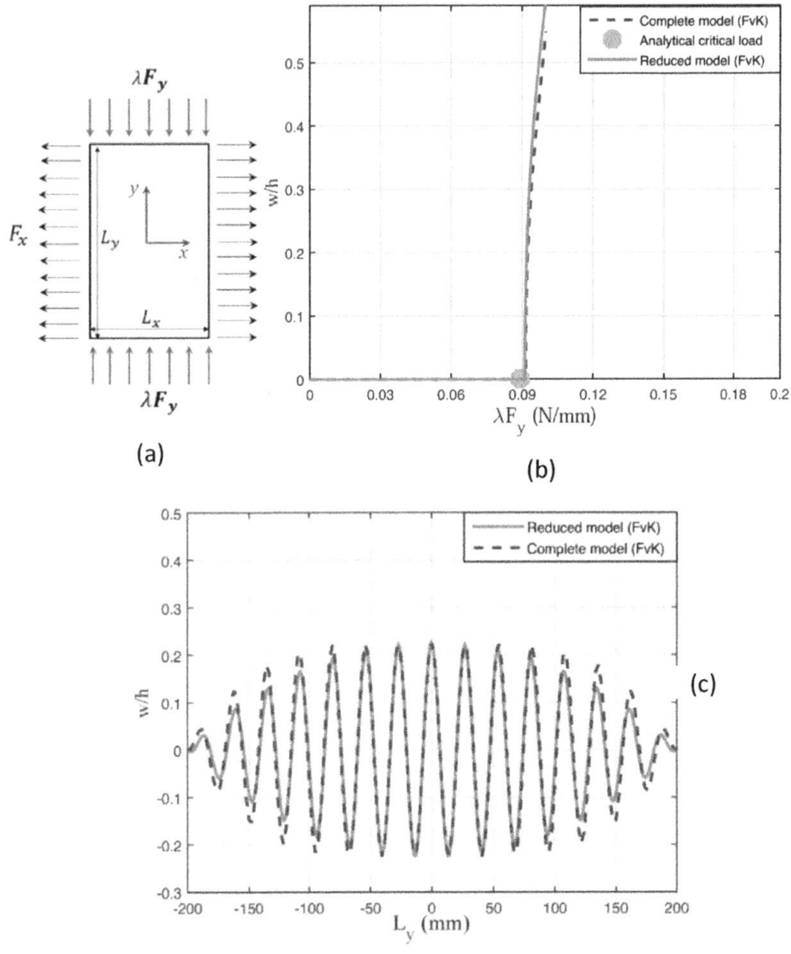

Figure 8.7. *(a) Wrinkling of a thin and elastic rectangular membrane under biaxial load. (b) Evolution of transversal displacement w/h as a function of the load λF_y for the FvK model and the analytical critical load. (c) Comparison of the complete and reduced models: deviation along the median line $(x = 0, y)$ for $\lambda = 0.092$*

Let us consider a thin rectangular membrane of width $L_x = 400$ mm, length $L_y = 120$ mm and thickness $h = 0.05$ mm, made from an isotropic, homogeneous and elastic material of Young's modulus $E = 70{,}000$ MPa and Poisson coefficient $\nu = 0.3$. This membrane is subjected to a shear load λF_x along the longest edges, with $F_x = 1$ N/mm (see Figure 8.8(a)). The membrane is subjected to the following boundary conditions: $v = w = 0$ along the longest edges and free on the two shortest edges; these boundary conditions are used for the complete model. For the boundary conditions used in the reduced model, we assumed that $v_0 = 0$; $w_1^R = 0$ and $w_1^l = 0$ on the two longest edges.

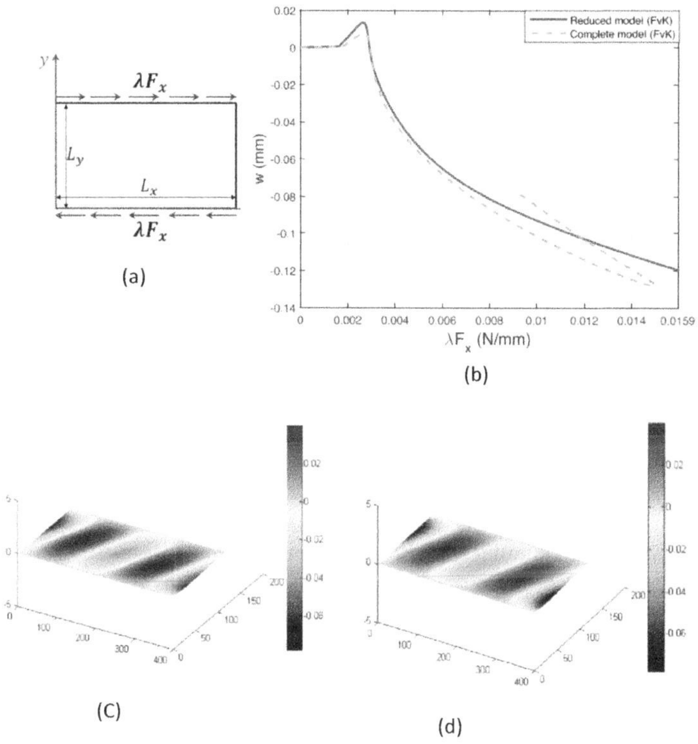

Figure 8.8. *(a) Wrinkling of a membrane subjected to shear. (b) Evolution of transversal displacement w as a function of the load λF_x for the FvK model. Membrane strained by the complete model (c) and reduced model (d) for $\lambda = 0.004$*

In this case, the direction of the chosen wave vector $q_x \boldsymbol{e}_x + q_y \boldsymbol{e}_y$ is parallel to the diagonal direction of the rectangle, and the wave number was estimated by a preliminary calculation using the industrial Abaqus code. The values obtained are $q_x = 0.021$ and $q_y = -q_x = -0.021$.

The domain occupied by the thin rectangular membrane is discretized by 47 nodes in the x direction and by 13 nodes in the y direction, or 3,666 dof for the complete model. For the reduced model, the domain is discretized by nine nodes along the x direction and by seven along the y direction, or 504 dof.

Figure 8.8(b) shows the load–displacement curve $(\lambda F_x, w)$ at the central node of the membrane for the complete and reduced models using the FvK theory. It can be noted in this figure that the solutions obtained by the complete and reduced models are in good agreement up to a large displacement $^{w}/_{h} \approx 2.5$, where the curve goes backwards. Such a return is often associated with a change of the wave number that is obviously taken into account only by the reduced model.

In Figures 8.8(c) and (d), the strained forms are compared. The reduced and complete models predict the same strained form with about two wrinkles. This sets the relevance of the reduced model. The numerical calculations of Wong and Pellegrino (2006a) found about 11 wrinkles as they considered an entirely different problem due to the prescribed transverse tension. Other recent and independent works show the strong influence of this transverse tension.

The third example is that of a classic annular membrane under traction, with radially oriented wrinkles that can only be taken into account by the new reduction technique involving a variable direction of the wave vector. Finally, we study another plate with a circular hole, but with a nonuniform pre-stress and hence with wrinkles of nonuniform amplitude.

We consider a thin annular membrane of inner radius $R_{int} = 400$ mm, and outer radius $R_{ext} = 1{,}000$ mm and thickness $h = 1$ mm, made from an elastic, homogeneous and isotropic material of Young's modulus $E = 69{,}000$ MPa and Poisson coefficient $\nu = 0.3$. This membrane is subjected to a traction force λF along the radial direction, with $F = 1$ N/mm as shown in Figure 8.8. For the complete model, we consider the following conditions: $w = 0$ on the inner boundary of radius R_{int}; $u = 0$; $v = 0$ and $w = 0$ on the outer boundary of radius R_{ext}. For the reduced model, we assume that $w_1^R = 0$; $w_1^I = 0$ on the outer boundary and $u_0 = 0$; $v_0 = 0$; $w_1^R = 0$ and $w_1^I = 0$ on the inner boundary. In this third test, the wave vector is chosen based on the geometry of the domain, so that wrinkles are orthogonal to the boundary: this leads to a phase $\varphi = q\theta$ and a wave vector $Q = \nabla\varphi = \frac{q}{R} e_\theta$, the chosen wave number $q = 13$ being based on a complete calculation.

For this third example, the domain occupied by the annular thin film is discretized in 23 nodes in $\boldsymbol{e}_r$ direction and in 160 nodes in $\boldsymbol{e}_\theta$ direction, or 22,080 dof for the complete model. For the reduced model, the domain is discretized in 19 nodes along the e_r direction and in 16 along the $\boldsymbol{e}_\theta$ direction, or 2,432 dof.

Figure 8.9(b) shows the load–displacement curve (λ, w) at the node of coordinates $(x = 700, y = 0)$.

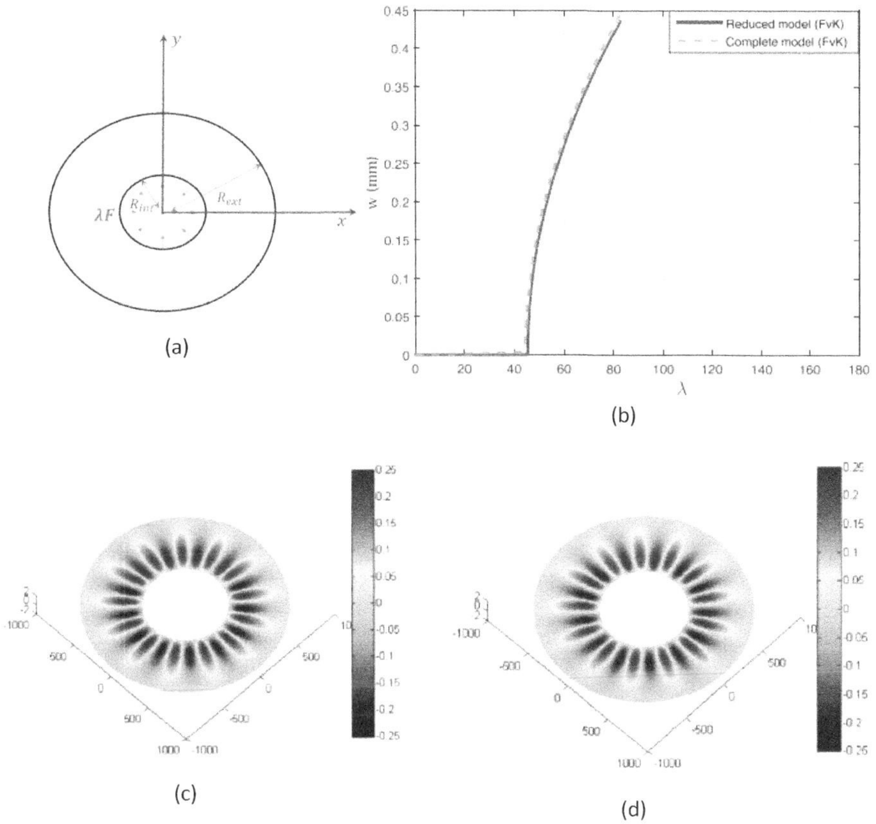

Figure 8.9. *(a) Wrinkling instabilities in a radially stretched annular thin film. (b) Evolution of transverse displacement w as a function of load λ for the FvK model at node $(x = 700, y = 0)$. Membrane strained by the complete model (c) and reduced model (d) for $\lambda = 49$*

It should be noted that complete and reduced models lead to the same response curves. The strained forms, presented in Figures 8.9(c) and (d), are also very close. This establishes that the new reduction procedure works equally well with a nonuniform wave vector.

Let us now examine the case of an instability by wrinkling in a disk pierced by a circular hole shifted under stretch.

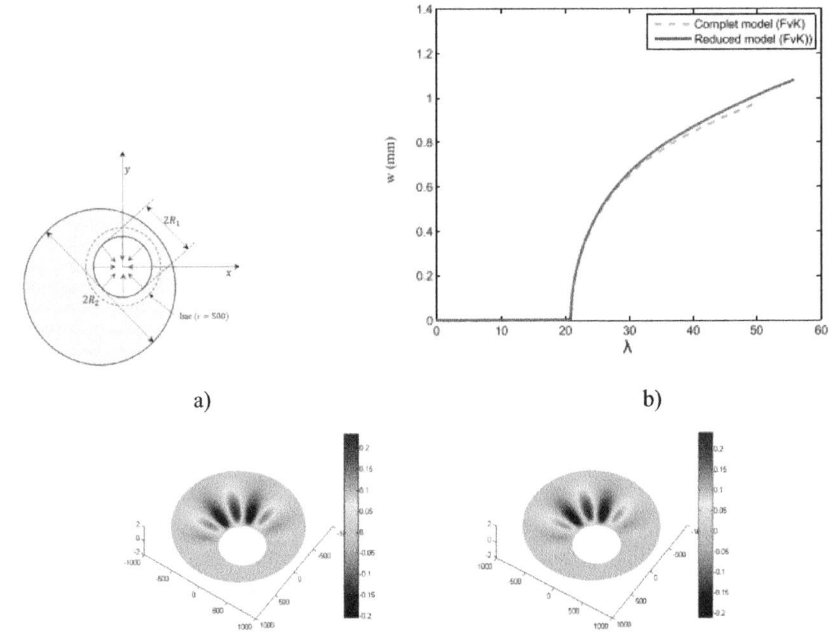

a) b)

c) Complete model for $\lambda = 21.6$ Reduced model for $\lambda = 21$

Figure 8.10. *(a) Wrinkling instabilities in a disk pierced by a circular hole shifted under stretch. (b) Evolution of the transverse displacement w as a function of the load* λF *for the FvK model. Membrane strained by the complete model (c) and reduced model (d) for* $\lambda = 21.6$

Let us consider a disk of radius $R_2 = 1{,}000$ mm and thickness $h = 1$ mm pierced by a shifted circular hole of radius $R_1 = 350$ mm made from an elastic, homogeneous and isotropic material of Young's modulus $E = 69{,}000$ MPa and Poisson coefficient $\nu = 0.3$. This membrane is subjected to a traction force λF along the radial direction, with $F = 1$ N mm as shown in Figure 8.10(a). The domain occupied by this thin film is discretized in 3,453 nodes, or 20,718 dof for the complete model. For the reduced model, the domain is discretized in 548 nodes or 4,384 dof. The adopted boundary conditions are similar to those in the previous example. In this last test, the wave vector is chosen based on the geometry of the domain, so that the wrinkles are orthogonal to the boundary of the hole. This leads to a phase $\varphi = q\theta$ and a wave vector $Q = \nabla\varphi = \frac{q}{R}\boldsymbol{e_\theta}$, and the chosen wave number $q = 6.8$ relies on a complete calculation.

Figure 8.10(b) presents the evolution of the transverse displacement w with respect to the load λ at the node of coordinates $(-353.49, -353.49)$. As shown by these figures, the complete and reduced models are in good agreement.

The strained forms obtained by the complete and reduced models for $\lambda = 21.6$, presented in Figure 8.10(c) are also very close. This establishes that the new reduction procedure works equally well with a nonuniform wave vector. These examples assume a response in the form of several modulated periodic patterns. This hypothesis is accurate in the vicinity of the bifurcation. When moving away from the bifurcation, interactions between modes are developing and the hypothesis becomes less accurate. This suggests the addition of other Fourier terms, but the reduced model must remain simpler than the complete model. Meshing has a significant influence on our reduced model. In this work, we conducted several tests in order to find a meshing that yields acceptable results.

8.7. Conclusion

We presented and discussed a model reduction method based on Fourier series with slowly varying coefficients. It can be used to define equations with macroscopic partial derivatives, with greatly reduced calculation costs. This reduction method was applied to buckling and wrinkling of thin membranes, but applications to fluid mechanics and other domains could be considered.

The weak point of this procedure is the necessity for a priori choice of the period and direction of oscillations. This reduction method should be improved to take into account the evolution of the wave vector.

8.8. References

Abdelmoula, R., Damil, N., Potier-Ferry, M. (1992). Influence of distributed and localized imperfections on the buckling of cylindrical shells. *International Journal of Solids and Structures*, 29, 1–25.

Allen, H.G. (1969). *Analysis and Design of Structural Sandwich Panels*. Pergamon, New York.

Attipou, K., Hu, H., Mohri, F., Potier-Ferry, M., Belouettar, S. (2015). Thermal wrinkling of thin membranes using a Fourier-related double scale approach. *Thin-Walled Structures*, 94, 532–544.

Audoly, B. and Boudaoud, A. (2008). Buckling of a stiff film bound to a compliant substrate–Part II: A global scenario for the formation of herringbone pattern. *Journal of the Mechanics and Physics of Solids*, 56, 2422–2443.

Banerjee, B., Shaw, A., Roy, D. (2009). The theory of Cosserat points applied to the analyses of wrinkled and slack membranes. *Computational Mechanics*, 43(3), 415–429.

Batoz, J.L., Chattopadhyay, A., Dhatt, G. (1976). Finite element large deflection analysis of shallow shells. *International Journal for Numerical Methods in Engineering*, 10(1), 39–58.

Ben Dhia, H. and Rateau, G. (2005). The Arlequin method as a flexible engineering design tool. *International Journal for Numerical Methods in Engineering*, 62, 1442–1462.

Boucif, M., Wesfreid, J.E., Guyon, E. (1991). Experimental study of wavelength selection in the elastic buckling instability of thin plates. *European Journal of Mechanics A/Solids*, 10, 641–661.

Budd, C.J. and Kuske, R. (2005). Localized periodic patterns for the non-symmetric generalized Swift–Hohenberg equation. *Physica D*, 208, 73–95.

Budd, C.J., Hunt, G.W., Kuske, R. (2001). Asymptotics of cellular buckling close to the Maxwell load. *Proceedings of the Royal Society of London*, A 457, 2935–2964.

Chen, X. and Hutchinson, J.W. (2004). Herringbone buckling patterns of compressed thin films on compliant substrates. *Journal of Applied Mechanics*, 71, 597–603.

Cochelin, B., Damil, N., Potier-Ferry, M. (1994). Asymptotic numerical methods and Padé approximants for nonlinear elastic structures. *International Journal for Numerical Methods in Engineering*, 37, 1187–1213.

Cochelin, B., Damil, N., Potier-Ferry, M. (2007). *Méthode asymptotique numérique*. Hermès-Lavoisier, Paris.

Cross, M.C. and Hohenberg, P.C. (1993). Pattern formation out of equilibrium. *Reviews of Modern Physics*, 65, 851–1112.

Cross, M.C., Daniels, P.G., Hohenberg, P.C., Siggia, E.D. (1983). Phase-winding solutions in a finite container above the convective threshold. *Journal of Fluid Mechanics*, 127, 155–183.

Damil, N. and Potier-Ferry, M. (1986). Wavelength selection in the postbuckling of a long rectangular plate. *International Journal of Solids and Structures*, 22, 511–526.

Damil, N. and Potier-Ferry, M. (1992). Amplitude equations for cellular instabilities. *Dynamics and Stability of Systems*, 7, 1–34.

Damil, N. and Potier-Ferry, M. (2006). A generalized continuum approach to describe instability pattern formation by a multiple scale analysis. *Comptes Rendus Mécanique*, 334, 674–678.

Damil, N. and Potier-Ferry, M. (2008). A generalized continuum approach to predict local buckling patterns of thin structures. *European Journal of Computational Mechanics*, 17, 945–956.

Damil, N. and Potier-Ferry, M. (2010). Influence of local wrinkling on membrane behavior: A new approach by the technique of slowly variable Fourier coefficients. *Journal of the Mechanics and Physics of Solids*, 58, 1139–1153.

Damil, N., Potier-Ferry, M., Hu, H. (2013). New nonlinear multi-scale models for wrinkled membranes. *Comptes Rendus Mécanique*, 341(8), 616–624.

Damil, N., Potier-Ferry, M., Hu, H. (2014). Membrane wrinkling revisited from a multi-scale point of view. *Advanced Modeling and Simulation in Engineering Sciences*, 1(1), 6.

Dawes, J.P.H. (2010). The emergence of a coherent structure for coherent structures: Localized states in nonlinear systems. *Philosophical Transactions of the Royal Society A*, 368, 3519–3534.

Daya, E.M., Jeblaoui, K., Potier-Ferry, M. (2003). A two scale method for modulated vibration modes of large, nearly repetitive, structures. *Comptes Rendus Mécanique*, 331, 443–448.

Drapier, S., Grandidier, J.C., Potier-Ferry, M. (2001). A structural approach of plastic microbuckling in long fiber composites: Comparison with theoretical and experimental results. *International Journal of Solids and Structures*, 38, 3877–3904.

Flores, F.G. and Oñate, E. (2011). Wrinkling and folding analysis of elastic membranes using an enhanced rotation-free thin shell triangular element. *Finite Elements in Analysis and Design*, 47(9), 982–990.

Forest, S. and Sab, K. (1998). Cosserat overall modeling of heterogeneous materials. *Mechanics Research Communications*, 25, 449–454.

Healey, T.J., Li, Q., Cheng, R.-B. (2013). Wrinkling behavior of highly stretched rectangular elastic films via parametric global bifurcation. *Journal of Nonlinear Science*, 23(5), 777–805.

Hoyle, R.B. (2006). *Pattern Formation. An Introduction to Methods*. Cambridge University Press, Cambridge.

Hu, H., Damil, N., Potier-Ferry, M. (2011). A bridging technique to analyze the influence of boundary conditions on instability patterns. *Journal of Computational Physics*, 230, 3753–3764.

Huang, Q., Hu, H., Yu, K., Potier-Ferry, M., Belouettar, S., Damil, N. (2015). Macroscopic simulation of membrane wrinkling for various loading cases. *International Journal of Solids and Structures*, 64, 246–258.

Hunt, G.W. and Wadee, M.A. (1998). Localization and mode interaction in sandwich structures. *Proceedings of the Royal Society A*, 454, 1197–1216.

Hunt, G.W., Da Silva, L.S., Manzocchi, G.M.E. (1988). Interactive buckling in sandwich structures. *Proceedings of the Royal Society A*, 417, 155–177.

Hunt, G.W., Peletier, M.A., Champneys, A.R., Woods, P.D., Wadee, M.A., Budd, C.J., Lord, G.J. (2000). Cellular buckling in long structures. *Nonlinear Dynamics*, 21, 3–29.

Iooss, G., Mielke, A., Demay, Y. (1989). Theory of steady Ginzburg–Landau equation in hydrodynamic stability problems. *European Journal of Mechanics – B/Fluids*, 8, 229–268.

Keller, H.B. (1987). Lectures on numerical methods in bifurcation problems. *Applied Mathematics*, 217, 50.

Khalil, S., Belaasilia, Y., Hamdaoui, A., Braikat, B., Damil, N., Potier-Ferry, M. (2020). A reduced-order modeling based on multi-scale method for wrinkles with variable orientations *International Journal of Solids and Structures*, 207, 89–103.

Kouznetsova, V., Geers, M.G.D., Brekelmans, W.A.M. (2004). Multi-scale second-order computational homogenization of multi-phase materials: A nested finite element solution strategy. *Computer Methods in Applied Mechanics and Engineering*, 193, 5525–5550.

Kyriakides, S., Arseculeratne, R., Perry, E.J., Liechti, K.M. (1995). On the compressive failure of fiber reinforced composites. *International Journal of Solids and Structures*, 32, 689–738.

Lecieux, Y. and Bouzidi, R. (2010). Experimental analysis on membrane wrinkling under biaxial load – Comparison with bifurcation analysis. *International Journal of Solids and Structures*, 47, 2459–2475.

Lecieux, Y. and Bouzidi, R. (2012). Numerical wrinkling prediction of thin hyperelastic structures by direct energy minimization. *Advances in Engineering Software*, 50, 57–68.

Leotoing, L., Drapier, S., Vautrin, A. (2002). Nonlinear interaction of geometrical and material properties in sandwich beam instabilities. *International Journal of Solids and Structures*, 39(13–14), 3717–3739.

Li, Q. and Healey, T.J. (2016). Stability boundaries for wrinkling in highly stretched elastic sheets. *Journal of the Mechanics and Physics of Solids*, 97, 260–274.

Liu, Y., Yu, K., Hu, H., Belouettar, S., Potier-Ferry, M., Damil, N. (2012). A new Fourier-related double scale analysis for instability phenomena in sandwich structures. *International Journal for Numerical Methods in Engineering*, 49, 3077–3088.

Mhada, K., Braikat, B., Hu, H., Damil, N., Potier-Ferry, M. (2012). About macroscopic models of instability pattern formation. *Internat. J. Solids Structures*, 49(21), 2978–2989.

Mhada, K., Braikat, B., Damil, N. (2013). A 2D Fourier double scale analysis of global-local instability interaction in sandwich structures. In *21ème congrès français de mécanique*, 26–30 August. Elsevier, Bordeaux.

Nayfeh, A.H. (1973). *Perturbation Methods*. John Wiley & Sons, New York.

Newell, A. and Whitehead, J. (1969). Finite band width finite amplitude convection. *Journal of Fluid Mechanics*, 38, 279–303.

Pomeau, Y. and Zaleski, S. (1981). Wavelength selection in one-dimensional cellular structures. *Journal de Physique*, 42, 515–528.

Potier-Ferry, M. (1983). Amplitude modulation, phase modulation and localization of buckling patterns. In *Collapse: The Buckling of Structure in Theory and Practice*, Thompson, J.M.T. and Hunt, J.W. (eds). Cambridge University Press, Cambridge.

Potier-Ferry, M., Mohri, F., Xu, F., Damil, N., Braikat, B., Mhada, K., Hu, H., Huang, Q., Nezamabadi, S. (2016). Cellular Instabilities analyzed by multi-scale Fourier series: A review. *Discrete and Continuous Dynamical Systems*, 9(2), 585–597.

Roddeman, D.G., Drukker, J., Oomens, C.W.J., Janssen, J.D. (1987). The wrinkling of thin membranes: Part I theory. *Journal of Applied Mechanics*, 54(4), 884–887.

Rossi, R., Lazzari, M., Vitaliani, R., Onate, E. (2005). Simulation of light-weight membrane structures by wrinkling model. *International Journal for Numerical Methods in Engineering*, 62, 2127–2153.

Sanchez-Palencia, E. (1980). *Non-Homogeneous Media and Vibration Theory*. Springer-Verlag, Heidelberg.

Schlichting, H. (1979). *Boundary-Layer Theory*, 7th edition. McGraw-Hill, New York.

Segel, L. (1969). Distant side walls cause slow amplitude modulation of cellular convection. *Journal of Fluid Mechanics*, 38, 203–224.

Swift, J.B. and Hohenberg, P.C. (1977). Hydrodynamic fluctuations at the convective instability. *Physical Review A*, 15, 319–328.

Trouffard, J., Cadou, J.M., Rio, G. (2010). Recherche de forme des gilets de sauvetage gonflables. *Mécanique & Industries*, 11, 117–122.

Waas, A.M. and Schultheisz, C.R. (1996). Compressive failure of composites, part II: Experimental studies. *Progress in Aerospace Sciences*, 32(1), 43–78.

Wadee, M.A. and Gardner, L. (2012). Cellular buckling from mode interaction in Ibeams under uniform bending. *Proceedings of the Royal Society A*, 468, 245– 268.

Wang, S., Song, J., Kim, D.H., Huang, Y., Rogers, J.A. (2008). Local versus global buckling of thin films on elastomeric substrates. *Applied Physics Letters*, 93(2), 023126.

Wang, C.G., Du, X.W., Tan, H.F., He, X.D. (2009). A new computational method for wrinkling analysis of gossamer space structures. *International Journal of Solids and Structures*, 46(6), 1516–1526.

Wang, T., Fu, C., Xu, F., Huo, Y., Potier-Ferry, M. (2019). On the wrinkling and restabilization of highly stretched sheets. *International Journal of Engineering Science*, 136, 1–16.

Wesfreid, J.E. and Zaleski, S. (eds) (1984). *Cellular Structures in Instabilities*. Springer-Verlag, Heidelberg.

Wong, W. and Pellegrino, S. (2006a). Wrinkled membranes III: Numerical simulations. *Journal of Mechanics of Materials and Structures*, 1(1), 63–95.

Wong, W. and Pellegrino, S. (2006b). Wrinkled membranes I: Experiments. *Journal of Mechanics of Materials and Structures*, 1(1), 3–25.

Wong, W. and Pellegrino, S. (2006c). Wrinkled membranes II: Analytical models. *Journal of Mechanics of Materials and Structures*, 1(1), 27–61.

9

Electrochemical Methods for the Study of Wall Transfer Phenomena

This chapter presents electrochemical methods and their use for the study of transfer phenomena in the vicinity of walls. The principle of the electrochemical method is based on mass transport in a flow in the vicinity of an electrode, where electrochemical reactions occur and the limiting diffusion current is measured. Depending on configuration, the method provides access to local and global mass transfer coefficients and to drag coefficients. The method can be used in laminar and turbulent flows. For the latter, various signal processing methods are presented so that intrinsic characteristics of the turbulence can be identified as accurately as possible based on the fluctuations of the limiting diffusion current. A significant part of the chapter is dedicated to electrochemical measurements in non-Newtonian fluids, rheological aspects and hydrodynamic drag reduction (DR) mechanisms in the case of long chain polymers and other additives.

9.1. Introduction

Due to the research conducted by V.G. Levich at the Frumkin Institute in Moscow and gathered in a seminal publication (Levich 1952), we know that mass transport in a flow in the vicinity of a wall is related to the electrochemical reactions taking place at the latter. Electrode potential or current control governs the electron exchanges with the species arriving at (or leaving) the wall. Therefore, in the general case, mass transport appears as one of the processes involved in the global current (or potential). If the rate of electron exchange becomes very high, then the mass input of reactive species becomes the limiting process and the current measurement contains

Chapter written by Claude DESLOUIS, Bernard TRIBOLLET, Abdellah ARHALIASS and Jack LEGRAND.

the information on the wall fluxes of these species, and therefore on the hydrodynamic and/or rheological properties of the studied fluid. The use of this property led to many studies of laminar and turbulent flows in Newtonian liquids and also to rheological studies in non-Newtonian liquids described below. The electrochemical technique is one of the non-intrusive methods used to measure the mass transfer coefficients and the wall shear rate. It is based on the determination of the limiting diffusion current delivered by a small probe (electrode) in contact with the liquid flow. By solving the convection-diffusion equation under steady state, Reiss and Hanratty (1962) proposed a solution linking the limiting diffusion current to the wall shear rate. This solution is only valid for high Péclet numbers when the axial diffusion term can be neglected. It is also known as the Lévêque (1928) solution.

Many works concerned the device involving a cylindrical space with rotation of one of the cylinders, plus an axial flow rate. The flow through such a device is known as Taylor–Couette–Poiseuille flow, which contains various flow regimes (Kreith 1968): laminar, laminar with Taylor vortices, turbulent and turbulent with vortices. This type of device was used for various applications due to high transfer coefficients that can be obtained and the possibility to vary two independent parameters (axial flow rate and rotation). These applications include the study of transfers in rotating machinery, heat exchangers or electrochemical reactors. Kataoka et al. (1977) extended to combined axial flow and rotation; the works of Mizushina (1971) concerned only rotation. These two articles were the first to use the electrochemical method to study the local mass transfer at the wall of the stationary outer cylinder. This configuration was studied for the design of an electrochemical reactor with coaxial cylindrical electrodes in which the rotation of the inner cylinder was combined with the forced axial flow of the electrolyte (Coeuret and Legrand 1981). The interesting aspect is the existence of vortex regimes that facilitate the local mixture of products generated at the electrodes or the transportation of a product from one electrode to another (case of certain organic electrosyntheses). This type of device has however been little studied for its application in electrochemistry; however, there are two particular cases, the flow of electrolyte by itself in the annular space (Ross and Wragg 1965) and the rotation by itself of the inner cylinder (rotating cylinder electrode) (Gabe and Robinson 1972), which were widely studied.

The study of mass transfer in a developing flow is important for the design of electrochemical reactors in which the electrodes are often located downstream of the electrolyte input. The analysis of the heat/mass development in the input region in pipes or annular spaces was widely studied. The analysis of the simultaneous development of mass and momentum transfers is very important for the control and prediction of the performances of electrochemical reactors. Ould-Rouis et al. (1989) applied an electrochemical method by inserting microelectrodes in a conducting inner cylinder made of nickel. This result has proved that the axial diffusion term in the diffusion–convection equation in the diffusion equation must be taken into account in order to numerically predict the mass transfer of electrochemical systems. It is possible to increase the heat or mass transfer by creating a vortex flow resulting from the application of a tangential component to the main axisymmetric axial motion of a fluid. Fluid rotation can be made at the input using a circular tangential input. This type of generator produces a decreasing vortex flow. The vortex motion is generated at the input, and then it freely decreases along the flow path and tends toward a fully developed axial flow far from the entry. Despite its simple design and its easy maintenance, few studies (Walsh and Wilson 1986; Legentilhomme and Legrand 1991) are dedicated to its application in electrochemistry. Moreover, mass transfer in a vortex flow is far less significant (Legentilhomme and Legrand 1990) than in the case of a fully developed axial flow. The importance of packing bed reactors in the chemical and electrochemical industry facilitated many works on mass transfer and flow modeling, the flow being assimilated to an axial dispersion reactor. The axial dispersion coefficients can be obtained using an electrochemical method based on experimental curves of residence time distribution (Legentilhomme et al. 1989). These coefficients, or their equivalent in number of perfectly stirred reactors in series, are necessary for the calculation of mass transfer coefficients during the flow through volume electrodes (Montillet et al. 1994).

The electrochemical method is also used for the study of transfer phenomena in microfluidic systems. The electrodiffusion microsensors prepared by photolithography have already been used for studying the structure of turbulence near the wall in flow channels (Deslouis et al. 2004) and their applications in microfluidic devices have been considered (Kjeang et al. 2007; Mouheb et al. 2011). Nowadays, it is possible to produce microfluidic devices with precisely shaped microelectrodes prepared by various microfabrication techniques. This offers the possibility to use such microelectrodes as electrochemical sensors for the study of flow near the wall in various microfluidic systems (e.g. in geometries of complex channels often used as micromixers; Hessel et al. 2005), or possibly in multiphase microfluidic systems dealing with the hydrodynamic behavior of bubbles and drops (Kreutzer et al. 2005).

9.2. Principles and definitions

Generally speaking, the current I flowing through an electrode in contact with an ionic conducting liquid has the form: $I = f(E, c_R, c_P, \vec{V})$, which is a function of various parameters, where E is the potential applied to the electrode, c_R and c_P wall concentrations of a reactant R and a product P, respectively, carried toward (or formed at) the electrode, and $\vec{V}$ is the velocity of the liquid in the vicinity of the electrode.

The various equations that formalize the expression of the current are (i) for the hydrodynamic part in the general case the Navier–Stokes equations, (ii) the equation governing the current-potential relation related to the interface electron transfer and (iii) the mass transport equation. The expression of current is in fine provided by the Fick equation, provided that the distribution of concentration near the wall is known.

The "electrochemical" equation may be more or less complex involving multiple stage mechanisms according to the possibility of interface modifications due to the dissolution of the electrode material or of deposit on its surface. For the transport phenomena considered in this article, we only consider the case of non-attackable electrodes, generally noble metals, gold, platinum or possibly carbonaceous materials.

The equation of mass transport to (or from) an electrode is written as:

$$\frac{\partial c}{\partial t} = D_M \Delta c - \vec{V}.\overrightarrow{grad.c} \quad [9.1]$$

D_M is the molecular diffusion coefficient of an electrochemical tracer involved in an oxido-reduction reaction and c its concentration:

$$ox + pe \rightleftarrows red \quad [9.2]$$

The *ox* and *red* species are therefore the reactants R or products P, and p is the number of electrons transferred in the wall reaction.

The currently used redox systems include particularly the following:

– the ferri-ferrocyanide couple ($Fe(CN)_6^{3-}/Fe(CN)_6^{4-}$);

– the triiodide-iodide couple (I_3^-/I^-);

– the benzoquinone/hydroquinone couple ($C_6H_4O_2/C_6H_6O_2$) in acidic medium;

– the O_2/OH^- couple.

The tracer couple is used in low concentration in the presence of a high-concentration support electrolyte (typically 1:100 ratio) in order to minimize the migration current so that only the diffusion–convection current is involved.

When the potential applied to the electrode exceeds a certain value with respect to a reference potential given by a reference electrode, the current becomes independent of the potential and a diffusion plateau is reached, corresponding to zero concentration at the wall of the reactant species. The control device employed is a potentiostat in a system with three electrodes: the working electrode provides current measurement, a reference electrode and a counter electrode. The study of mass transport is conducted for this value of the current referred to as limiting diffusion current.

The study of the mass transfer to a surface may concern the global surface consisting of a metallic electrode, but it can also be performed locally, by insertion of microelectrodes. In the latter case, two situations can be identified:

1) The microelectrode is inserted in the metallic electrode and applied the same potential while being isolated from it by a very small insulating space, so that the measurement of the current in the "conductive wall" offers information about the local flux J and therefore on the local mass transfer coefficient k_c with $k_C = {}^{J}/_{C_\infty}$ and $I_L = pFSJ$, where F is the Faraday's number (F = 96485 C mol^{-1}), S is the electrode surface, I_L is the limiting diffusion current and c_∞ is the tracer concentration in the solution.

2) The microelectrode is inserted in the surface consisting of an inert material ("microelectrode in inert wall") and offers information on the local properties of the flow. These two situations are schematically presented in Figure 9.1.

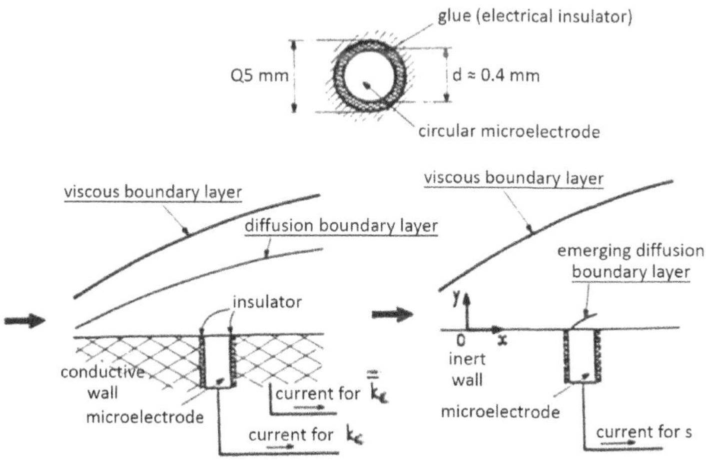

Figure 9.1. *Insertion of microelectrodes (Elements of electrochemical engineering, Coeuret F and Storck A. (1963) © Technique & Documentation – Lavoisier 1993)*

9.3. Results in Newtonian fluids

9.3.1. *Laminar flow and average turbulence quantities*

9.3.1.1. *Rotating disk*

The electrochemists interested in the study of interface kinetic processes are aware of the essential role of mass transport, particularly when there is no clearly limiting process. There is a flow system for which transport control is known, namely in the vicinity of a rotating disk. Introduced by Levich (1952), it has the remarkable property of having a diffusion layer of uniform thickness. This property is the direct result of the uniformity of the hydrodynamic layer, as shown by Cochran (1934). In case of mass transport control, the expression of the limiting diffusion current takes the form $I \propto \omega^{\frac{1}{2}}$, where ω is the angular rotation velocity of the electrode.

9.3.1.2. *Local friction*

The works of Levich (1952) were followed by various research works focusing on turbulent flows by means of microelectrodes (microdisks, rectangular microelectrodes, of submillimetric dimensions typically of the order of 100 μm) that can be used to collect local information.

Indeed, the integration of equation [9.1] in steady state ($\frac{\partial c}{\partial t} = 0$) leads to the local distribution of concentration of the electroactive species $c(y)$, where y is the coordinate normal to the wall. In a liquid, the mass diffusion coefficient is typically with three orders of magnitude below the kinematic viscosity ν of the solution so that the concentration profile in the diffusion layer is fully developed in a region where the velocity profile is characterized by a linear gradient α, $\alpha = \left.\frac{\partial V_x}{\partial y}\right|_{y=0}$, $V_x\,(y) \sim \alpha y$, where x is the coordinate along the flow direction.

The current I_L results from Fick's equation:

$$I_L = -pFSD_M \left.\frac{\partial c}{\partial y}\right|_{y=0} \qquad [9.3]$$

The integration of equation [9.1] reveals the dependence of c as a function of α. The first works in this field were conducted by T. J. Hanratty's team at the University of Illinois Urbana Champaign in the 1960s. Pioneering works on thermal transfer had been conducted (Lévêque 1928; Ling 1963).

The expression of current on a rectangular microelectrode of width L and length l along the flow direction has the following form (Reiss and Hanratty 1962):

$$I_L = 0.8075\ p\ F\ Lc_\infty\ l^{2/3} D_M^{2/3}\ \ \alpha^{1/3} \qquad [9.4]$$

Also, on a circular microelectrode of diameter d:

$$I_L = 0.6776\ p\ F\ c_\infty\ d^{5/3} D_M^{2/3} \alpha^{1/3} \qquad [9.5]$$

The microelectrode plays the role of a local drag sensor τ, where the shear stress is $= \mu\alpha$, μ being the dynamic viscosity, hence $\alpha\ \propto I^3$.

9.3.1.3. *Turbulence diffusivity*

As shown in Figure 9.1, the measurements at the conductive wall imply that the diffusion boundary layer, of thickness δ_c, develops on the large size transfer surface compared to those of microelectrodes. This boundary layer is not interrupted on the microelectrodes, provided that the insulating parts between the microelectrodes and the transfer surface are sufficiently thin.

The average current obtained on the microelectrodes in the conductive wall is linked to the local mass transfer coefficient $\overline{k_c}$:

$$\overline{k_c} = \frac{I_L}{pFSC_\infty} \qquad [9.6]$$

where I_L is the time average of the limiting diffusion current. The global coefficient of mass transfer between the fluid and the transfer surface, of area S, is given by:

$$\overline{K_c} = \frac{1}{S} \int \overline{k_c}\, ds \qquad [9.7]$$

The section below shows the component of fluctuating velocity that predominates for the electrode in the conductive wall and what information it brings on local turbulence.

In turbulent flow, the velocity vector involves the average value along x and three fluctuating components:

$$V = \overline{V_x}(y) + v'_x + v'_y + v'_z$$

On the other hand, in the electrolytes, the Schmidt number (Sc =ν/D_M) is of the order of 1,000, which means that the diffusion boundary layer is much thinner than the hydrodynamic boundary layer.

Consequently, the velocity field can be written as a Taylor series expansion near the wall:

$$\overline{V_x}(y) = \overline{\alpha_x}\, y\, v'_x = \alpha'_x\, y\, (\alpha'_x = \left.\frac{\partial v'_x}{\partial y}\right|_{y=0})$$
$$v'_z = \alpha'_z\, y\, (\alpha'_z = \left.\frac{\partial v'_z}{\partial y}\right|_{y=0}) \qquad [9.8]$$

Our objective is to determine the expression of v'_y.

The continuity equation of the fluid is written as:

$$\frac{\partial\left(\overline{V_x} + v'_x\right)}{\partial x} + \frac{\partial v'_y}{\partial y} + \frac{\partial v'_z}{\partial z} = 0 \qquad [9.9]$$

The regime being established $\frac{\partial(\overline{V_x})}{\partial x} = 0$; therefore, it can be deduced:

$$\left.\frac{\partial v'_y}{\partial y}\right|_{y=0} = [-y(\alpha'_x + \alpha'_z)]|_{y=0} = 0 \qquad [9.10]$$

The Taylor series expansion of velocity v'_y is written as:

$$v'_y = v'_y\big|_{y=0} + y\left.\frac{\partial v'_y}{\partial y}\right|_{y=0} + \frac{y^2}{2}\frac{\partial^2 v'_y}{\partial y^2} + \ldots\ldots$$

The first two terms of the expansion being equal to zero, we get:

$$v'_y = \alpha'_y(x, y, z)y^2 \qquad [9.11]$$

In equations [9.8] and [9.11], the terms α'_x, α'_y and α'_z represent fluctuating wall velocity gradients. Since the component v'_y varies as a function of y^2, it can be neglected compared to v'_x and v'_z for a microelectrode with inert wall, while it becomes dominant for a conductive wall.

Introducing the decomposition $c_j = \overline{c_j} + c'_j$, the diffusion–convection equation is written as:

$$\frac{\partial c'_j}{\partial t} + \left(\overline{V_x} + v'_x\right)\frac{\partial\left(\overline{C_j} + c'_j\right)}{\partial x} + v'_y\frac{\partial\left(\overline{C_j} + c'_j\right)}{\partial y} + v'_z\frac{\partial\left(\overline{C_j} + c'_j\right)}{\partial z}$$
$$= D_{Mj}\Delta_\ell\left(\overline{c_j} + c'_j\right) \qquad [9.12]$$

Writing:

$$\left(\overline{V_x} + v'_x\right)\frac{\partial\overline{C_j}}{\partial x} + v'_y\frac{\partial\overline{C_j}}{\partial y} + v'_z\frac{\partial\overline{C_j}}{\partial z} = \vec{V}\,\overrightarrow{grad\,\overline{c_j}} \qquad [9.13]$$

Also, adding the zero term (equation [9.9]) corresponding to the continuity equation of the fluid, equation [9.12] becomes:

$$\frac{\partial c'_j}{\partial t} + \vec{V}\,\overrightarrow{grad\,\overline{c_j}} + \overline{V_x}\frac{\partial c'_j}{\partial x} + v'_x\frac{\partial c'_j}{\partial x} + v'_y\frac{\partial c'_j}{\partial y} + v'_z\frac{\partial c'_j}{\partial z}$$
$$+ c'_j\left(\frac{\partial v'_x}{\partial x} + \frac{\partial v'_y}{\partial y} + \frac{\partial v'_z}{\partial z}\right) = D_{Mj}\Delta_\ell\left(\overline{c_j} + c'_j\right) \qquad [9.14]$$

Finally, the equation of convective diffusion is written as:

$$\frac{\partial c'_j}{\partial t} + \vec{V}\,\overrightarrow{grad\,\overline{c_j}} + \overline{V_x}\frac{\partial c'_j}{\partial x} + div\left(c'_j v'_i\right) = D_{Mj}\Delta_\ell\left(\overline{c_j} + c'_j\right) \qquad [9.15]$$

The index i corresponds to three directions x, y, z. Equation [9.15] is decomposed into a stationary equation:

$$V_x\frac{\partial\overline{c_j}}{\partial x} + div\left(\overline{c'_j v'_i}\right) = D_{Mj}\Delta_\ell\,\overline{c_j} \qquad [9.16]$$

and a non-stationary equation:

$$\frac{\partial c'_j}{\partial t} + \overline{V_x}\frac{\partial c'_j}{\partial x} + v'_i\,\overrightarrow{grad\,\overline{c_j}} + div\left(c'_j v'_i - \overline{c'_j v'_i}\right) = D_{Mj}\Delta_\ell\,c'_j \qquad [9.17]$$

The quadratic term $\overline{c'_j v'_i}$ differs from zero and it establishes the correlation between the velocity field and concentration. It can be defined by introducing a coefficient of turbulent diffusion, D_{ti}, in the direction i:

$$\overline{c'_j v'_i} = D_{ti} \frac{\partial \overline{C}_j}{\partial x_i} \quad [9.18]$$

For example, $\overline{c'_j v'_y}$ corresponds to the average time of production of instantaneous concentration along the component v'_y:

$$\overline{c'_j v'_y} = D_{ty} \frac{\partial \overline{C}_j}{\partial y} \quad [9.19]$$

The coefficient of turbulent diffusion can be expressed as a function of the dimensionless distance to the wall, $y^+ = V^* y/\nu$, where V^* the drag velocity is defined as $V^* = \sqrt{\tau/\rho}$.

$$\frac{D_{ty}}{\nu} = a_n y^{+m} \quad [9.20]$$

The fluctuations of the limiting diffusion current collected on a microelectrode in the conductive wall are linked to the fluctuating gradient normal to the wall, α'_y.

According to equation [9.17], neglecting the variations in the z and x directions, because the diffusion boundary layer is developed above microelectrodes, it can be written with: $v'_y = \alpha'_y y^2$

$$\frac{\partial c'_j}{\partial t} + \alpha'_y y^2 \frac{\partial \overline{C}_j}{\partial y} + \frac{\partial}{\partial y}\left(c'_j \alpha'_y y^2 - \overline{c'_j v'_y}\right) = D_{Mj} \frac{\partial^2 c'_j}{\partial y^2} \quad [9.21]$$

Similarly, the stationary equation (equation [9.16]) becomes:

$$\frac{\partial}{\partial y}\left(\overline{c'_j v'_y}\right) = D_{Mj} \frac{\partial^2 \overline{C}_j}{\partial y^2} \quad [9.22]$$

Using [9.19], equation [9.22] is written as:

$$\frac{\partial}{\partial y}\left[D_{ty} \frac{\partial \overline{C}_j}{\partial y} + D_{Mj} \frac{\partial \overline{C}_j}{\partial y}\right] = 0 \quad [9.23]$$

Also, using expression [9.20], we get:

$$\frac{\partial}{\partial y}\left[\left(a_n {y^+}^m + \frac{1}{Sc}\right)\frac{\partial \overline{C_j}}{\partial y}\right] = 0 \qquad [9.24]$$

Equation [9.24] is solved using the following boundary conditions:

$\overline{C_j} = 0$ in $y= 0$ $\overline{c_j} = c_\infty$ when $y \to \infty$, we get:

$$\overline{k_c} = V^* \frac{m}{\pi} sin\frac{\pi}{m}\, a_n^{1/m}\, Sc^{\frac{1-m}{m}} \qquad [9.25]$$

The values of m and a_n being determined from experimental measurements, the coefficient of turbulent diffusion, D_{ty}, can be calculated from equation [9.20].

Based on measurements on the rotating disk (Deslouis et al. 1980), it was shown that the exponent $n = 3$ with $a_n = 4.45 \times 10^{-4}$ is more compatible with the experimental data than $m = 4$. This relation is applicable in the viscous sublayer $y^+ \leq 5$.

NOTE.– The electrochemical method can be used to determine, at the same place, the local coefficients of mass transfer (k_c) and momentum (α_x), and therefore, it provides information on the analogy between the various transfers.

9.3.1.4. *From scalar measurement to vector measurement, split electrodes*

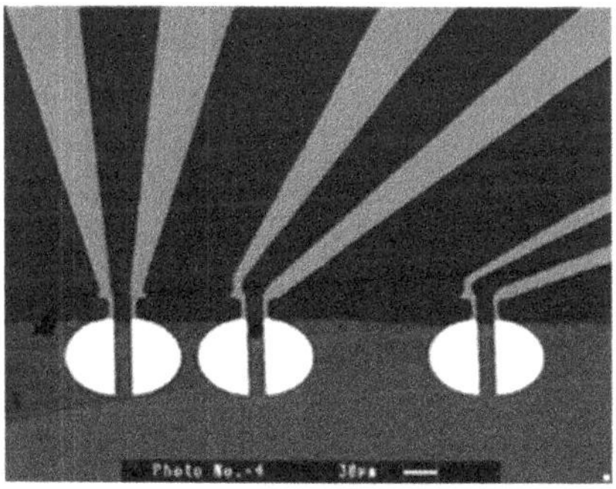

Figure 9.2. *Scanning electron microscope image of split circular electrodes prepared by photolithography. The golden tracks covered by protective resin appear in gray (Deslouis et al. 2004)*

Similarly to its thermal equivalent, the electrochemical microelectrode is a drag sensor that provides only scalar information. Microelectrodes coupled, for example, in chevron arrangements (Sirkar and Hanratty 1970) or split (Wein and Sobolik 1987) were proposed to obtain vector information providing access to longitudinal and transverse components of the flow. Bisegmented microelectrodes (Deslouis et al. 2004) or trisegmented microelectrodes (Baléras 1995) were made by photolithography (Figures 9.2 and 9.3).

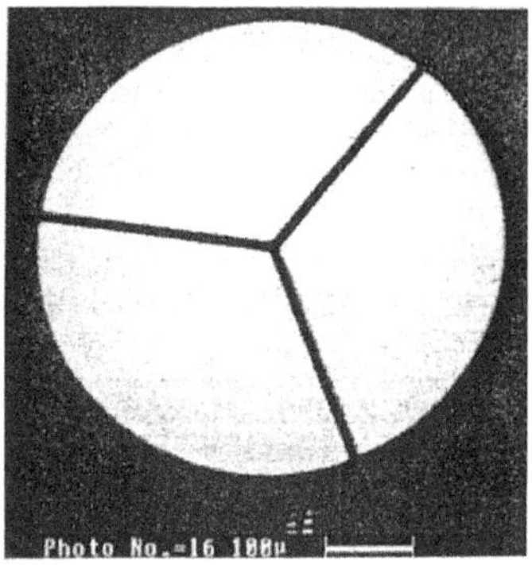

Figure 9.3. *Trisegmented probe obtained by photolithography on a flexible kapton film holder (Baléras 1995)*

9.3.2. *Access to time depending hydrodynamic quantities*

This section is dedicated to the use of the measurement of time-dependent fluctuations of the diffusion–convection current to obtain the temporal fluctuations of the velocity gradient. The necessity of this operation is explained by the role played by the diffusion layer as low-frequency filter. The statistical processing can then be directly performed on the returned signal $\alpha(t)$. Another path consists of conducting the statistical study on the current fluctuations in order to obtain the velocity fluctuations by integrating the role of the filter of the diffusion layer.

9.3.2.1. *Calculations of $\alpha(t)$ from $I(t)$*

For very high Péclet numbers in steady state for a circular probe, and by neglecting the axial diffusion term, mass transfer is written in the form of an analytical solution (Lévêque 1928), given by equation [9.20], which leads to the average velocity gradient:

$$\alpha_x = 1.556 k_c^3 d / D_{Mj}^2 \qquad [9.26]$$

In the presence of unsteady flows, the quasi-stationary solution deviates from the actual response of the probe because of the frequency response of the concentration sub-layer. To correct the calculations obtained from the quasi-stationary solution, various methods were proposed.

9.3.2.2. *Correction of the Lévêque solution*

Sobolik et al. (1987) presented a transient correction of the Lévêque solution based on the quasi-stationary solution and the response of the probe in the transient regime. This method can be applied to large fluctuations of wall shear rate but it is limited to the cases of negligible axial mass transfer (high Péclet numbers). They proposed the following expression to take into account temporal variations:

$$\alpha_c(t) = \alpha(t) + \frac{2}{3} t_0 \left(\frac{\partial \alpha}{\partial t}\right) \qquad [9.27]$$

where $\alpha_c(t)$ is the value of the shear rate corrected with respect to the probe inertia. The values of $\alpha(t)$ are calculated using equation [9.26]. t_0 is a characteristic time related to operating conditions, which can be determined from the temporal response to a polarization step (Wein 1981):

$$t_0 = 0.426\, d^{2/3} D_M^{-1/3} \;\; \alpha(t)^{-2/3} \qquad [9.28]$$

9.3.2.3. *Reverse method*

Mao and Hanratty (1991) developed a reverse method to solve the reverse problem of the convection–diffusion equation for a single probe using the Beck's algorithm (Beck et al. 1985). In their approach, the axial mass transfer was neglected and the objective was the study of flows with high fluctuations of wall shear rate. This method was adopted (Maquinghen 1999) in a three-dimensional case taking into account the axial diffusion for rectangular probes. Rehimi et al. (2006) used the same methodology in a two-dimensional case by improving the stability of the speed of the algorithm.

When the length of the electrochemical probe is small compared to the characteristic length of the variations of the wall shear rate, we can assume a homogeneous wall shear rate on the surface of the microelectrode. In this case, the diffusion–convection equation becomes:

$$\frac{\partial C}{\partial t} + \alpha(t)\, y\, \frac{\partial C}{\partial x} = D\left(\frac{\partial^2 C}{\partial x^2} + \frac{\partial^2 C}{\partial y^2}\right) \qquad [9.29]$$

If f is a characteristic frequency of the flow, l is the width of the probe, c_∞ is the concentration in solution and $\bar{\alpha}$ is the temporal mean of the wall shear gradient, the dimensionless form of equation [9.29] can be written as follows:

$$f^* \, \overline{Pe}^{-2/3} \frac{\partial C}{\partial t^*} + \alpha^* \, y^* \frac{\partial C}{\partial x^*} = \overline{Pe}^{-2/3} \left(\frac{\partial^2 C}{\partial x^{*2}} + \frac{\partial^2 C}{\partial y^{*2}} \right) \quad [9.30]$$

with:

$$x^* = \frac{x}{\ell};\ y^* = \frac{y}{\delta_c};\ t^* = t\,f;\ C = \frac{c}{c_\infty};\ \alpha^* = \frac{S}{\bar{S}};\ \delta_c = \left(\frac{\ell\,D}{\bar{S}}\right)^{1/3} f^* = \frac{f\,\ell^2}{D}$$

and $\overline{Pe} = \frac{\bar{\alpha}\ell^2}{D}$ the average Péclet number.

A finite volume method was used (Rehimi et al. 2006) for the discretization of equation [9.30]. An unconditionally stable implicit approach with an upwind diagram was used to solve the equation. Reverse methods rely on the minimization of the difference between the experimental results and those found using a direct model describing the phenomena (convection–diffusion equation, equation [9.30]) in order to determine unknown parameters or functions. The method relies on the use of the sequential estimation algorithm (Beck et al. 1985) to find the wall shear rate $\alpha^*_{exp}(t^*)$. In the sequential estimation method, $\alpha^*_{exp}(t^* \leq t^*_{i-1})$ is assumed to be known. At the time t^*_i, we define $\hat{\alpha}^*_{num}(t^*_i)$ as the first estimation of $\alpha^*_{exp}(t^*_i)$. $\hat{\alpha}^*_{num}(t^*_i)$ must be a good estimation of $\alpha^*_{exp}(t^*_i)$. A Taylor series expansion truncated at the order of the first estimation makes it possible to write the experimental Sherwood number as follows:

$$Sh_{exp}(t^*_i) = \hat{S}h_{num}(t^*_i) + \left(\alpha^*_{exp}(t^*_i) - \hat{\alpha}^*_{num}(t^*_i)\right) \left(\frac{\partial \hat{S}h_{num}(t^*_i)}{\partial S^*}\right)_{\hat{\alpha}^*_{num}(t^*_i)} \quad [9.31]$$

Based on [9.31], the iterative sequential process of inversion is given by:

$$\alpha^*_{exp}(t^*_i) = \hat{\alpha}^*_{num}(t^*_i) + \frac{\left(Sh_{exp}(t^*_i) - \widehat{Sh}_{num}(t^*_i)\right)}{\left(\frac{\partial \widehat{Sh}_{num}(t^*_i)}{\partial S^*}\right)_{\hat{\alpha}^*_{num}(t^*_i)}} \quad [9.32]$$

$\hat{\alpha}^*_{num}(t^*_i)$ is the value of the gradient numerically estimated by equation [9.30]. $\frac{\partial \hat{S}h_{num}(t^*_i)}{\partial \alpha^*}$ is numerically estimated by the following equation:

$$\left(\frac{\partial \hat{S}h_{num}(t^*_i)}{\partial \alpha^*}\right)_{\hat{\alpha}^*_{num}} = \frac{\hat{S}h_{num}\left(t^*_i, (1+\varepsilon)\ \hat{S}^*_{num}\right) \ - \ \hat{S}h_{num}\left(t^*_i, (1-\varepsilon)\ \hat{\alpha}^*_{num}\right)}{2\ \varepsilon\ \hat{\alpha}^*_{num}} \qquad [9.33]$$

According to Beck et al. (1985), ε is chosen equal to 10^{-4} in order to obtain a linear evolution between $Sh(t^*_i)$ and $\alpha^*\ (t^*_i)$. The new value of the wall velocity gradient $\alpha^*_{num}(t^*_i)$ is injected in the model in order to update the numerical value of the Sherwood number at the time t^*_i. After this operation, the time is incremented and the operation is repeated. The first estimation of $\hat{\alpha}^*_{num}(t^* = t^*_0 = 0)$ is calculated from equation [9.30]. The wall shear rate is calculated with the following expression:

$$\alpha^*_{num}(t^*_i) = \hat{\alpha}^*_{num}(t^*_i) + \frac{\sum_{k=1}^{r}\left(Sh_{exp}\ \ (t^*_{i+k}) - \hat{S}h_{num}(t^*_{i+k})\right)\cdot\left(\frac{\partial \hat{S}h_{num}}{\partial S^*}\right)_{\hat{\alpha}^*_{num}}}{\sum_{k=1}^{r}\left(\left(\frac{\partial \hat{S}h_{num}}{\partial S^*}\right)_{\hat{\alpha}^*_{num}}\right)^2} \qquad [9.34]$$

The estimation algorithm is the following:

i) Calculation of the field of concentration $C(x^*, y^*, t_i{}^*)$ from the equation [9.30] with $\hat{\alpha}^*_{num}(t^*_i) = \alpha^*_{num}(t^*_{i-1})$ or $\hat{\alpha}^*(t_i) = 2\ \alpha^*_{num}(t^*_{i-1}) - \alpha^*_{num}(t^*_{i-2})$.

ii) Calculation of $\left(\frac{\partial \hat{S}h_{num}(t^*_i)}{\partial S^*}\right)_{\alpha^*=\hat{\alpha}(t^*_i)}$ with the equation [9.33] by solving the equation [9.30] with t^*_i for the shear rate $(1+\varepsilon)\ \hat{\alpha}^*_{num}(t^*_i)$ that can be used to calculate $S\hat{h}_{num}\left(t^*_i, (1+\varepsilon)\ \hat{\alpha}^*_{num}(t_i)\right)$ with: $Sh(t^*) = \overline{Pe}^{\frac{1}{3}}\ \int_0^1 \left(\frac{\partial C}{\partial y^*}\right)_{y*=0} dx^*$ and the same operation is repeated for $S\hat{h}_{num}\left(t^*_i, (1-\varepsilon)\ \hat{\alpha}^*_{num}(t^*_i)\right)$.

iii) Calculation of $\alpha^*_{num}(t^*_i)$ with the equation [5.32] and recalculation of $S\hat{h}_{num}(t^*_i)$. Due to a good stability of the algorithm, we have $\left|\left|Sh(t^*_i)_{num}(t^*_i)_{exp}\right| < \left|Sh(t^*_{i-1})_{num}(t^*_{i-1})_{exp}\right|\right|$.

iv) Time incrementation and repetition of operations from i. When $\left|Sh(t^*_k)_{num}(t^*_k)_{exp}\right| \leq \left|\varepsilon_{precision}\right|$, the algorithm converges and we obtain

$\alpha^*_{exp}\ (t_k^*) \approx \alpha^*_{num}(t_k^*)$ for $t^* \geq t_k^*$. The difference between the exact value of the wall shear rate and the solution obtained by the inverse method automatically decreases for each new time step and rapidly converges to the expected accuracy.

9.3.2.4. *Transfer function – spectral analysis*

The hydrodynamic fluctuations are filtered through the diffusion boundary layer, which can be characterized by an impulse response, $h(t)$.

$\alpha'(t) \rightarrow \boxed{h(t)} \rightarrow I'(t)$

In the time domain, α' and I' are related by the convolution product (Max 1981): $I'(t) = h(t) * \alpha'(t)$. Hence, the determination of $\alpha'(t)$ from $I'(t)$ is simple only in the quasi-stationary case (see section 9.3.2.1). This relation extended to the autocorrelation functions is written as:

$$C_{\tilde{\iota}\tilde{\iota}'(\tau)} = C_{\tilde{\alpha}\tilde{\alpha}'(\tau)} * C_{hh(\tau)}$$

The Fourier transforms of the autocorrelation functions are the power spectral densities (PSD) (Max 1981) $W_{\tilde{\iota}\tilde{\iota}'(f)}$ and $W_{\tilde{\alpha}\tilde{\alpha}'(f)}$ and the convolution product becomes a simple product in the frequency domain.

$$W_{\tilde{\iota}\tilde{\iota}'(f)} = [H(f)]^2 \times W_{\tilde{\alpha}\tilde{\alpha}'(f)} \qquad [9.35]$$

The expression $H(f)$ is obtained from the resolution of the non-stationary convective–diffusion equation by considering a sinusoidal perturbation of low amplitude compared to the average flow. This equation was numerically solved for a periodic flow and experimentally confirmed on a pulsed flow in annular cavity or on a sine modulated rotating disk (Ambari et al. 1986; Nakoryakov et al. 1986; Deslouis et al. 1990).

For a microdisk and the dimensionless frequency:

$$\omega^* = 2\pi f \left(\frac{d^2}{D_M \bar{\alpha}^2}\right)^{1/3} \qquad [9.36]$$

For $\omega^* \leq 6$:

$$\left|\frac{\mathrm{H}(\omega^*)}{\mathrm{H}(0)}\right| = \left(1 + 0.049\ \omega^{*2} + 0.0006\ \omega^{*4}\right)^{-0,5} \qquad [9.37]$$

for $\omega^* \geq 6$:

$$\frac{H(\omega^*)}{H(0)} = \frac{4.416}{i\omega^*} - \frac{5.3}{(i\omega^*)^{3/2}} \qquad [9.38]$$

and:

$$|H(0)|^2 = \frac{1}{9}\frac{\overline{I_L^2}}{\overline{S_x^2}} \qquad [9.39]$$

For a rectangular microelectrode of width l along the flow direction, the expression [9.36] can be used by replacing d with l. In the expression [9.37], the coefficients 0.049 and 0.0006 are, respectively, replaced by 0.056 and 0.00126. Similarly, in the expression [5.38], the coefficients 4.416 and 5.3 are, respectively, replaced by 3.715 and 3.99.

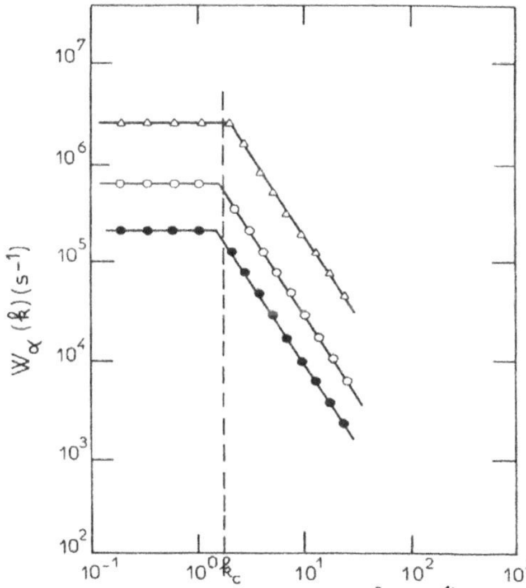

Figure 9.4. *PSD of fluctuations of the wall velocity gradient as a function of the wave number. L = 200 μm; Sc = 1130. Re =11250 (●); 15,800 (O); 24,490 (△) (Deslouis et al. 1993)*

The use of equations [9.35], [9.36], [9.37] and [9.38] for a rectangular microelectrode placed in a nearly 2D geometry rectangular channel made it possible to plot the turbulent velocity gradient spectra. As an example, Figure 9.4 shows spectra of $W_\alpha(k)$ as a function of the wave number $k = 2\pi f/V$ (V channel flow velocity) for various Reynolds numbers (Deslouis et al. 1993). These spectra correctly describe the inertial domain that appears as the linear section of the high frequency domain of slope -1.65 ± 0.1.

9.3.2.5. *Spatial correlations*

Using split probes represented in Figure 9.2, we can access correlation lengths of large structures of turbulent flow by placing several sensors at the interface, particularly following the longitudinal and transversal components (Deslouis et al. 2004). To this end, the probes are positioned in the rectangular channel, the line of separation between the two half-disks placed along the flow direction (x).

Hence, the sum of currents of the two half-disks ($i_1 + i_2$) provides the axial component α_x (respectively, the component u_x in the diffusion layer) and their difference ($i_1 - i_2$) yields the transverse component α_z (respectively, the component u_z fluctuating around the zero average value for this flow). The spatial correlation functions were obtained by comparing the current fluctuations measured at various transversal positions Z.

$$R(z) = \frac{\overline{\tilde{\imath}(Z,t)}\ \overline{\tilde{\imath}(Z+z,t)}}{\sqrt{\overline{\tilde{\imath}^2(Z,t)}}\ \sqrt{\overline{\tilde{\imath}^2(Z+z,t)}}} \qquad [9.40]$$

Under these experimental conditions (size of microelectrodes, Reynolds number), it was shown that the cut-off frequency of the transfer function H(f) is higher than that of the hydrodynamic spectrum of the turbulence and that H(f) has a constant value throughout the frequency domain. Hence:

$$R(z) = \frac{\overline{\tilde{\imath}(Z,t)}\ \overline{\tilde{\imath}(Z+z,t)}}{\overline{\tilde{\imath}^2(Z,t)}} \cong \frac{\overline{\tilde{\alpha}(Z,t)}\ \overline{\tilde{\alpha}(Z+z,t)}}{\tilde{\alpha}^2(Z,t)} \qquad [9.41]$$

$$g(z) = \frac{\overline{\tilde{\alpha}_x(Z,t)}\ \overline{\tilde{\alpha}_x(Z+z,t)}}{\tilde{\alpha}_x{}^2(Z,t)} \qquad [9.42]$$

Also, the longitudinal correlation coefficient $f(z)$ of the transversal velocity component (Figure 9.5):

$$f(z) = \frac{\overline{\tilde{\alpha}_z(Z,t)}\ \overline{\tilde{\alpha}_z(Z+z,t)}}{\tilde{\alpha}_z{}^2(Z,t)} \qquad [9.43]$$

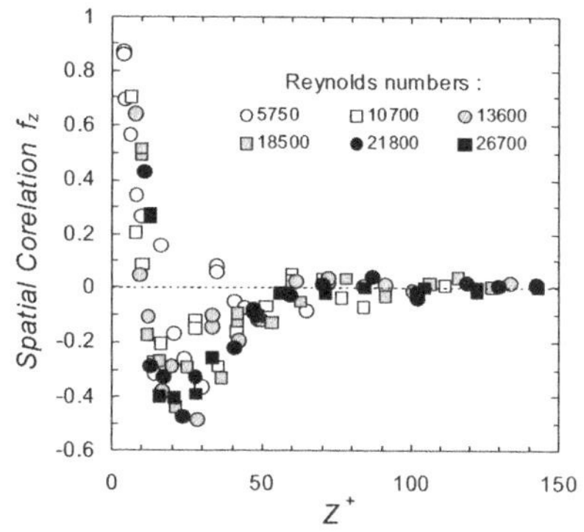

Figure 9.5. *Longitudinal correlation of the transversal component of the wall shear rate (Deslouis et al. 2004)*

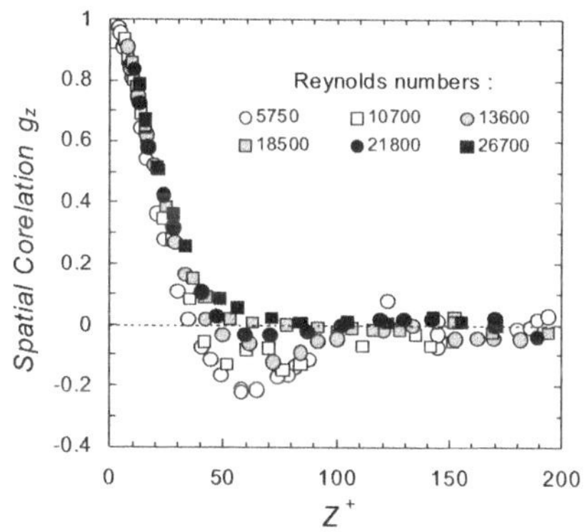

Figure 9.6. *Lateral correlation of the axial component of the wall shear rate (Deslouis et al. 2004)*

This leads to the two spatial correlation functions: the lateral correlation coefficient g(z) of the axial velocity component (Figure 9.6).

9.4. Electrochemical measurements in non-Newtonian media

Most works dedicated to the study of transport phenomena using electrochemistry concern Newtonian liquids, for which the relation between stress and strain is proportional, the proportionality coefficient being the dynamic viscosity μ ($\tau = \mu \frac{\partial V}{\partial x}$). These are essentially aqueous media. This proportionality does not exist for non-Newtonian media. This concerns either aqueous media in which additives are introduced (high polymer solutions, surfactant solutions, etc.), or non-aqueous media whose structuring yields various rheological laws (threshold liquids, etc.).

As it will be also seen, certain media are incompatible with the use of electrochemical methods due to the introduction of redox couples and also above-mentioned associated support electrolytes which, particularly given their high concentration (of the order of mole/L), are susceptible to change the nature of the medium or its structure, thus leading it to losing its non-Newtonian character. Other redox couples and/or support media can then be proposed.

9.4.1. *Long chain linear polymer solutions*

9.4.1.1. *Shear-thinning behavior*

Concentrated solutions of linear chain high polymers have a shear-thinning behavior characterized by a rheological law of the following form:

$$\tau_{xy} = K\left(\frac{\partial v_x}{\partial y}\right)^n \qquad [9.44]$$

Here, τ_{xy} is the shear stress along x, $\frac{\partial v_x}{\partial y}$ is the shear velocity along x, K is the consistency and n is the flow index (with $n \leqslant 1$). When the polymer concentration decreases, the index n tends to 1 and K tends to the dynamic viscosity μ. We find again the Newtonian rheological behavior and also the transport law for the rotating disk $I \propto \Omega^{1/2}$. The flow on a rotating disk for the rheological law [9.44] was calculated (Mitschka and Ulbrecht 1965). Hansford and Litt (1968) and then Greif and Paterson (1973), with a corrective taking into account the radial diffusion–convection, have used the results of the velocity field to calculate the mass flux and have obtained a relation of the following form:

$$J(R) = g(K, n, R, c_\infty, D_M)\Omega^{\frac{1}{1+n}} \qquad [9.45]$$

R denotes the disk radius, and the other parameters of the function g are defined above.

Hansford and Litt (1968) verified experimentally the dependence of the mass flux as a function of the angular velocity of the disk by measuring the rate of dissolution by weighing benzoic acid or β-naphtol pellets in aqueous solution of carboxymethylcellulose or polyethyleneoxide (PEO). The advantage of the electrochemical method compared to this technique is to provide a better accuracy and measure instantaneous values of the mass flux. The same verification was brought by Greif et al. (1972) using the reduction of O_2 in NaCl medium; other authors used the reduction of potassium ferricyanide into ferrocyanide (Shulman et al. 1972) in NaCl medium and (Deslouis et al. 1973) in KCl medium in the presence of PEO of average mass 4×10^6. Several authors reported departures from this law at low velocity with values below those predicted by relation [9.45] and at high velocity with a current drop and the appearance of instabilities. The behavior at low velocity is characterized by the presence of a cusp (Figure 9.7; Deslouis and Tribollet 1975): from centrifugal above this point, it becomes centripetal below it, elastic forces being predominant. The behavior at high velocity will be analyzed in the next section.

9.4.1.2. *Unstable regime: coil ⇄ stretch transition*

In Figure 9.7, the behavior at high velocity clearly shows a current drop for a velocity v_{c1} with the emergence of instabilities, the return to stable behavior taking place at a lower velocity v_{c2}, following a hysteresis loop. For velocities below that of the hysteresis loop and above that of the cusp, we observe a linear domain of slope $\frac{1}{1+n}$ in double logarithmic coordinates.

Such a behavior at high velocity for solutions of high polymers with linear chain of PEO type was observed for other types of flows: the flow through a hole in a thin wall (Piau 1974) or the flow in an extrusion capillary (Kataoka and Ueda 1971). Similar behaviors on the heat transfer to a cylinder positioned transversally to the flow (Smith et al. 1967; James and Acosta 1970; Piau 1980) or the mass transfer (Ambari et al. 1984a, 1984b, 1985b) have been observed.

These various flows and also that of the rotating disk share the property of possessing an elongation component. The proposed interpretation relies on the capacity of this type of strain to unwind polymer coils initially at rest (Flory 1953): this demonstrated that for a dilute solution, the transition, referred to as coil ⇄ stretch, takes place if the elongation stress γ and the terminal relaxation time (i.e. for a complete unwinding) τ_R are linked by the relation: $\gamma \times \tau_R > 1$ (Peterlin 1966; de Gennes 1974).

This process is accompanied by an increase in viscosity, the contribution of the elongation viscosity being typically two orders of magnitude greater than the shear viscosity. For example, in the case of the rotating disk, it results in a notable thickening of the hydrodynamic layer measured by flow birefringence, showing also the appearance of a birefringence characteristic for a structure anisotropy corresponding to the presence of elongated objects (stretched chains) parallel to the disk plane (Deslouis et al. 1981a). The relation between γ and τ_R strictly applies only to dilute conditions. The transition with hysteresis was observed both for the rotating disk and the flow around the cylinder.

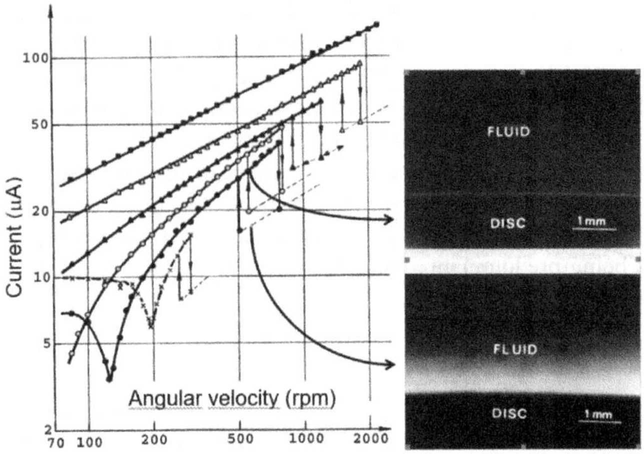

Figure 9.7. *Limiting diffusion current i_d for the reduction of ferricyanide (10^{-3} M) as a function of the rotational velocity Ω of an electrode with rotating disk for various concentrations of a KCl (M) solution containing* polyethyleneoxide *("polyox WSR 301" of average molecular weight 4 × 10^6): (■) 0 ppm; (△) 2,000 ppm; (▲) 5,000 ppm; (O) 6,500 ppm; (●) 7,000 ppm; (×) 8,000 ppm (Deslouis and Tribollet 1975)*

COMMENT ON FIGURE 9.7.– *Images of flow birefringence taken at the same rotational velocity of the disk in the hysteresis cycle: the upper image corresponds to the stable condition, and there is no birefringence; the lower image taken in unstable condition shows a strong birefringence with thickening of the hydrodynamic layer (Deslouis et al. 1981a).*

More precisely, in the case of the rotating disk, the elongation stress was identified at the outer boundary of the hydrodynamic layer based on the velocity gradient tensor of the dimensionless components of the field of velocities F_1, G_1, H_1 (Cochran 1934). This tensor has only one real eigenvalue $\gamma = H_1'$. The equation of continuity leads to $H_1' = -2F_1$, hence $\gamma = -2\,\Omega\,F_1(\xi)$ reaching the maximal value of 0.36 for $\xi = 1$, that is, at the outer boundary of the hydrodynamic layer (Ambari et al. 1985b). The characteristic time associated with the polymer is Zimm time τ_Z:

$$\tau_Z = \frac{\eta_s R_F^3}{k_B T} \qquad [9.46]$$

η_s is the dynamic viscosity of the solvent, R_F is the Flory radius of the macromolecular coil and k_B is Boltzmann constant.

The extrapolation at zero polymer concentration yields values of critical Ω, and therefore of γ^1 of the order of 3–4 × 10–^{2} s, an order of magnitude consistent with the calculated values of τ_Z near 10^{-2} s according to [9.46].

Similarly, in the case of a cylinder placed transversally to the uniform flow, the current drop is associated with the appearance of instabilities (Figure 9.8).

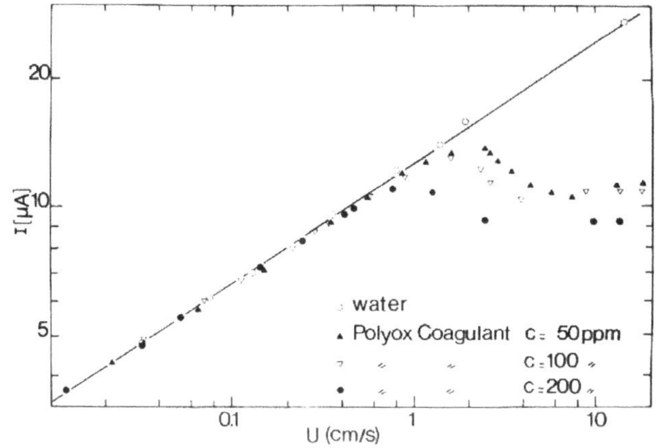

Figure 9.8. *Limiting diffusion current I as a function of the flow velocity transversal to a cylinder (platinum wire of diameter 50 μm) for the reduction of ferricyanide (10^{-3} M) in PEO solution (50, 100 and 200 ppm) (Ambari et al. 1984a)*

The flow near upstream of cylinder involves a pure elongation gradient deduced from the velocity field (Lamb 1945), whose maximum value is:

$$\alpha_{yy'} = -\alpha_{xx'} = 0.77\frac{V}{\alpha d} \qquad [9.47]$$

V is the velocity of the upstream uniform flow, d is the diameter of the cylinder and α is of the order of one at transition (Figure 9.9). According to Figure 9.10, macromolecules subjected to this elongational gradient undergo a contraction along xx' and an elongation along yy'.

As in the case of the rotating disk, the values of γ^1 calculated from the critical elongational gradient yield for the three considered concentrations in PEO, 2 × 10^{-2} s, 2 × 10^{-2} s and 6 × 10^{-2} s. Here, we have the same order of magnitude as the Zimm time of 10^{-2} s.

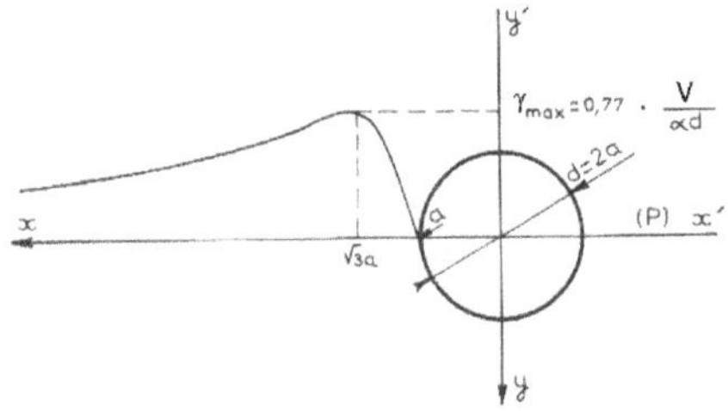

Figure 9.9. *Variation of the elongation gradient in the cylinder upstream region near the plane of symmetry (Ambari et al. 1984b)*

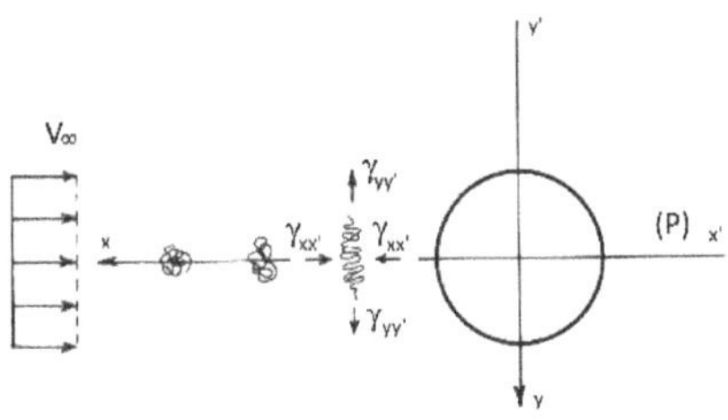

Figure 9.10. *Visualization of the strain of a macromolecule convected toward the cylinder (Ambari et al. 1984b)*

9.4.1.3. *Reduction of hydrodynamic drag*

The works on DR by macromolecular compounds and other additives were initiated in the United States at the end of World War Two. The first published works date back to 1949 and are due to Toms (1949), which explains why this phenomenon is sometimes referred to as the Toms effect.

Research skyrocketed in the 1970s and 1980s due to the enthusiasm generated by potential applications. The studied compounds were linear high polymers either synthetic (PEO and polyacrylamide (PAA) in aqueous media, and polystyrene sulfonate (PSS) in organic medium) or natural as carboxymethylcellulose (CMC) guar gum, or micellar solutions, or rodlet-shaped suspensions (e.g. asbestos). A non-exhaustive list includes the following works: (Fortuna and Hanratty 1972; Hanratty et al. 1977; Oldaker and Tiederman 1977; Gyr and Bewersdorf 1995).

The common characteristics are a maximum of DR that can reach 80% at polymer concentrations of several dozen ppm, reaching lower values for higher molar mass. The maximal DR regime is however higher than the extrapolation of the laminar regime. Finally, a degradation of the effect is observed by chain splitting if turbulence is maintained in time.

The mechanisms explained are as follows:

– The shear-thinning effect is rapidly excluded since at the very low concentrations used, the solutions are quasi-Newtonian.

– The direct interaction of turbulent structures and particularly the turbulent bursts that involve elongation components with polymer chains at molecular levels is the privileged mechanism and it highlights the coil–stretch process. The assumption is that these elongation stresses last long enough to generate chain unwinding. Two alternatives are proposed.

According to Lumley (1973), under the effect of elongation stresses, the chains may fully stretch out of the viscous sub-layer with an increase in viscosity (due to the contribution of elongation viscosity) being able to dampen dissipative scales.

de Gennes (1986) postulates the role of elasticity with alternating elongations and shears during which molecules store elastic energy. Similar to Lumley (1973), there must be a match between a characteristic time of the turbulence and Zimm time (coil-stretch analogue).

There is a simultaneous reduction in the transport of heat and mass (Virk and Suraiya 1977). Deslouis et al. (1981b) inserted microelectrodes on a rotating disk and measured the mass flux reduction in conductive wall (see section 5.2), and then on microelectrode in insulating wall to measure the local DR in PEO solutions. At maximum DR, the flux measured in turbulent flow is near the extrapolation of laminar regime.

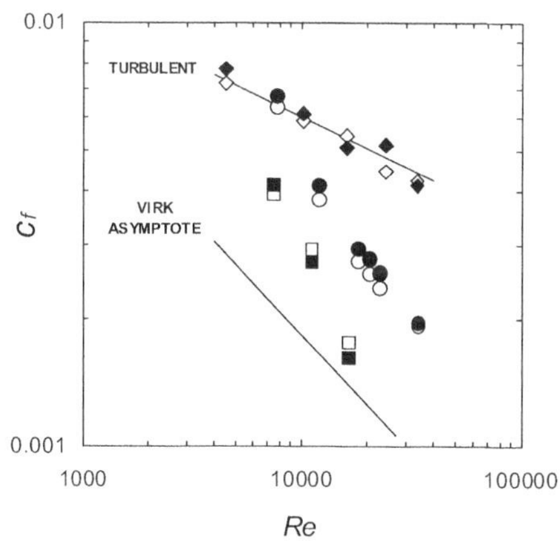

Figure 9.11. *Comparison of drag data measured by electrochemistry (solid symbols) and by pressure loss (hollow symbols) in water (rhombi) and for 20 ppm PAA solutions (circles) and 50 ppm PEO (squares) (Deslouis et al. 2004)*

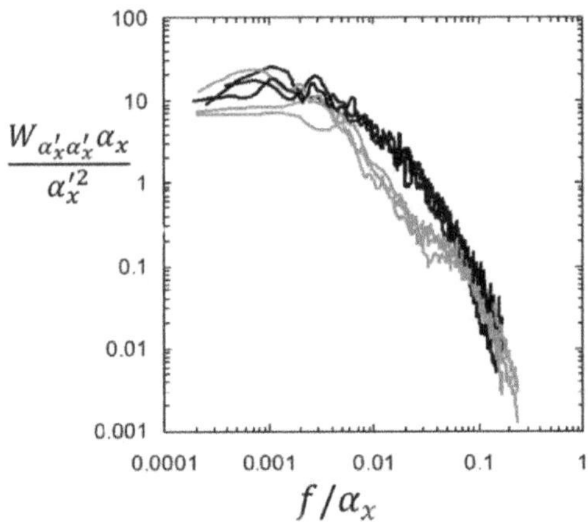

Figure 9.12. *PSD spectra normalized by the average shear of fluctuations along the flow direction. PSD for water (black lines) correspond to Re = 10,700, 18,500 and 26,700. The spectra for a (100 ppm) PEO solution (gray lines) correspond to Re = 9,040, 17,200 and 35,700 (with DR values of 62%, 72% and 63%, respectively) (Deslouis et al. 2004)*

The use of sensors fabricated by photolithography leads to an excellent agreement, without resorting to a calibration of microelectrodes, between electrochemically measured values of drag.

$c_f = (2\ \alpha\mu)/(\rho\ V^2)$ and those obtained by pressure loss $c_f = (H\ \Delta P)/(L\rho\ V^2)$ both in water and in polymer solutions (Figure 9.11).

The PSD shown for water in Figure 9.4 are modified in the presence of polymer by decreasing both the cut-off frequency and the amplitude of wall velocity fluctuations. This is all the more so for the component of the transversal velocity. Using shear velocity as a scale parameter for frequency, PSD for polymer solutions show an attenuation in a median domain around $f/\alpha_x \sim 0.01$ (Figure 9.12). The slight rebound associated with it corresponds to the time of relaxation of polymer chains of the order of several milliseconds (Ambari et al. 1984).

Spatial correlation functions such as those presented in Figures 9.5 and 9.6 are also affected in the presence of polymer. These functions are used to provide a description of large coherent structures as pairs of vortices operating as counter-rotating rollers, in which the fluid is subjected to strong elongation stresses, which facilitates polymer stretch (Gyr 1984). The dimensionless values of the space between the threads of these large structures ("streak spacing") $\lambda+(\lambda+ = \lambda u^*/\nu)$ corresponding to twice the minimum observed for gz are quasi-constant for water ($\lambda+\sim100$). They increase linearly up to a DR ~50%, following the law $\lambda+ = 100 +$

1.9 DR (DR in %) (Oldaker and Tiederman 1977), to reach a value λ+~350 at maximum of DR in our experiments (Deslouis et al. 2004).

9.4.2. *Surfactant solutions*

Surfactant solutions are another class of drag reducers that has been long known (White 1967) but less studied because of their physicochemical characteristics.

DR by high linear polymers shares many characteristics with surfactants: the maximum rate of DR is even higher than that by polymers. The law governing streak spacing is identical to that of polymers.

From a physicochemical viewpoint, linear long-chain polymers and surfactants are very different, surfactants being small molecules with polar heads and hydrophobic tails. In aqueous medium, they form small-size spherical micelles (beyond a critical concentration cmc1; see Figure 9.13) that cannot give the hydrodynamic effects observed with high polymers. At increasing concentration, they form beyond a second critical concentration cmc2 worm-like micelles. On the other hand, there are particular surfactants concerning the counter-ion nature that lead to the formation of worm-like micelles before the formation of spherical micelles at low concentrations. This is the case of a cationic surfactant, alkyltrimethylammonium, the alkyl chain in C16 or C18, the counter-ion being the para-cholorobenzoate (PCB). This type of counter-ion that has a benzene ring facilitates the worm-like micelle structuring. Measurements using an Atomic Force Microscopy (AFM) confirmed this structure, estimating that the characteristic lengths of these micelles were of the order of several dozen nm and the thread diameter was twice the length of the alkyl chain (Chaal et al. 2006) (Figure 9.14).

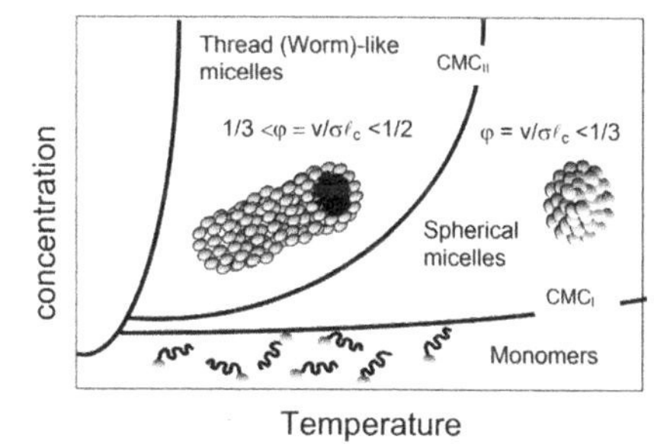

Figure 9.13. *Schematic phase diagram (concentration/temperature) of a surfactant that can aggregate in spherical or worm-like micelles*

It was observed that the behavior using the electrochemical technique was similar to that found for high polymers in laminar flow on a rotating disk ("Levich" behavior ($I_L \propto \Omega^{1/2}$) for low-concentration solutions (about 1–2 mM L^{-1}) and from a critical velocity, current drop, instabilities and hysteresis) (Boutoudj et al. 2002) (Figure 9.15). The ferri/ferrocyanide couple could not be used as supporting electrolyte, because the associated anions replace the PCB in the micelles and destroy the worm-like structure. Oxygen reduction was used without supporting electrolyte.

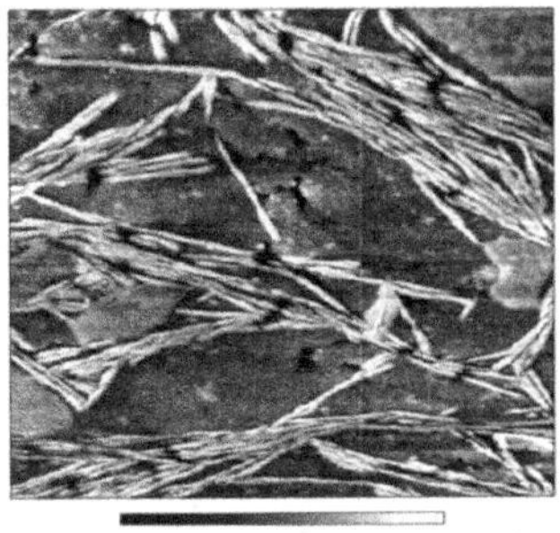

Figure 9.14. *AFM image (2 μm × 2 μm) in phase mode of a micellar solution deposited on a mica support coated with a gold film: octodecyltrimethylammonium and parachlorobenzoate solution (Chaal et al. 2006). For a color version of this figure, see www.iste.co.uk/saouab/mechanics.zip*

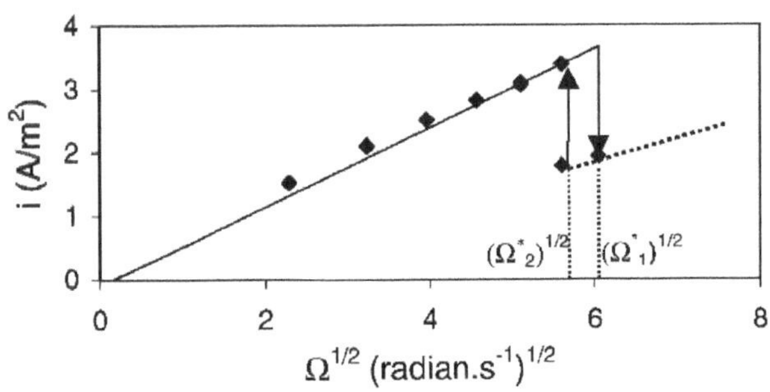

Figure 9.15. *Limiting diffusion current of oxygen reduction as a function of the square root of rotational velocity of a disk in a 1.28 mM L^{-1} surfactant solution (Arquad S50) composed of hexadecyltrimethylammonium (12%) and octadecyltrimethylammonium (82%) in the presence of PCB (3.75 mM L^{-1}) (Boutoudj et al. 2002)*

Using measurements of electrochemical noise (Chaal et al. 2004) showed that at transition and in an instability regime, the increase in conductivity noise was the reason for the destruction of worm-like micelles. These micelles can be reconstructed in calmer regions of the flow, unlike the high polymers, for which degradation is irreversible. The measurements were conducted with another redox system, the reduction of methyl viologen that does not affect the worm-like form of the micelle because of the positive charge of ox and red species. Local measurements of drag in turbulent flow with microelectrodes in inert wall were conducted in rectangular channels (Boutoudj et al. 2008) or on rotating disk (Boutoudj et al. 2015) (see Figure 9.16).

Given the reversible character of their degradation in turbulent flow, the use of these micellar solutions was considered for preventing flow-induced localized corrosion (Schmitt 2001; Deslouis 2003). This was confirmed on copper microelectrodes inserted in a rotating cage, initially covered with an oxide layer that was not pulled off in the presence of a micellar solution (S50/salicylate ion) (Chaal et al. 2007). A subsequent study on platinum microelectrodes inserted in various points of coupons of the rotating cage made it possible to compare the importance of shears in relation with erosion–corrosion and to confirm the efficiency of DR by micelles (Chaal et al. 2009).

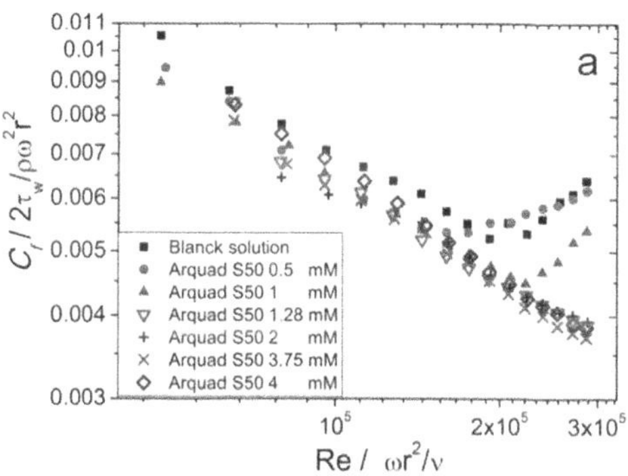

Figure 9.16. *Drag factor C_f on gold microelectrodes (∅ 200 μm) inserted in inert wall of a rotating disk at 3.9 cm from the rotational center. C_f is deduced from the shear stress τ_w calculated from the limiting diffusion current of oxygen reduction. Surfactant solutions Arquad S50 (see Figure 9.16) at various concentrations/PCB (3.75 mM L^{-1}). Drag reduction reaches 40% for Re = 3 × 10^5 (Boutoudj et al. 2015). For a color version of this figure, see www.iste.co.uk/saouab/mechanics.zip*

The mechanism referred to for explaining the viscoelastic behavior observed with micellar solutions cannot be simply deduced from the "coil-stretch" process of high polymers. An analogy was proposed (de Gennes 1986; Doi and Edwards 1989) by using the reptation model of semi-diluted solutions (case of micellar solutions of several 10^3 ppm). These solutions form an entangled network on which an elongation flow generates the reptation-based alignment of micelle, releasing it from the network. The reptation time t_{rep} is defined by: $Lm^2 = D_{rep} \times t_{rep}$.

Lm is the length of the micelle, and D_{rep} is the reptation coefficient. Moreover, micelles undergo statistical splits whose characteristic time t_{sc} between two splits is defined by the curvilinear length lc along which the micelle moved during the disentanglement: $lc^2 = D_{rep} \times t_{sc}$.

It can be shown that the global process (time for a micelle to lose the memory of a previous configuration) is defined by a single time (Cates 1987): $t^* = (t_{sc} \times t_{rep})^{1/2}$. Viscoelastic effects are expected when the elongation gradient γ exceeds t^{*-1}.

9.4.3. *Electrochemical measurements in other viscoelastic aqueous or organic media*

9.4.3.1. *Supramolecular compounds*

Supramolecular polymers are aggregates of small molecules associated by non-covalent bonds. It was shown that they can present a viscoelastic behavior at low concentration in non-polar organic solvents (Bellot and Bouteiller 2008). In particular, bis-urea, 2,4-bis(2-ethylhexylureido) toluene ("EHUT") can be assembled in two supramolecular structures, either along threads by hydrogen bonds or in more rigid tubes. The EHUT presents a DR in octane or toluene, measured in turbulent Couette flow. Though these media are less accessible to electrochemical measurements due to their very low conductivity and to the limited number of available redox couples, Aoudia et al. (2015) conducted an electrochemical study in laminar flow using a rotating disk to extend the results found on linear high polymers and on surfactants in aqueous medium. To do this, the medium (toluene) was rendered conductive using a ionic liquid, tris(pentafluoroethyle) trifluorophosphate ("aph4-cph12") and the ferrocene/ferrocenium system was chosen as redox couple (Duffy and Bond 2006). The three techniques – electrochemistry, viscosimetry and light diffusion – were jointly conducted:

– Light diffusion measurements indicate an entangled network of pseudo mesh of the order of 4 nm.

– Viscosimetry measurements (cone-plane and plane-plane) yielded characteristic responses of viscoelasticity of 8 and 25 s^{-1} for aph4-cph12 concentrations of 0.03 and 0.04 M L^{-1}, respectively.

– The limiting diffusion current as a function of the angular velocity of the disk shows a "Levich" behavior up to a velocity where a break occurs and instabilities emerge (Figure 9.17). The critical velocities of 30 and 50 s^{-1} yield an agreement on the order of magnitude with the above viscosimetric data.

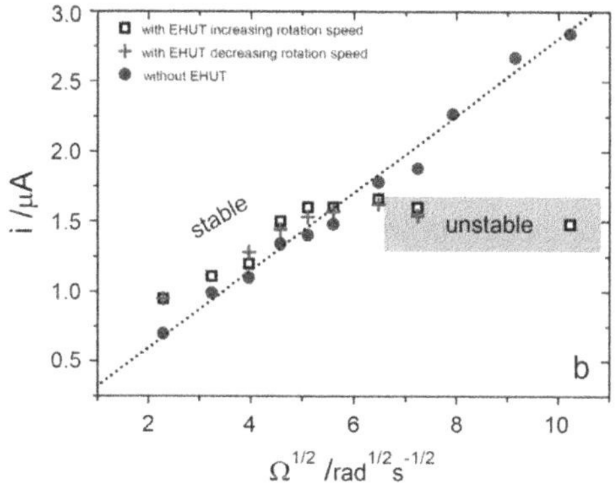

Figure 9.17. *Limiting diffusion current as a function of $\Omega^{1/2}$ for the oxidation of ferrocene in toluene medium. Supporting electrolyte aph4-cph12 0.04 M L^{-1}. (●) solution without EHUT (Aoudia et al. 2015). For a color version of this figure, see www.iste.co.uk/saouab/mechanics.zip*

9.4.3.2. *Comb-like amphiphilic copolymers*

These compounds could advantageously combine the properties of surfactants and of linear polymers for use in DR. Such a compound was synthesized (80C12) (Dutertre et al. 2015): it consists of a polystyrene chain on which quaternary ammonium groups were grafted (Figure 9.18).

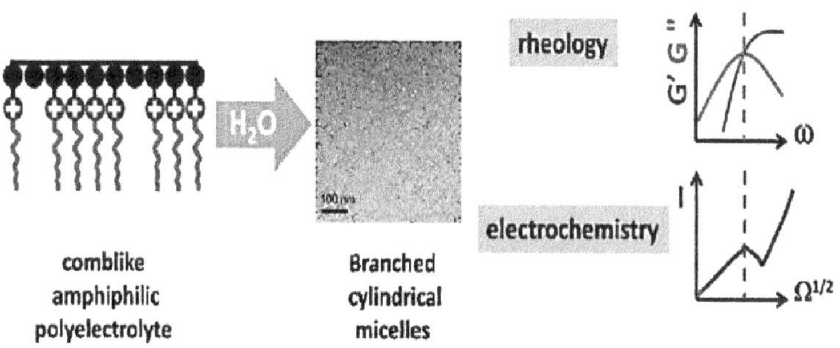

Figure 9.18. *Diagram showing the structure of comb-like copolymer 80C12 formed of a polystyrene frame on which quaternary ammonia are grafted. The image obtained in cryoTEM shows the structure in branched cylindrical micelles. The image on the right shows the identity of the inverse of characteristic times found by electrochemistry and the rheological measurement at the intersection of loss and storage modules. Adapted from Talantikite et al. (2017). For a color version of this figure, see www.iste.co.uk/saouab/mechanics.zip*

This polymer self-assembles in water to form cylindrical aggregates. Electrochemical measurements could be conducted in water in KCl supporting electrolyte and by using methyl viologen as redox tracer. They were compared to measurements of light diffusion and cryo-TEM to characterize their structure and to rheological measurements (Talantikite et al. 2017). The relaxation rate of the basic compound, too long for electrochemical measurements, can be reduced by adding a non-ionic PEO (Brij C12E10).

The electrochemical measurements on rotating disk show, similar to linear high polymers, surfactants or supramolecular compounds, a transition at critical velocity with limiting current drop and the appearance of instabilities. Here, once more the characteristic times given by rheology or by electrochemistry are in a comparable order of magnitude.

9.5. Conclusion

This chapter focused on showing the various uses of electrochemical methods to analyze wall transfer phenomena. The interest in these methods is justified by their non-intrusive character and their ability to provide accurate information on momentum and mass transfer, and by analogy heat transfer. To our knowledge, these methods are the only ones giving access to inaccessible (or difficult to access) zones to other techniques, even non-intrusive such as Laser Doppler Anemometry (LDA) for rotating systems or in internal parts of reactors. A significant part of the chapter is dedicated to the use of electrochemical methods for the analysis of the rheological behavior of polymer solutions, particularly for DR, domain to which Prof. A. Ambari brought a major contribution.

9.6. References

Achia, B.U. and Thompson, D.W. (1977). Structure of the turbulent boundary layer in drag-reducing pipe flow. *J. Fluid Mech.*, 81, 439–464.

Ambari, A., Deslouis, C., Tribollet, B. (1984a). Mass transfer to a cylindrical wire in dilute polymer solutions. *Int. Comm. Heat Mass Transfer*, 11, 353–360.

Ambari, A., Deslouis, C. Tribollet, B. (1984b). Coil-stretch transition of macromolecules in laminar flow around a small cylinder. *Chem. Eng. Commun.*, 29, 63–78.

Ambari, A., Deslouis, C., Tribollet B. (1985a). Coil-stretch transition of dilute polymer solutions in elongational laminar flows. *Physicochemical Hydrodynamics*, 6, 815–827.

Ambari, A., Deslouis, C., Tribollet, B. (1985b). Mass transfer in elongational flows of dilute polymer solutions. In *The Influence of Polymer Additives on Velocity and Temperature Fields: IUTAM Symposium Essen 1984*, Gampert, B. (ed.). Springer, Berlin, Heidelberg.

Ambari, A., Deslouis, C., Tribollet, B. (1986). Frequency response of the mass transfer rate in a modulated flow at electrochemical probes. *Int. J. Heat Mass Transfer*, 29, 35–45.

Aoudia K., Brinis N., Chaal L., Saidani B., Benyahia L., Chassenieux C., Deslouis C. (2015). Electrochemical characterization of viscoelastic solutions of supramolecular polymers in phosphonium-based ionic liquids. *Journal of Electroanalytical Chemistry*, 744, 101–109.

Baléras, F. (1995). Un capteur électrochimique pour la mesure locale d'un champ de vitesse tridimensionnel. Réalisation par des techniques microélectroniques et application à l'étude de l'écoulement dans une cellule à jet immerge. PhD Thesis, Université de Paris-Sud, Orsay.

Beck, J.V., Blackwell, B., St-Clair, C.R. (1985). *Inverse Heat Conduction: Ill-Posed Problems*. Wiley, New York.

Bellot, M. and Bouteiller, L. (2008). Thermodynamic description of bis-urea self-assembly: Competition between two supramolecular polymers. *Langmuir*, 24, 14176–14182.

Boutoudj, M.S., Ouibrahim, A., Deslouis, C. (2002). Mass transfer in a laminar elongational flow of a drag reducing surfactant. *J. Non-Newtonian Fluid Mech.*, 103, 141–148.

Boutoudj, M.S., Ouibrahim, A., Barbeu, F., Deslouis, C., Martemianov, S. (2008). Local shear stress measurements with microelectrodes in turbulent flow of drag reducing surfactant solutions. *Chemical Engineering and Processing*, 47, 793–798.

Boutoudj, M.S., Ouibrahim, A., Deslouis, C. (2015). Mass transfer in elongational laminar and turbulent flows of drag reducing solutions of quaternary ammonium surfactants. Influence of the counter-ion to surfactant concentrations ratio. *Chemical Engineering and Processing*, 93, 34–43.

Cates, M.E. (1987). Reptation of living polymers: Dynamics of entangled polymers in the presence of reversible chain-scission reactions. *Macromolecules*, 20, 2289–2296.

Chaal, L., Boutoudj, M.S., Ouibrahim, A., Saidani, B., Nogueira R.P., Deslouis, C. (2004). Mass transfer and conductivity instabilities in surfactant solutions submitted to a laminar elongational flow. *J. Non-Newtonian Fluid Mech.*, 121, 143–155.

Chaal, L., Pillier, F., Saidani, B., Joiret, S., Pailleret, A., Deslouis, C. (2006). Characterization of counterion and surface on micelle formation using tapping mode atomic force microscopy in air. *J. Phys. Chem. B*, 110, 21710–21718.

Chaal, L., Deslouis, C., Pailleret, A., Saidani, B. (2007). On the mitigation of erosion-corrosion of copper by a drag-reducing cationic surfactant in turbulent flow conditions using a rotating cage. *Electrochim. Acta*, 52, 7786–7795.

Chaal, L., Albinet, B., Deslouis, C., Al-Janabi, Y.T., Pailleret, A., Saidani, B., Schmitt, G. (2009). Wall shear stress mapping in the rotating cage geometry and evaluation of drag reduction efficiency using an electrochemical method. *Corrosion Science*, 51, 1809–1816.

Cochran, W.G. (1934). The flow due to a rotating disk. *Proc. Camb. Phil/Soc.*, 30, 365–375.

Coeuret, F. and Legrand, J. (1981). Mass transfer at the electrodes of concentric cylindrical reactors combining axial flow and rotation of the inner cylinder. *Electrochim. Acta*, 26, 865–872.

Coeuret, F. and Storck, A. (1993). *Eléments de génie électrochimique*. Technique et Documentation, Lavoisier, Paris.

Deslouis, C. (2003). Microscopic aspects of surfactants action on flow induced corrosion. *Electrochim. Acta*, 48, 3279–3288.

Deslouis, C. and Tribollet, B. (1975). Etude du transport de matière dans des solutions concentrées de polyoxyéthylène à l'aide d'une electrode à disque tournant. *Journal de Chimie Physique*, 2, 224–228.

Deslouis, C., Epelboin, I., Tribollet, B., Viet, L. (1973). Etude du transport de matière dans un électrolyte à la surface d'une électrode tournante à disque ou à anneau en présence d'un haut polymère. *C. R. Acad. Sc.*, Paris, 277(Série C), 353–356.

Deslouis, C., Tribollet, B., Viet, L. (1980). Local and overall mass transfer rates to a rotating disk in turbulent and transition flows. *Electrochimica Acta*, 25, 1027–1032.

Deslouis, C., Tribollet, B., Cressely, R., Hocquart, R. (1981a). Electrochemical and optical studies of a stable to unstable transition in a polymer solution flowing close to a rotating disc. *Rheol. Acta*, 20, 29–34.

Deslouis, C., Tribollet, B., Viet, L. (1981b). Réductions du transport de matière et du frottement hydrodynamique en présence de macromolécules au voisinage de microélectrodes placées sur un disque tournant. *Electrochimica Acta*, 26, 1457–1466.

Deslouis, C., Gil, O., Tribollet, B. (1990). Frequency response of electrochemical sensors to hydrodynamic fluctuations. *J. Fluid Mech.*, 215, 85–100.

Deslouis, C., Huet, F., Robin, S., Tribollet, B. (1993). Spectral analysis of wall turbulence with photolithography devised electrochemical probes. *Int. J. Heat Mass Transfer*, 36, 823–829.

Deslouis, C., Tribollet, B., Tihon, J. (2004). Near-wall turbulence in drag-reducing flows investigated by the Photolithography-electrochemical probes. *Journal of Non-Newtonian Fluid Mechanics*, 123(2–3), 141–150.

Doi, M. and Edwards, S. F. (1989). *The Theory of Polymer Dynamics*. Clarendon, Oxford.

Donohue, G.L., Tiederman, W.G., Reischman M.M. (1972). Flow visualization of the near-wall region in a drag-reducing channel flow. *J. Fluid Mechanics*, 56, 559–575.

Duffy, N.W. and Bond, A.M. (2006). Macroelectrode voltammetry in toluene using a phosphonium-phosphate ionic liquid as the supporting electrolyte. *Electrochem. Commun.*, 8, 892–898.

Dutertre, F., Gaillard, C., Chassenieux, C., Nicolai, T. (2015). Branched wormlike micelles formed by self-assembled comblike amphiphilic copolyelectrolytes. *Macromolecules*, 48, 7604–7612.

Eckelman, L.D., Fortuna, G., Hanratty, T.J. (1977). Drag reduction and the wavelength of flow-oriented wall eddies. *Nature Phys. Sci.*, 236, 94–96.

Flory, P.J. (1953). *Principles of Polymer Chemistry*. Cornell University Press, Ithaca, New York.

Fortuna, G. and Hanratty, T.J. (1972). The influence of drag-reducing polymers on turbulence in the viscous sublayer. *J. Fluid Mechanics*, 53(3), 575–586.

Gabe, D.R. and Robinson, D.J. (1972). Mass transfer in a rotating cylinder cell–I. Laminar flow. *Electrochimica Acta*, 17, 1121–1127.

de Gennes, P.G. (1974). Coil-stretch transition of dilute flexible polymers under ultrahigh velocity gradients. *J. Chem. Phys.*, 60, 5030–5042.

de Gennes, P.G. (1986). Towards a scaling theory of drag reduction. *Physica*, 140A, 9–25.

Greif, R. and Paterson, J.A. (1973). Mass transfer to a rotating disk in a non-Newtonian fluid. *Physics of Fluids*, 16, 1816.

Greif, R., Kappesser, R., Cornet, I. (1972). Diffusivity of oxygen in a non-Newtonian saline solution. *Int. J. Heat Mass Transfer*, 15, 593–597.

Gyr, A. (1984). Direct evidence that drag reduction is an effect of the elongation of the polymer molecules. In *Proceedings of the Third International Conference on Drag Reduction*. University of Bristol, Bristol.

Gyr, A. and Bewersdorf, H.-W. (1995). *Drag Reduction in Turbulent Flows by Additives*. Kuwer, London.

Hanratty, T.J., Chorn, L.G., Hatziavramidis, D.T. (1977). Turbulent fluctuations in the viscous wall region for Newtonian and drag reducing fluids. *Phys. Fluids*, 20, 112–119.

Hansford, S. and Litt, M. (1968). Mass transport from a rotating disk into power-law liquids. *Chem. Engng., Sci.*, 23, 849–864.

Hessel, V., Lowe, H., Schonfeld, F. (2005). Micromixers–A review on passive and active mixing principles. *Chem. Eng. Sci.*, 60, 2479–2501.

James, D.F. and Acosta, A.J. (1970). The laminar flow of dilute polymer solutions around circular cylinders. *J. Fluid Mech.*, 42, 269–288.

Kataoka, T. and Ueda, S.H. (1971). Flow instability of high molecular weight linear polyethylene. *Rheol. Acta*, 10, 446–447.

Kataoka, K., Doi, H., Komai, T. (1977). Heat/mass transfer in Taylor vortex flow with constant axial flow rates. *Int. J. Heat Mass Transfer*, 20, 57–63.

Kjeang, E., Roesch, B., McKechnie, J., Harrington, D.A., Djilali, N., Sinton, D. (2007). Integrated electrochemical velocimetry for microfluidic devices. *Microfluid Nanofluid*, 3, 403–416.

Kreith, F. (1968). Convection heat transfer in rotating systems. In *Advances in Heat Transfer*, Irvine Jr., T.F. and Hartnett, J.P. (eds). Academic Press, New York, 129–251.

Kreutzer, M.T., Kapteijn, F., Moulijn, J.A., Heiszwolf, J.J. (2005). Multiphase monolith reactors: Chemical reaction engineering of segmented flow in microchannels. *Chem. Eng. Sci.*, 60, 5895–5916.

Lamb, B.M. (1945). *Hydrodynamics*, 6th edition. Dover, New York.

Legentilhomme, P. and Legrand, J. (1990). Overall mass transfer in swirling decaying flow in annular electrochemical cells. *J. Appl. Electrochem.*, 20, 216–222.

Legentilhomme, P. and Legrand, J. (1991). Developing mass transfer for annular swirling decaying flow induced by means of a tangential inlet. *Int. J. Heat Mass Transfer*, 34, 1281–1291.

Legentilhomme, P., Legrand, J., Comiti, J. (1989). Axial dispersion in electrolyte flow through anisotropic packed beds. *Journal of Applied Electrochemistry*, 19, 263–270.

Lévêque, M.A. (1928). Les lois de transmission de la chaleur par convection. *Ann. Mines*, 13, 381–412.

Levich, V.G. (1952). *Physico-Chemical Hydrodynamics*. Prentice Hall, New York.

Ling, S.C. (1963). Heat transfer from a small isothermal spanwise strip on an insulated boundary. *Trans. ASME C: J. Heat Transf.*, 85, 230–235.

Lumley, J.L. (1973). Drag reduction in turbulent flow by polymer additives. *Annu. Rev. Fluid Mech.*, 7, 263–290.

Mao, Z. and Hanratty, T.J. (1991). Analysis of wall shear stress probes in large amplitude unsteady flows. *Int. J. Heat Mass Transfer*, 34, 281–290.

Maquinghen, T. (1999). Métrologie tridimensionnelle instationnaire à l'aide de la méthode polarographique. Thesis, Université de Valenciennes et du Hainaut-Cambrésis, Valenciennes.

Max, J. (1981). *Méthodes et Techniques de Traitement de Signal 1*, 3rd edition. Masson, Paris.

Mitchell, J.E. and Hanratty, T.J. (1966). A study of turbulence at a wall using an electrochemical wall shear stress meter. *J. Fluid Mech*, 26, 199–221.

Mitschka, P. and Ulbrecht, J. (1965). Nicht-Newtonsche Flüssigkeiten IV. Strömung nicht-Newtonscher Flüssigkeiten Ostwald-de-Waeleschen Typs in der Umgebung rotierender Drehkegel und Scheiben. *Coll. Czechoslov. Chem. Comm.*, 30, 2511–2526.

Mizushina, T. (1971). The electrochemical method in transport phenomena. In *Advances in Heat Transfer*, Irvine Jr., T.F. and Hartnett, J.P. (eds). Academic Press, New York, 87–161.

Montillet, A., Comiti, J., Legrand, J. (1994). Application of metallic foams in electrochemical reactors of the filter-press type. Part II: Mass transfer performance. *Journal of Applied Electrochemistry*, 24, 384–389.

Mouheb, N.A., Montillet, A., Solliec, C., Havlica, J., Legentilhomme, P., Comiti, J., Tihon, J. (2011). Flow characterization in T-shaped and cross-shaped micromixers. *Microfluid Nanofluid*, 10, 1185–1197.

Nakoryakov, V., Budukov, A.P., Kashinsky, O.N., Geshev, P.I. (1986). *Electrodiffusion Method of Investigation into the Local Structure of Turbulent Flows*. VG Gasenko, Novosibirsk.

Oldaker, K. and Tiederman, W.G. (1977). Spatial structure of the viscous sublayer in drag-reducing channel flows. *Phys. Fluids*, 20, 133–144.

Ould-Rouis, M., Nouar, C., Salem, A., Legrand, J., Legentilhomme, P. (1989). Hydrodynamics and mass transfer of the laminar flow in the entrance region of an annular electrochemical reactor. *Int. Chem. Eng. Symp. Series*, 112, 17–28.

Peterlin, A. (1966). Hydrodynamics of macromolecules in a velocity field with longitudinal gradient. *J. Polym. Sci. Part B*, 4, 287–291.

Piau, J.M. (1974). Comportement des tubes de pression totale dans les solutions diluées de polyoxyéthylène. *C. R. Acad. Sci., Paris*, 278B, 447–449.

Piau, J.M. (1980). A characteristic length for dilute drag reducing polymer solutions. Measurements using hot film anemometers. *Rheol. Acta*, 19, 724–730.

Rehimi, F., Aloui, F., Ben Nasrallah, S., Doubliez, L., Legrand, J. (2006). Inverse method for electrodiffusional diagnostics of flows. *Int. J. Heat Mass Transfer*, 45, 1242–1254.

Reiss, L.P. and Hanratty, T.J. (1962). Measurement of instantaneous rates of mass transfer to a small sink on a wall. *AIChE J.*, 8(2), 245–247.

Ross, T.K. and Wragg, A.A. (1965). Electrochemical mass transfer studies in annuli. *Electrochim. Acta*, 10, 1093–1106.

Schmitt, G. (2001). Drag reduction by corrosion inhibitors – A neglected option for mitigation of flow induced localized corrosion. *Mater. Corros.*, 52, 329–343.

Shulman, Z.P., Pokryvailo, N.A., Kordonskii, V.I., Kaberdina, E.B. (1972). Rheodynamics and mass transfer in rotating flows of anomalous viscous fluids. *Progress in Heat and Mass Transfer*, 5, 177–185.

Sirkar, K.K. and Hanratty, T.J. (1970). The limiting behaviour of the turbulent transverse velocity component close to a wall. *J. Fluid Mech.*, 216(2), 234–245.

Smith, K.A., Merill, E.W., Mickley, H.S., Virk, P.S. (1967). Anomalous pitot tube and hot film measurements in dilute polymer solutions. *Chem. Eng. Sci.*, 22, 619–626.

Sobolik, V., Wein, O. Cermak, J. (1987). Simultaneous measurement of film thickness and wall shear stress in wavy flow of non-Newtonian liquids. *Collection Czechoslovak Chem. Comun.*, 52, 913–928.

Talantikite, M., Aoudia, K., Benyahia, L., Chaal, L., Chassenieux, C., Deslouis, C., Gaillard, C., Saidani, B. (2017). Structural, viscoelastic and electrochemical characteristics of self-assembled amphiphilic comb-like copolymers in aqueous solutions. *J. Phys. Chem. B*, 121, 867–875.

Toms, B.A. (1949). Some observations on the flow of linear polymer solutions through straight tubes at large Reynolds numbers. *Proc. First Congress on Rheology*, 2, 135–141.

Virk, P.S. and Suraiya, T. (1977). Mass transfer at maximum drag reduction. In *Proc. 2nd Int. Conf. on Drag Reduction*. BHRA Fluid Engineering, Cranfield.

Walsh, F.C. and Wilson, G. (1986). The electrolytic removal of gold from spent electroplating liquors. *Trans. IMF*, 64, 55–61.

Wein, O. (1981). On the transient Leveque's problem with an application in electrochemistry. *Collection Czechoslovak Chem. Comun.*, 46, 3209–3220.

Wein, O. and Sobolik, V. (1987). Theory of direction sensitive probes for electrodiffusion measurements of wall velocity gradients. *Coll. Czech. Commun.*, 52, 2169–2180.

White, A. (1967). Flow characteristics of complex soap systems. *Nature*, 214, 585–586.

10

Simulation of the Resin Infusion Process using the Multilayer Approach: An Industrial Case Application

The resin infusion process is widely used to elaborate large size and complex-shaped composite structures. It is an economic process requiring a mold composed of a half rigid part and a flexible bag. Moreover, the use of a flexible bag as counter-mold makes it possible to fabricate parts that cannot be molded in a rigid mold. However, its success, which relies mainly on the impregnation of the entire fibrous preform without leaving dry zones, requires the control of the front flow progression inside the mold. This chapter proposes an industrial application of the resin infusion simulation using the control volume finite element method (CVFEM). The resulting simulation tool can be used to predict the scenario of mold filling to guarantee its success. First of all, the numerical model developed is experimentally validated on the basis of infusion of parts with simple geometry. Then, its robustness is tested for simulating the infusion of an industrial demonstrator with a complex geometry representative for aeronautic nacelles. The results of this simulation were compared with the actual infusion and showed a good concordance.

10.1. Introduction

The family of resin infusion processes through a fibrous medium covers a wide range of processes of long fiber composite fabrication, such as the vacuum-assisted resin transfer molding (VARTM), vacuum infusion (VI or VARI) (Modi et al.

Chapter written by Raounak LOUDAD, Abdelghani SAOUAB and Pierre BEAUCHENE.

For a color version of all of the figures in this chapter, see www.iste.co.uk/saouab/mechanics.zip.

2009), liquid resin infusion (LRI) (Celle et al. 2008), resin injection under flexible tooling (RIFT) (Williams et al. 1996), vacuum bag resin transfer molding (VBRTM) (Kang et al. 2001) and Seemann composites resin infusion molding process (SCRIMP) (Han et al. 2000). These processes consist of placing the fibrous preform between a half rigid mold and a flexible bag, as shown in Figure 10.1.

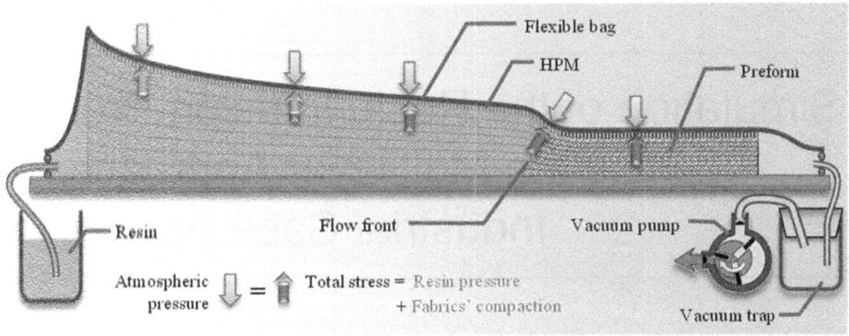

Figure 10.1. *Schematic representation of the resin infusion process*

By generating a vacuum inside the mold, the flexible bag compacts the preform under atmospheric pressure. Once the preform is compacted, the inlet valves are open and resin infuses the mold. The inlet pressure being limited to the atmospheric pressure, the use of a distribution medium (DM), which is a high permeability net (also referred to as high permeability media (HPM)), is required to accelerate the mold filling, particularly for the infusion of large size parts. Placing the DM between the preform and the vacuum bag, the resin flows through it at a higher velocity and impregnates the preform throughout its thickness. Consequently, the simulation of this process requires taking into account planar flows and through-thickness flows. Moreover, the hydro-mechanical coupling, that is, the mutual influence between the fibrous medium compressibility and the resin flow, must be taken into consideration in order to accurately describe the resin flow.

The resin infusion process was widely studied in the literature. Many researchers simulated this process mainly by using the CVFEM for its simplicity. The thickness of the infused parts is generally very small compared to other dimensions. This is why some researchers (Calado and Advani 1996; Chen et al. 2004; Yoon and Dolan 2008; Simacek and Advani 2016) took advantage of the shell-like geometry of the infused parts to model this process using a 2D surface mesh. However, the out-of-plane flow involved by the use of DM combined with the plane flow leads to a 3D flow that cannot be modeled in 2D. Other researchers have therefore modeled this infusion process in 3D (Han et al. 2000; Kang et al. 2001; Kang and Lee 2004; Dong 2006; Yang et al. 2016), which is more accurate, but requires a much longer computation time. Young (2013) developed a two-layer model, one modeling the planar flow in the DM, and the other the planar flow in the preform.

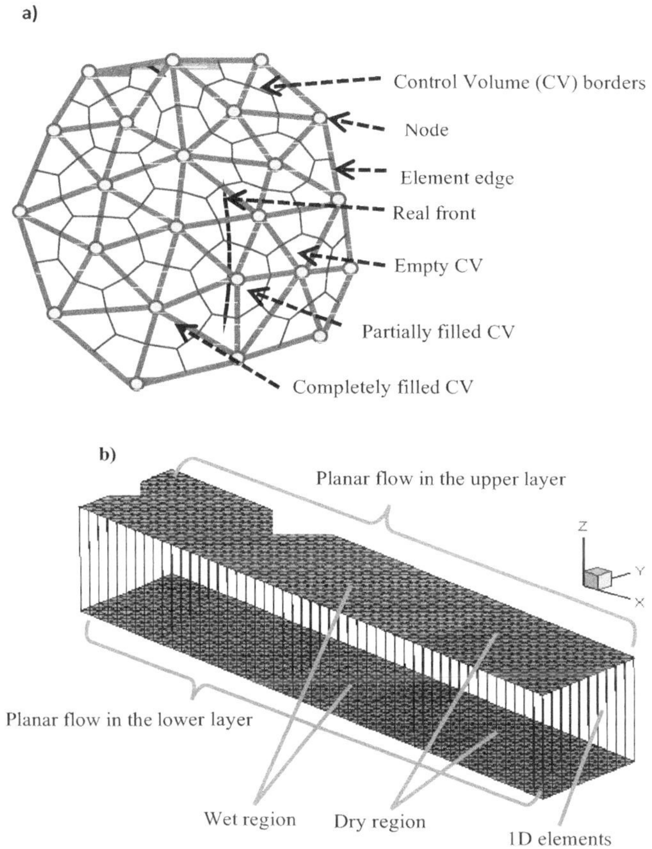

Figure 10.2. *CVFEM computation domain according to (a) standard 2D approach and (b) multilayer approach*

Similarly, we developed a multilayer CVFEM model in a previous work (Loudad et al. 2017), in which we modeled each layer superimposed in 2D½ (2D with one degree of freedom on the thickness) and connected these layers through the thickness by 1D elements, which model the out-of-plane flow, as shown in Figure 10.2. These layers may have different materials and orientations.

In a previous work (Loudad et al. 2017), we have experimentally validated this multilayer CVFEM model in the case of a plane plate and a ply drop infusion. In this chapter, we focus on the study of the robustness of this model. We present an experimental validation of this approach based on the infusion of an omega-shaped stiffener case. We have also tested this simulation approach in an industrial context. This test was conducted as part of an infusion of a large size industrial demonstrator, of complex shape, and which involves many singularities.

10.2. Mathematical modeling

The flow of resin through a deformable fibrous medium was widely studied in the literature (Tari et al. 1998; Hsiao et al. 2000; Mathur et al. 2001; Ouahbi et al. 2007; Song and Youn 2008; Park and Saouab 2009). It is governed by the combination of Darcy law and the equation of mass balance in a deformable representative volume element (RVE). Assuming that the velocity of the solid phase is negligible compared to that of the fluid phase, this equation can be written as follows (Park and Saouab 2009):

$$\nabla \cdot \left(\frac{-\overline{\overline{K}}}{\mu} \overline{\nabla P} \right) = \frac{1}{V_f} \frac{\partial V_f}{\partial t} \qquad [10.1]$$

where V_f is the fiber volume fraction, $\overline{\overline{K}}$ is the permeability tensor of the medium, μ is the viscosity of the resin and P is the pressure.

During the process, the fibrous medium is subjected to the ambient pressure applied by the flexible bag and to the internal pressure applied by resin or gas. For an incompressible solid phase, this equilibrium is described by Terzaghi's (1943) theorem. Assuming that the out-of-plane shear is negligible, this equilibrium can be considered in 1D (through the thickness) and expressed as follows:

$$\sigma = P_{atm} - P \qquad [10.2]$$

where σ is the compressibility stress and P_{atm} is the atmospheric pressure.

The compressibility of the fibrous preform has an impact on its thickness and its fiber volume fraction. This leads to a modification of the permeability of this medium and, consequently, of the fluid pressure. The resulting hydro-mechanical coupling of this mutual influence between the preform strain and the resin flow is described by the governing equation [10.1] and the constitutive laws [10.3]:

$$\begin{cases} V_f = F_{comp}(P_{atm} - P) \\ K_{i,j} = F_{perm}(V_f) \dots \dots \end{cases} \qquad [10.3]$$

These relations F_{comp} and F_{perm} describe the behavior of fibrous media in terms of compressibility and permeability, namely, the variation of the fiber volume fraction and that of the permeability as a function of the compressibility of the fibrous medium. Consequently, systems [10.1]–[10.3] will be associated with boundary conditions that are a pressure imposed to the injection threshold and the vent.

10.3. Numerical approach

The governing equation [10.1] is discretized using Green's formula and transforms into equilibrium of fluxes at the boundaries of control volume:

$$\int_{CV} \nabla \cdot \left(\frac{-\overline{\overline{K}}}{\mu} \overline{\nabla P} \right) dV = \int_{CS} \left(\frac{-\overline{\overline{K}}}{\mu} \overline{\nabla P} \right) \cdot \bar{n}\, dS = \int_{CV} \frac{1}{V_f} \frac{\partial V_f}{\partial P} \frac{\partial P}{\partial t} dV \qquad [10.4]$$

Then, the fluxes at the boundaries of control volume are calculated for 1D and 2D elements (Figure 10.3, equation [10.5]):

$$\int_{CS} \left(\frac{-\overline{\overline{K}}}{\mu} \overline{\nabla P} \right) \cdot \bar{n}\, dS = \sum_{2D\, elements} \sum_{nodes} H \cdot \|\overline{ac}\| \cdot \left(\frac{-\overline{\overline{K}}}{\mu} \overline{\nabla P} \right) \cdot \overline{n_{ac}} + \sum_{1D\, elements} \sum_{nodes} S_{cv} \cdot \left(\frac{-\overline{\overline{K}}}{\mu} \overline{\nabla P} \right) \cdot \overline{n_z} \qquad [10.5]$$

where H is the thickness of the 2D element, $\|\overline{ac}\|$ is the distance between the median points a and c (Figure 10.3) and $\overline{n_{ac}}$ is the external normal vector on this segment. S_{cv} is the surface of the control volume and $\overline{n_z}$ the external normal vector through the thickness.

Using the linear interpolation in the elements, the pressure gradient in equation [10.5] can be expressed in terms of node coordinates and nodal pressures:

$$P = \overline{\nabla P} \cdot \bar{X} + P_0 \;\rightarrow\; \overline{\nabla P} = f(P_i; \bar{X}_\iota) \qquad [10.6]$$

where $\bar{X}$ are the coordinates, P_0 is the pressure at the origin, P_i is the pressure at node i and $\bar{X}_\iota$ is the coordinates of node l. The factorization of these nodal pressures leads to the calculation of the local matrix of the element.

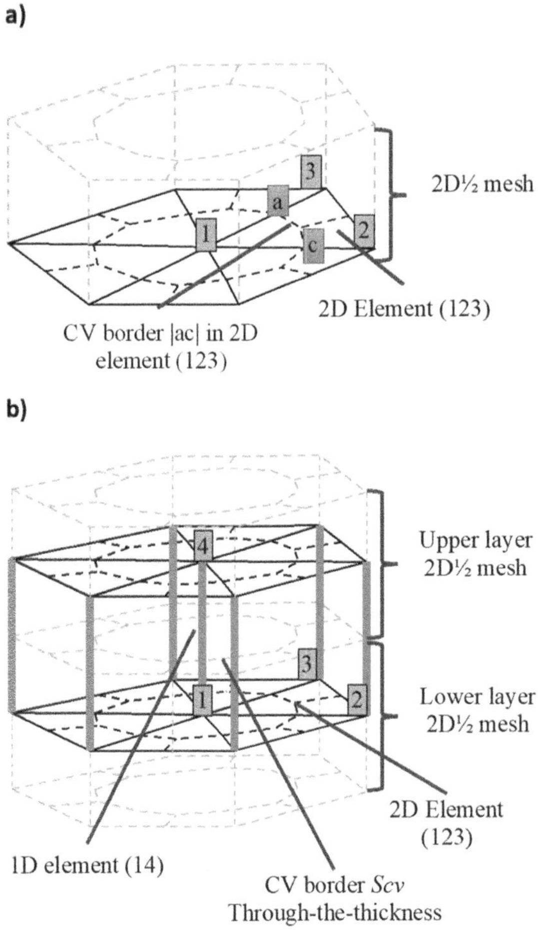

Figure 10.3. *Edges of control volume in (a) 2D elements and (b) 1D elements*

The proposed multilayer approach can be used to accurately simulate planar and out-of-plane flows more rapidly than the 3D modeling, as the size of the resulting global matrix in this approach is below that of the equivalent 3D modeling. Indeed, it is composed of less than half of the number of coefficients. Moreover, unlike the 3D modeling in which the layers of various materials share the same nodes at the interface between them, no nodes are shared by the layers in the multilayer approach, since the layers are linked by these 1D elements. This makes it possible to avoid the approximation of properties at the nodes of the interface. However, in the multilayer approach, the 1D elements must be parallel to the direction of out-of-plane permeability and the mesh of the surface parallel to the plane of planar permeability. Likewise, the regions of layers that are connected through thickness by 1D elements must be similarly meshed.

The source term of equation [10.4] is evaluated according to an implicit scheme, depending on the current nodal pressure P_i, the nodal pressure at the previous time step P_i^{t-1} and the time step Δt as follows:

$$\int_{CV} \frac{1}{V_f}\frac{\partial V_f}{\partial P}\frac{\partial P}{\partial t} dV = H \cdot S_{cv} \cdot \frac{1}{V_f}\frac{\partial V_f}{\partial P}\frac{P_i - P_i^{t-1}}{\Delta t} \qquad [10.7]$$

10.4. Analytical modeling of 1D VARI

During a right infusion using the VARI process in the main direction of the preform x (Figure 10.4), the governing equation is written as:

$$\frac{\partial}{\partial x}\left(\frac{K}{\mu}\frac{\partial P}{\partial x}\right) = -\frac{1}{V_f}\frac{\partial V_f}{\partial t}$$

$$x = 0 : P = P_{atm} \qquad [10.8]$$

$$x = L(t) : P = 0$$

where $L(t)$ is the position of the fluid front.

The permeability and compressibility behaviors chosen in this study are modeled by the following power laws:

$$K = a{V_f}^b$$

$$\sigma = c{V_f}^d \qquad [10.9]$$

Combining these behavior laws, we obtain:

$$K = a\left(\frac{\sigma}{c}\right)^{\frac{b}{d}} \qquad [10.10]$$

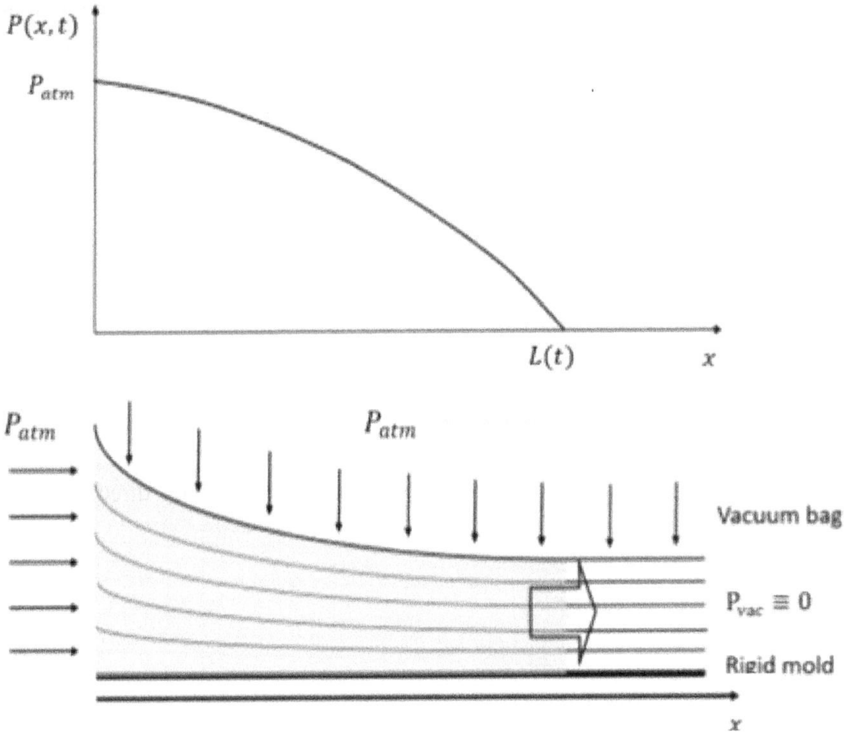

Figure 10.4. *Hydro-mechanical coupling during 1D VARI process*

The differentiation of the equilibrium relation [10.2] yields:

$$\sigma = p_{atm} - p \;\Rightarrow\; \frac{\partial P}{\partial x} = -\frac{\partial \sigma}{\partial x} \qquad [10.11]$$

Expressing the source term as a function of compressibility σ, we get:

$$\frac{1}{V_f}\frac{\partial V_f}{\partial t} = \frac{1}{d}\frac{1}{\sigma}\frac{\partial \sigma}{\partial t} \qquad [10.12]$$

Using equations [10.8] to [10.12], we obtain the governing equation as a function only of compressibility:

$$\frac{\partial}{\partial x}\left[\frac{a}{\mu}\left(\frac{\sigma}{c}\right)^{\frac{b}{d}}\left(-\frac{\partial \sigma}{\partial x}\right)\right] = -\frac{1}{d}\frac{1}{\sigma}\frac{\partial \sigma}{\partial t} \qquad [10.13]$$

The simplification of the previous equation yields:

$$\frac{a \cdot d^2}{\mu \cdot (b+d) \cdot c^{\frac{b}{d}}} \frac{\partial^2}{\partial x^2}\left(\sigma^{\frac{b+d}{d}}\right) = \frac{1}{\sigma}\frac{\partial \sigma}{\partial t} \quad [10.14]$$

Introducing the new variable S_f, we get:

$$S_f = \sigma^{\frac{b+d}{d}} \quad [10.15]$$

The governing equation is written as follows:

$$\frac{a \cdot d}{\mu \cdot c^{\frac{b}{d}}} \frac{\partial^2}{\partial x^2}(S_f) = \frac{1}{S_f}\frac{\partial S_f}{\partial t}$$

$$x = 0 : P = P_{atm};\ \sigma = 0;\ S_f = 0 \quad [10.16]$$

$$x = L(t): P = 0;\ \sigma = P_{atm};\ S_f = P_{atm}^{\ \frac{b+d}{d}}$$

The exact solution of the previous PDE is the following (Park and Saouab 2009):

$$S_f = M\left\{\left(\frac{x}{L}\right) - \left(\frac{x}{L}\right)^2\right\} + P_{atm}^{\frac{b+d}{d}}\left(\frac{x}{L}\right) \quad [10.17]$$

where: $M = \dfrac{P_{atm}^{\frac{b+d}{d}}}{(b+d)\left\{1-\left(\frac{P_{atm}}{c}\right)^{\frac{1}{d}}\right\}+1}$

This is how we find the analytical expressions of various sought-for quantities (compressibility, pressure, fiber volume fraction, position of the front and height of the preform):

$$\sigma(x,t) = \left[M\left\{\left(\frac{x}{L}\right) - \left(\frac{x}{L}\right)^2\right\} + P_{atm}^{\frac{b+d}{d}}\left(\frac{x}{L}\right)\right]^{\frac{d}{b+d}} \quad [10.18]$$

$$P(x,t) = P_{atm} - \sigma(x,t) \quad [10.19]$$

$$V_f(x,t) = \left(\frac{P_{atm}}{c}\right)^{\frac{1}{d}} \left[\frac{\left\{\left(\frac{x}{L}\right) - \left(\frac{x}{L}\right)^2\right\}}{(b+d)\left\{1 - \left(\frac{P_{atm}}{c}\right)^{\frac{1}{d}}\right\} + 1} + \left(\frac{x}{L}\right) \right]^{\frac{d}{b+d}} \quad [10.20]$$

$$L(t) = \sqrt{\frac{2 \cdot a \cdot d \cdot P_{atm}^{\frac{b+d}{d}} \cdot t}{\mu \cdot c^{\frac{b}{d}} \cdot \left[(b+d)\left\{1 - \left(\frac{P_{atm}}{c}\right)^{\frac{1}{d}}\right\} + 1\right]}} \quad [10.21]$$

$$H(x,t) = \frac{H_0(x)\, V_{f0}(x)}{V_f(x,t)} \quad [10.22]$$

H_0 denotes the initial height of the preform and V_{f0} is the initial fiber volume fraction.

10.5. Model validation

For the validation of the proposed numerical model, we will present in the following section a comparison of the numerical results with the analytical and experimental results.

10.5.1. *Analytical/numerical comparison*

The case proposed for this comparison is presented in Figure 10.5 and the data related to the materials employed are the following (Park and Saouab 2009):

– permeability: $K = aV_f^b$ with $a = 5.607 \times 10^{-12}$ m^2 and $b = -4.397$;

– compressibility: $\sigma = c\, V_f{}^d$ with $c = 3.5886 \times 10^7$ Pa and $d = 10.309$;

– and resin viscosity: $\mu = 0.1$ Pa · s.

Figure 10.6 presents the pressure fields at the end of infusion and the kinetics of the front. There is a very good agreement between the analytical and numerical curves of the pressure field. On the other hand, there is a slight difference between the numerical and analytical kinetics of the front. This difference is due to the pressure gradient deviation at the front. Indeed, this pressure gradient determines the front velocity, which explains the sensitivity of the front position kinetics to this value.

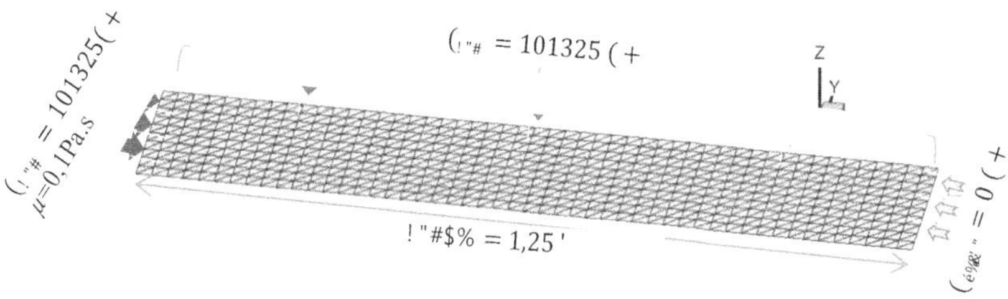

Figure 10.5. *1D VARI case study*

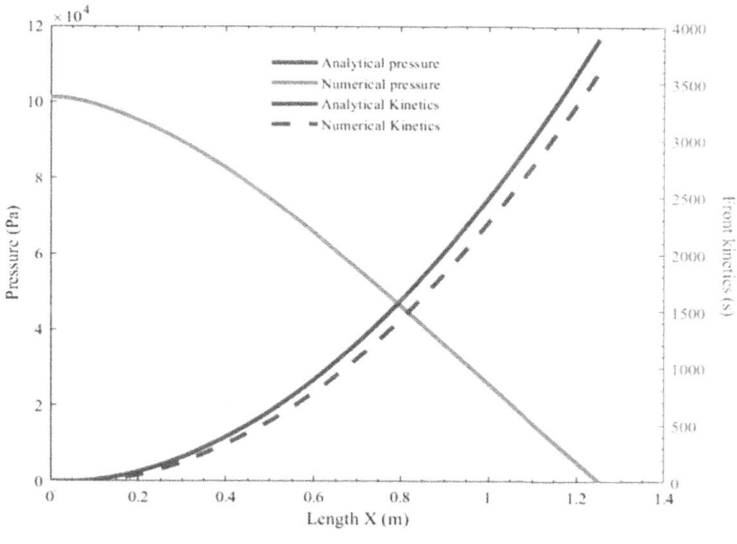

Figure 10.6. *Comparison of pressure fields and of the analytical/numerical front kinetics*

Figure 10.7 describes the evolution of thickness and of fiber volume fraction during infusion. It can be noted that there is a slight deviation between the numerical and analytical results of TVF and thickness at the beginning of the infusion. This deviation is due to the power law that yields a zero TVF at the point of injection, where the pressure is equal to the atmospheric pressure and therefore compressibility at this point is zero.

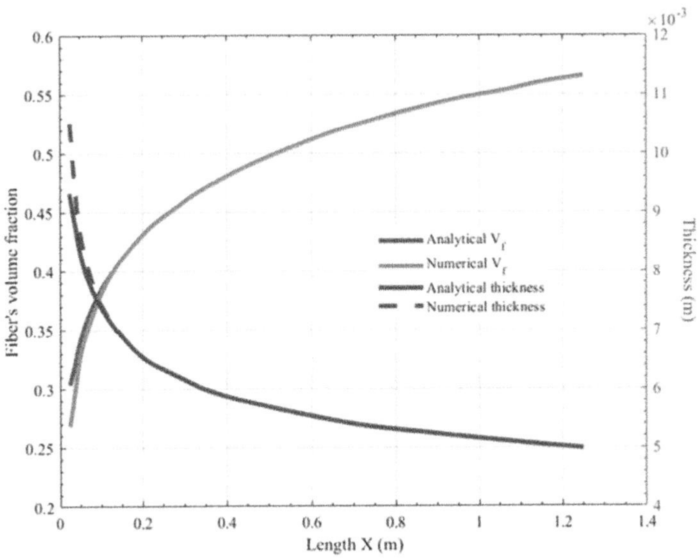

Figure 10.7. *Field of thickness and fiber volume fraction: analytical/numerical comparison*

10.5.2. *Experimental validation*

10.5.2.1. *Characterization of materials*

In order to simulate the infusion process, we need permeability and compressibility behavior laws of infused materials, as shown by equations [10.1]–[10.3]. The preform used in this study is a carbon fiber fabric. The preform and the DM used in this experimental validation were characterized in Loudad et al. (2017).

10.5.2.1.1. Compressibility

The preform has a nonlinear viscoelastic behavior and a plastic behavior due to layer nesting. Moreover, its compressibility depends on whether it is dry or soaked (Loudad et al. 2017; George et al. 2019).

In this study, we chose to model compressibility with a nonlinear elastic model. In order to measure the compressibility of the preform under conditions similar to the infusion, we performed, during the measurement, loadings similar to those to which the preform is subjected during infusion. Indeed, before the infusion, three pre-compacting cycles are performed. Hence, using the universal testing machine, three similar cycles were conducted on a sample placed in a rigid tank. This sample is then lubricated and unloaded. The compressibility values measured during this unloading are assumed to be the closest possible to the compressibility during infusion (Figure 10.8).

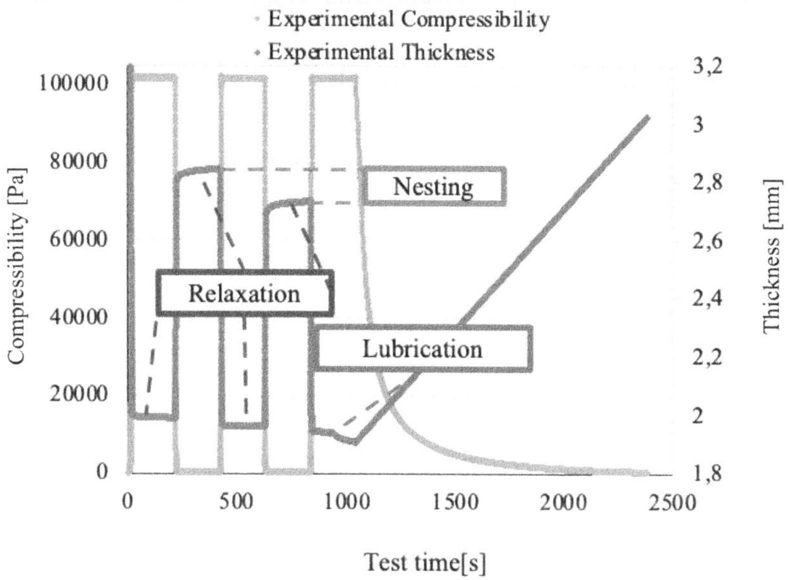

Figure 10.8. *Compressibility and thickness of the four-layer preform after the infusion protocol*

These experimental data are modeled by the Grimsley model (Grimsley et al. 2004) expressed in [10.23]:

$$H = \frac{A\sigma + B}{C\sigma + 1} \qquad [10.23]$$

with $A = 7.12 \times 10^{-7}\ \mathrm{m} \cdot \mathrm{Pa}^{-1}$; $B = 3.03$ mm; $C = 3.77 \times 10^{-4}\ \mathrm{Pa}^{-1}$.

The compressibility of the DM was also measured with the universal testing machine during unloading. It is modeled by the power model expressed in [10.24]:

$$H = A\sigma^{B} \qquad [10.24]$$

with $A = 1.1407$ mm; $B = -0.054$.

10.5.2.1.2. Permeability

Permeability was characterized using air injection benches presented in Figure 10.9 (Loudad et al. 2017).

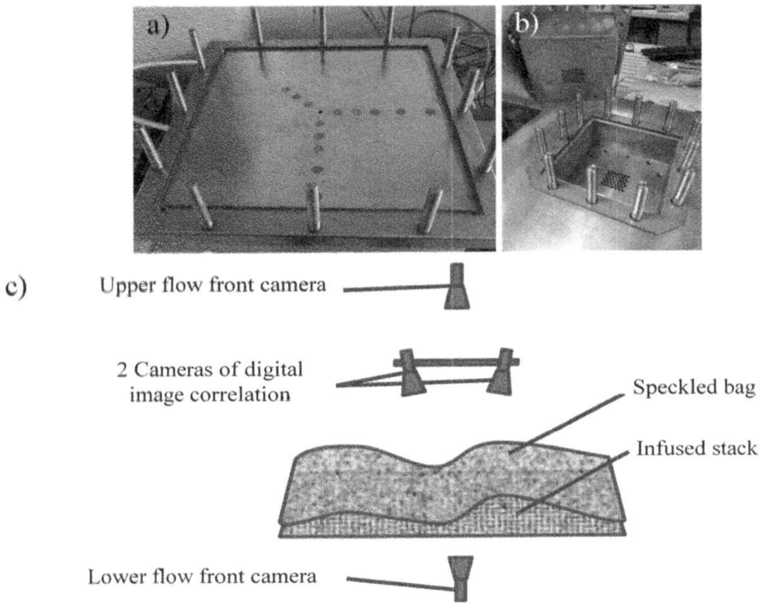

Figure 10.9. *Experimental set-up: (a) Planar permeability bench; (b) transversal permeability bench; (c) infusion recording device (front and DIC monitoring)*

For each fiber volume fraction, the measurement of permeability involves setting an air velocity and collecting the pressure values at several positions. Many researchers (Kim et al. 2002, 2003; Kim and Daniel 2005; Xiao et al. 2015; Agogue et al. 2018) used the air flux to measure the permeability of fiber reinforcement. Using air requires respecting the Knudsen condition in order to remain under continuum conditions and avoid the Klinkenberg slippage effect (Ziarani and Aguilera 2012), ensure a laminar regime at the Reynolds number of the air flux and take into account the air compressibility (Scholz et al. 2007). These conditions have been satisfied during the measurements. The preform permeability tensor was measured and modeled by the power law [10.25]:

$$K_1 = 1.41 \times 10^{-12}\, V_f^{-5.73};\ K_2 = 4.90 \times 10^{-14}\, V_f^{-8.04};\ K_3 = 2.74 \times 10^{-15}\, V_f^{-9.40} \qquad [10.25]$$

where K_1 and K_2 are the values of planar permeability and K_3 is the through-thickness permeability. The permeability of the DM was determined by reverse analysis based on experimental perfusions and considered constant in the planar direction:

$$K_{DMI} = 1.6 \times 10^{-9}\ \text{m}^2 \quad [10.26]$$

As for the through-thickness permeability of the DM, given that it is very high compared to the K_3 permeability of the preform in the same direction, its influence is negligible (Song and Youn 2008) and could be modeled by a high permeability, such as 1 m². Indeed, as these materials are placed in series in the direction of the thickness, the equivalent permeability of the stack is close to the lowest permeability.

10.5.2.2. *Infusion of a stiffened plate*

In this case of validation, the infusion of a plate containing an Omega stiffener is studied. This study makes it possible to evaluate the multilayer approach when a dry zone is created on the stiffener.

10.5.2.2.1. Case study data

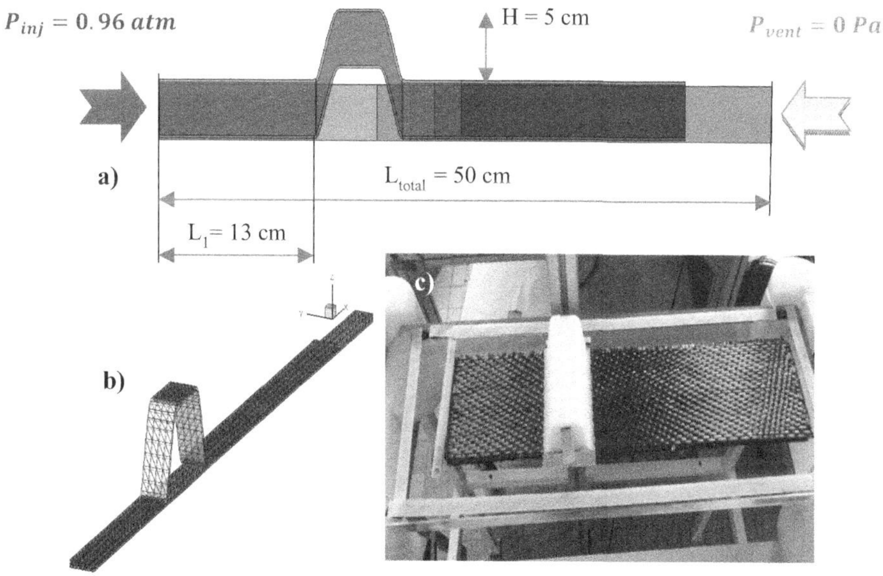

Figure 10.10. *Stiffened plate of the study: (a) case data; (b) numerical mesh; (c) experimental position of the stiffener on the two lower layers*

The infused pack studied in this case is composed of four layers of 50 cm along the characterized preform, two positioned under the stiffener, two positioned on the stiffener and one DM that covers all these layers (Figure 10.10). This infusion is made with silicone oil of 1 Pa·s viscosity. The input pressure is 0.96×10^5Pa, as the free surface of the oil is 30 cm under the mold. The mesh used in this simulation (Figure 10.10(b)) is composed of three layers: the upper layer models the DM, the intermediate layer models the preform above the stiffener and the lower layer models the preform below the stiffener. These surface meshes are linked in the direction of the thickener by 1D elements. The number of 2D elements composing the mesh is 5,456 and that of 1D elements is 1,995.

10.5.2.2.2. Thickness

The thickness of the stiffened plate is measured at the level of the vacuum bag using the numerical image correlation device (Sutton et al. 1983) (Figure 10.11(c)). These results correspond to the height and the shape of the vacuum bag, which includes the collapse of the vacuum bag in the DM as an effect of the difference between the fluid pressure and the external pressure (Simacek and Advani 2007). This results in many fluctuations in the measured data. In order to accurately measure the height of DM, we have taken the mean height of the spikes.

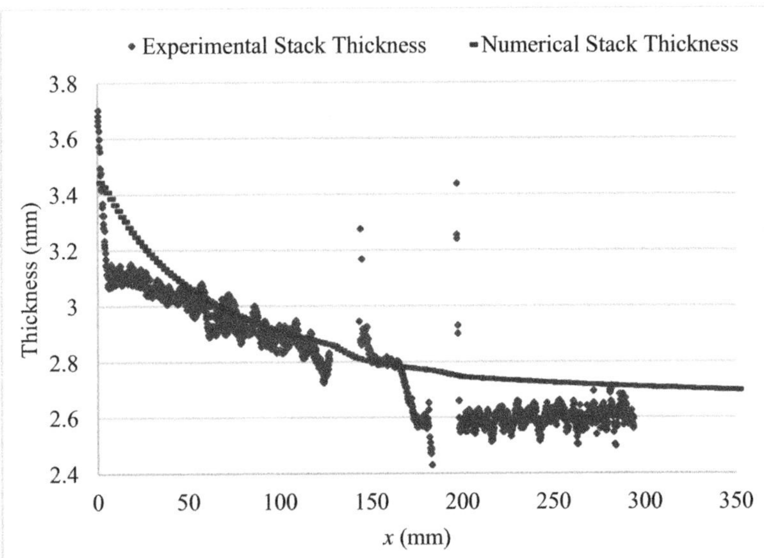

Figure 10.11. *Experimental and numerical distribution of thickness at the moment of filling the mold*

Figure 10.11 presents the experimental and numerical distributions of the thickness of the infused stack at the moment of filling the mold, the axis x

representing the position along the infused plate. The experimental results are limited to the camera vision field of 29 cm. Figure 10.11 shows that, near the input line, there is a significant deviation between the numerical thickness and the experimental thickness (about 0.3 mm). This is due to the high sensitivity of the compressibility to low values and the input pressure uncertainty (Loudad et al. 2017). Then, the numerical curve corresponds to experimental data up to the stiffener position, where another difference between the numerical and experimental results can be noted. This deviation is due to the lower quality of the DIC measurement at the height of the stiffener, as the calibration was made on the height of the plate.

10.5.2.2.3. Kinetics of the flow fronts

The kinetics of the flow front is recorded on both sides of the mold using two cameras (Loudad et al. 2017) (Figure 10.9(c)). The position of the flow front in the DM is measured at each moment, as well as the positions of the saturated and non-saturated flow fronts at the bottom of the preform, on the transparent side of the rigid mold. These experimental results are compared to the numerical results in the DM and the preform. As the numerical model does not take into account the double scale of the preform porosity, only one numerical flow front is calculated in the preform.

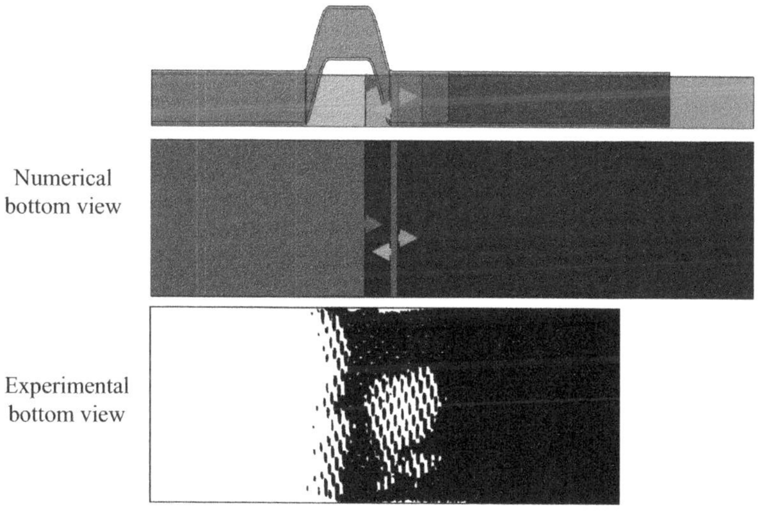

Figure 10.12. *Flow fronts under the stiffener (bottom view)*

Due to the high permeability of DM, the layers above the stiffener are filled before those below it. Then, a backward flow appears under the stiffener, as illustrated in Figure 10.12, with a forward flow under the stiffener represented in red, the front of the backward flow under the stiffener in green and the front of the forward flow after the stiffener in blue.

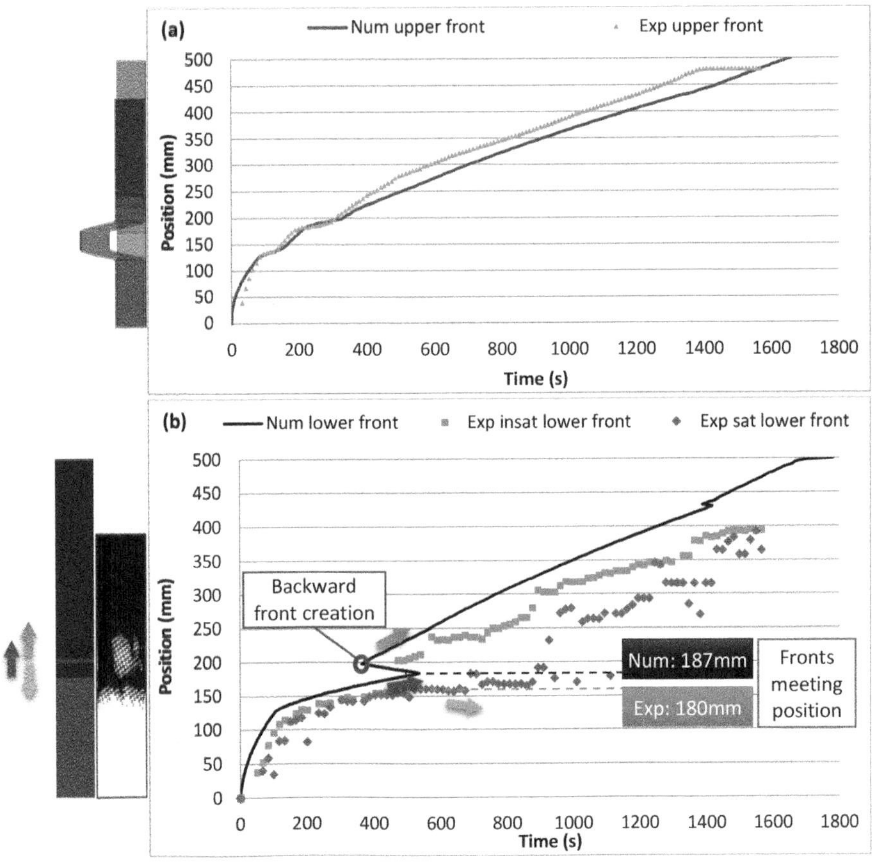

Figure 10.13. *Kinetics of the flow front in the case of the stiffener infusion in the (a) DM (b) preform*

Figure 10.13(a) presents the kinetics of the flow front in the DM (the image on the left of the y axis is scaled as a function of the position). A good correspondence can be noted between the numerical and experimental results up to the right side of the stiffener (t ≈ 300 s, position ≈ 20 cm). Then, there is a delay of about 2 min between these curves. After that, the flow continues both experimentally and numerically with the same velocity, which can be deduced from the fact that the two curves are parallel.

Figure 10.13(b) presents the kinetics of the flow front on the preform side (the images on the left of the y axis are scaled as a function of the position). This figure shows that the numerical flow is slightly faster than the experimental flow.

Similarly, the difference between the moment of creation of the numerical and experimental recoil front is 80 s. A good correspondence between the meeting position of the numerical and experimental fronts can be noted (with a difference of 7 mm).

10.5.2.3. *Other experimental validations*

Besides the infusion of the stiffened board, the multilayer model was validated in a previous study (Loudad et al. 2017) in the case of infusion of a flat plate and plate with one straight step for various stacks of ply layers.

10.6. Industrial application

After having experimentally validated the multilayer CVFEM approach in three configurations, each with its own singularity, the focus of this section is on the LRI process and its simulation of an industrial demonstrator that involves these three forms of singularity. This demonstrator is representative for the aeronautic nacelles (Figure 10.14). It was designed and manufactured by SAFRAN Nacelles in order to generate the manufacturing difficulties encountered in the aircraft industry. In order to evaluate the efficiency of the model in an industrial context, the simulation of this infusion was performed without having access to the infusion results collected during the manufacturing.

10.6.1. *Presentation of the demonstrator*

The demonstrator is a double-curved half-cylinder. Its diameter is 2.5 m and its width is 1.5 m. It is composed of three longitudinal stiffeners of various sections: an IPN section, a J section and an Omega section with a flattened area. It also contains two axial omega stiffeners, a rectangular stiffener and a thicker rectangular area (Figure 10.14). The demonstrator was infused in a convex mold with a DM covering the entire preform. The resin input lines are placed in the middle of the mold and the evacuation lines (vents) are placed on the edges of the mold, as shown in Figure 10.15.

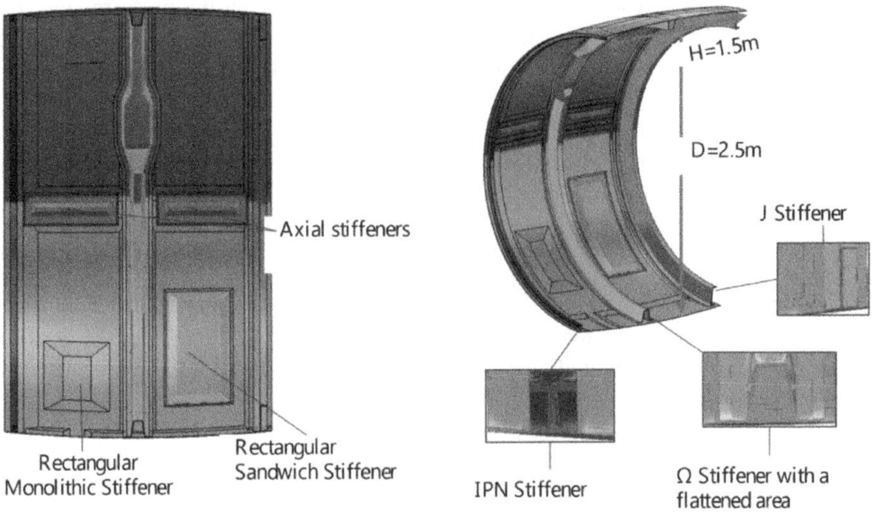

Figure 10.14. *Geometry of the demonstrator and its components*

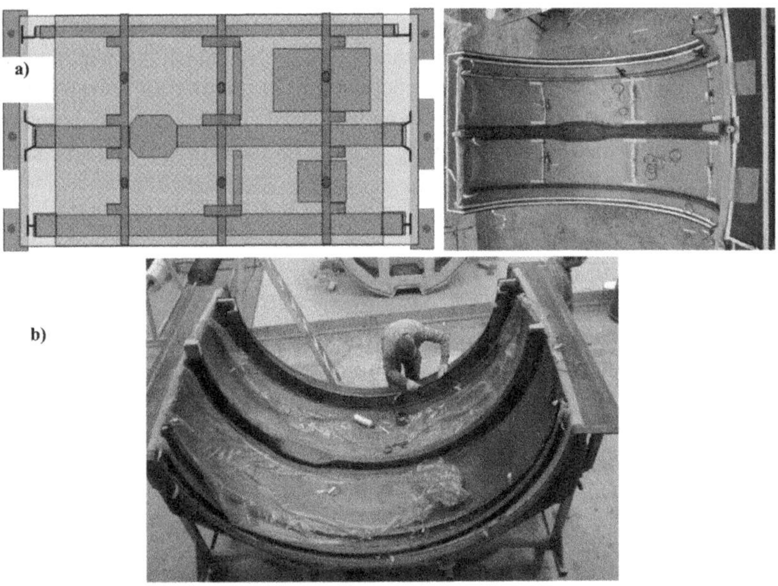

Figure 10.15. *(a) Experimental positions of the input lines. (b) Presentation of the demonstrator*

The material of the infused preform is carbon fiber reinforcement. This preform and the DM used have properties similar to those of the materials characterized in section 10.5.2. The input pressure is 1 atm and the output pressure is 0 Pa. The resin used in this case is an epoxy resin. In the simulation, we considered a constant viscosity of 0.1 Pa·s as the time required to fill the mold is <10 min.

10.6.2. *Geometry meshing*

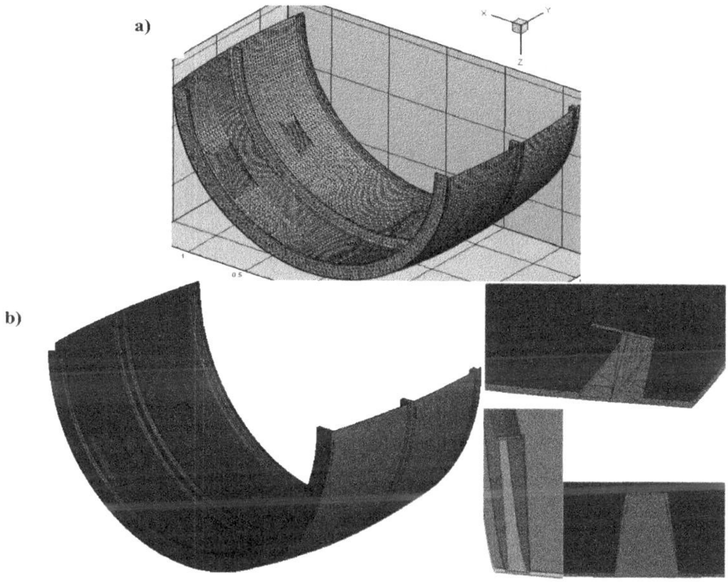

Figure 10.16. *(a) Structure meshing. (b) Material of the layers (DM in blue, preform in red)*

The structure is discretized by a surface meshing for the DM, a surface meshing for the cylindrical preform (including the thickest zone), and a surface meshing for each stiffener according to the multilayer method (Figure 10.16). These surfaces are related by 1D elements that model the through-thickness flow. To make this connection with 1D elements, all the surfaces are similarly meshed so that each node has its through-thickness conjugate. For this purpose, the surfaces are meshed with triangular elements using a structured meshing method. This structure mesh is composed of 23,741 nodes, 45,529 2D elements and 12,188 1D elements.

10.6.3. *Comparison of numerical results and demonstrator infusion*

Figure 10.17 presents numerical distributions of the pressure in the DM and the preform thickness. The filling time error due to neglecting gravity is verified on the basis of the works of Yoon et al. (2005). At the moment of the filling, a difference of 1 cm was noted between the flow front position with and without considering the gravity. On both sides of the injection line, the pressure and the thickness have a symmetrical profile, they have significant values in the vicinity of the injection threshold and they decrease along the x direction toward the edge of the mold, where the vents are positioned.

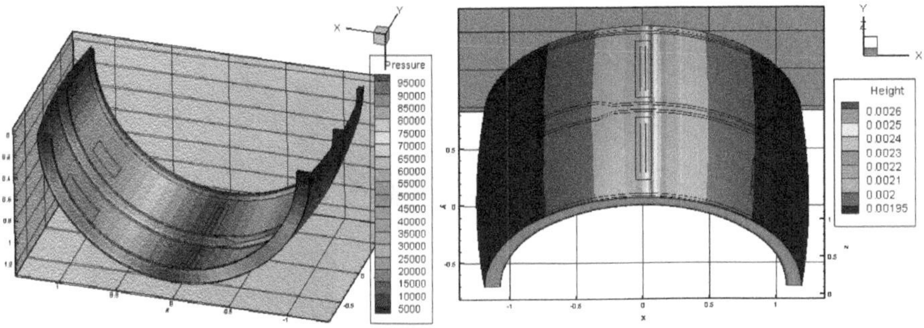

Figure 10.17. *Numerical results during the filling of the mold, on the left: pressure field in the DM in Pa; on the right: thickness of the preform in m*

Figure 10.18 shows the position of the flow front at several instants as the process unfolds. It can be noted on this figure that the flow front traps certain areas representative for the dry areas formed in the preform at the end of the mold filling operation. However, as the infusion model is based on a single-phase approach, the (simulated) flow continues until the mold is entirely filled. However, the position of these areas representative of macro vacuums (or macro porosities) can be identified by analyzing the kinetics of the flow front.

The analysis of the flow front kinetics presented in Figure 10.19 can be used to identify the location of the dry areas, even if the simulation is based on a single-phase modeling approach. The simulation also shows that the fluid soaks the preform under the longitudinal stiffeners from the sides. This leads to a sharp "V" flow front. This type of filling may drive the formation of air bubbles.

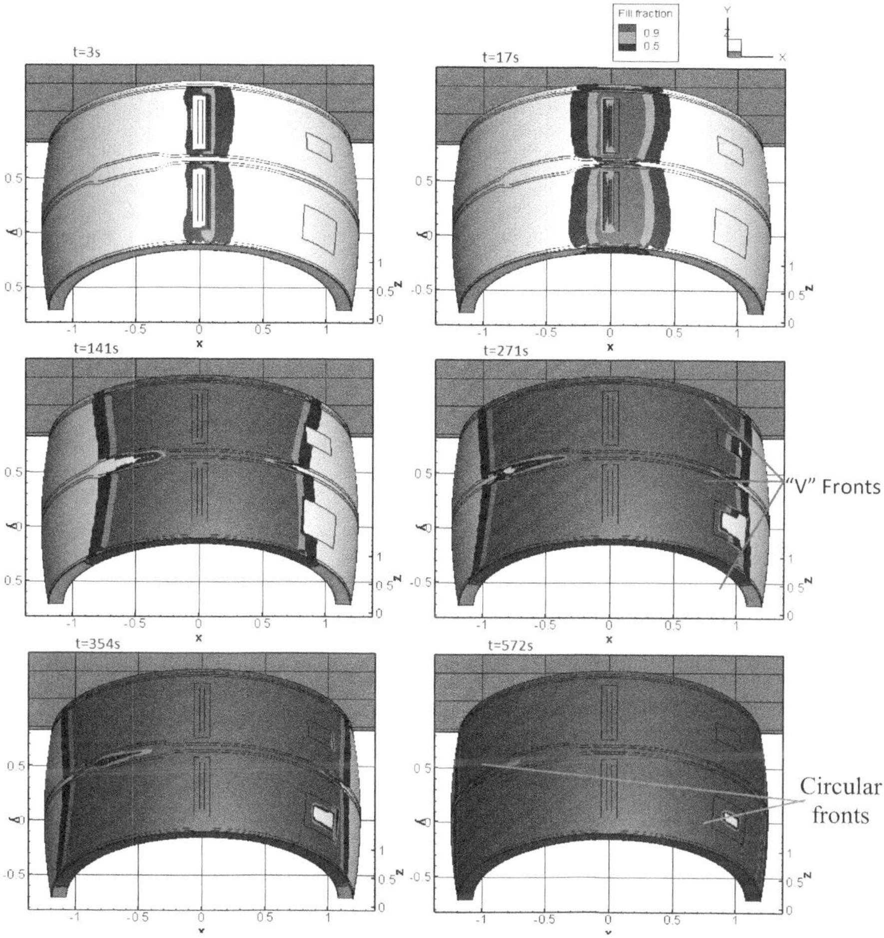

Figure 10.18. *Mold filling at various instants*

After having performed the simulation without access to experimental data, the experimental positions of defects (dry areas) obtained during the infusion of the demonstrator were compared with the simulation-based predictions. This comparison between the numerical and experimental results is presented in Figure 10.20. We can conclude that the predictions based on the analysis of the flow front kinetics obtained with the multilayer CVFEM approach are correct, even though a constant resin viscosity was employed in the simulation.

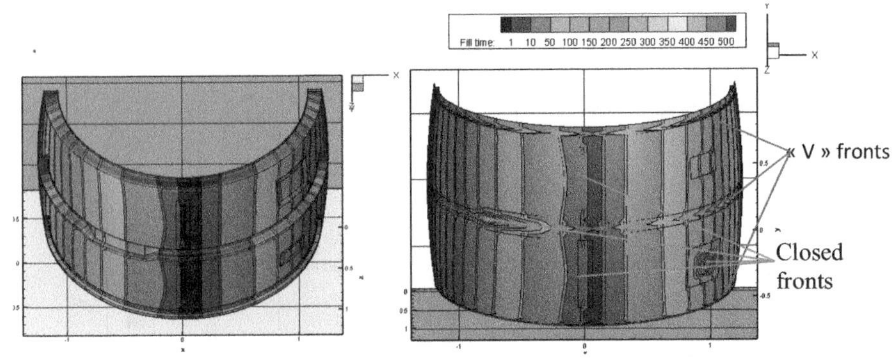

Figure 10.19. *Kinetics of the flow front; on the left: upper view; on the right: lower view*

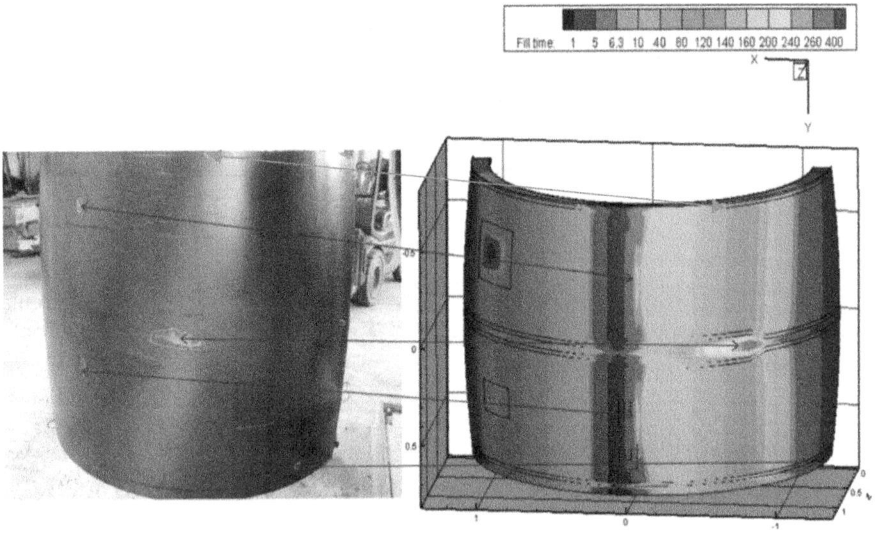

Figure 10.20. *Experimental dry areas and their positions in the kinetics of the numerical flow front*

10.7. Conclusion

This chapter presented an application of the multilayer CVFEM approach in an industrial context. In the first part, the approach is validated in the context of a comparison with various analytical results and that of an experimental case studying the infusion of a stiffened panel. These validations were performed using the behavior laws of the permeability and compressibility of infused materials resulting from our experimental measurements. Then, we simulated the infusion of a complex industrial structure of large size (2.5 m in diameter) that features the main singularities encountered in aircraft parts.

In the three cases, the comparisons between the numerical, analytical results and the experimental measurements confirm the validity of the proposed approach and prove its capacity to simulate the infusion of complex-shaped parts. Indeed, it can be used to describe the flow front progression during the infusion and predict the final geometric characteristics of the part developed, namely, its thickness and the fiber volume fraction.

Moreover, the analysis of the results of the simulation of the two experimental cases was used to accurately determine the position of the dry areas after filling the mold. However, the description of the creation and the displacement of dry areas requires a two-phase modeling. Likewise, the bubble formation and transport should be taken into account for a more complete modeling. These phenomena will be implemented in the multilayer CVFEM model in future research.

10.8. References

Agogue, R., Chebil, N., Deleglise-Lagardere, M., Beauchêne, P., Park, C.H. (2018). Efficient permeability measurement and numerical simulation of the resin flow in low permeability preform fabricated by automated dry fiber placement. *Applied Composite Materials*, 25, 1169–1182.

Calado, V.M.A. and Advani, S.G. (1996). Effective average permeability of multi-layer preforms in resin transfer molding. *Compos. Sci. Technol.*, 56, 519–531.

Celle, P., Drapier, S., Bergheau, J.-M. (2008). Numerical modelling of liquid infusion into fibrous media undergoing compaction. *Eur. J. Mech. – A/Solids*, 27, 647–661.

Chen, R., Dong, C., Liang, Z., Zhang, C., Wang, B. (2004). Flow modeling and simulation for vacuum assisted resin transfer molding process with the equivalent permeability method. *Polym. Compos*, 25, 146–164.

Dong, C. (J.). (2006). Development of a process model for the vacuum assisted resin transfer molding simulation by the response surface method. *Compos Part A Appl. Sci. Manuf.*, 37, 1316–1324.

George, A., Hannibal, P., Morgan, M., Hoagland, D., Stapleton, E. (2019). Compressibility measurement of composite reinforcements for flow simulation of vacuum infusion. *Polym. Compos*, 40, 961–973.

Grimsley, B.W., Cano, R.J., Hubert, P., Loos, A.C., Kellen, C.B., Jensen, B.J. (2004). Preform characterization in VARTM process model development. In *36th International SAMPE (Society for the Advancement of Material and Process Engineering) Technical Conference*, San Diego.

Han, K., Jiang, S., Zhang, C., Wang, B. (2000). Flow modeling and simulation of SCRIMP for composites manufacturing. *Compos Part A Appl. Sci. Manuf.*, 31, 79–86.

Hsiao, K.T., Mathur, R., Advani, S.G., Gillespie, Jr., J.W., Fink, B.K. (2000). A closed form solution for flow during the vacuum assisted resin transfer molding process. *J. Manuf. Sci. Eng. ASME*, 122, 463–475.

Kang, M.K. and Lee, W.I. (2004). A dual-scale analysis of macroscopic resin flow in vacuum assisted resin transfer molding. *Polym. Compos*, 25, 510–520.

Kang, M.K., Lee, W.I., Hahn, H.T. (2001). Analysis of vacuum bag resin transfer molding process. *Compos Part A Appl. Sci. Manuf.*, 32, 1553–1560.

Kim, S.K. and Daniel, I.M. (2005). Transient gas flow technique for inspection of fiber preforms in resin transfer molding. *Compos Part A Appl. Sci. Manuf.*, 36, 1694–1699.

Kim, S.K., Opperer, J.G., Daniel, I.M. (2002). Determination of permeability of fibrous medium considering inertial effects. *Int. Commun. Heat Mass Transf.*, 29, 879–885.

Kim, S.K., Opperer, J.G., Kim, D.-H., Daniel, I.M. (2003). Determination of in-plane permeability of fiber preforms by the gas flow method using pressure measurements. *Polym. Compos*, 24, 34–44.

Loudad, R., Saouab, A., Beauchene, P., Agogue, R., Desjoyeaux, B. (2017). Numerical modeling of vacuum-assisted resin transfer molding using multilayer approach. *J Compos Mater.*, 51, 3441–3452.

Mathur, R., Heider, D., Hoffmann, C. Advani, S.G., Gillespie, Jr., J.W., Fink, B.K. (2001). Flow front measurements and model validation in the vacuum assisted resin transfer molding process. *Polym. Compos*, 22, 477–490.

Modi, D., Johnson, M., Long, A., Rudd, C. (2009). Analysis of pressure profile and flow progression in the vacuum infusion process. *Compos Sci. Technol.*, 69, 1458–1464.

Ouahbi, T., Saouab, A., Bréard, J., Ouagne, P., Chatel, S. (2007). Modelling of hydro-mechanical coupling in infusion processes. *Compos Part A Appl. Sci. Manuf.*, 38, 1646–1654.

Park, C.H. and Saouab, A. (2009). Analytical modeling of composite molding by resin infusion with flexible tooling: VARI and RFI processes. *J. Compos Mater.*, 43, 1877–1900.

Scholz, S., Gillespie, Jr., J.W., Heider, D. (2007). Measurement of transverse permeability using gaseous and liquid flow. *Compos Part A Appl. Sci. Manuf.*, 38, 2034–2040.

Simacek, P. and Advani, S.G. (2007). Modeling resin flow and fiber tow saturation induced by distribution media collapse in VARTM. *Compos Sci. Technol.*, 67, 2757–2769.

Simacek, P. and Advani, S.G. (2016). "Equivalent" permeability and flow in compliant porous media. *Compos Part A Appl. Sci. Manuf.*, 80, 107–110.

Song, Y.S. and Youn, J.R. (2008). Modeling of resin infusion in vacuum assisted resin transfer molding. *Polym. Compos*, 29, 390–395.

Sutton, M., Wolters, W., Peters, W., Ranson, W., McNeill, S. (1983). Determination of displacements using an improved digital correlation method. *Image and Vision Computing*, 1, 133–139.

Tari, M.J., Imbert, J.P., Lin, M.Y, Lavine, A.S., Hahn, H.T. (1998). Analysis of resin transfer molding with high permeability layers. *J. Manuf. Sci. Eng.*, 120, 609–616.

Terzaghi, K. (1943). *Theoretical Soil Mechanics*. John Wiley & Sons, New York.

Williams, C., Summerscales, J., Grove, S. (1996). Resin infusion under flexible tooling (RIFT): A review. *Compos Part A Appl. Sci. Manuf.*, 27, 517–524.

Xiao, X., Hu, J., Hua, T, Zeng, X., Long, A. (2015). Through-thickness air permeability of woven fabric under low pressure compression. *Text Res. J.*, 85, 1732–1742.

Yang, B., Tang, Q., Wang, S, Jin, T., Bi, F. (2016). Three-dimensional numerical simulation of the filling stage in resin infusion process. *J. Compos Mater.*, 50, 4171–4186.

Yoon, M.K. and Dolan, D.F. (2008). Homogenous modeling of VARTM processes with hybrid layered media. *J. Compos Mater.*, 42, 805–824.

Yoon, M.-K., Baidoo, J., Gillespie, Jr., J.W., Heider, D. (2005). Vacuum assisted resin transfer molding (VARTM) process incorporating gravitational effects: A closed-form solution. *J. Compos Mater.*, 39, 2227–2242.

Young, W.-B. (2013). A two-layer model for the simulation of the VARTM process with resin distribution layer. *Appl. Compos Mater.*, 20, 1305–1319.

Ziarani, A.S. and Aguilera, R. (2012). Knudsen's permeability correction for tight porous media. *Transp. Porous Media*, 91, 239–260.

List of Authors

Azita AHMADI-SENICHAULT
École nationale supérieure
des Arts et Métiers
Bordeaux
France

Saïd ANISS
Université Hassan II
Casablanca
Morocco

Abdellah ARHALIASS
Laboratoire GEnie des Procédés
Environnement – Agroalimentaire
(GEPEA)
UMR 6144, CNRS
Nantes Université
Saint-Nazaire
France

Mouh ASSOUL
Université Hassan II
Casablanca
Morocco

Pierre BEAUCHENE
Office National des Études et
Recherches Aérospatiales
Chatillon
France

Mabrouk BENHAMOU
Université Moulay Ismail
Meknès
Morocco

Mohamed Khaled BOURBATACHE
Institut National des sciences
appliquées
Rennes
France

Stéphane CHAMPMARTIN
École nationale supérieure
des Arts et Métiers
Angers
France

Marie-Catherine CHARRIER-MOJTABI
Université Paul Sabatier Toulouse III
France

Olivier CRUMEYROLLE
Université Le Havre Normandie
France

Noureddine DAMIL
Université Hassan II
Casablanca
Morocco

Claude DESLOUIS
Laboratoire Interfaces et Systèmes Electrochimiques (LISE)
UMR 8235, CNRS
Sorbonne Université
Paris
France

Mustapha HELLOU
Institut National des sciences appliquées
Rennes
France

Fayçal KELAI
Université Le Havre Normandie
France

Noureddine LATRACHE
Université Bretagne Occidentale
Brest
France

Jack LEGRAND
Laboratoire GEnie des Procédés Environnement - Agroalimentaire (GEPEA)
UMR 6144, CNRS
Nantes Université
Saint-Nazaire
France

Franck LOMINÉ
Institut National des sciences appliquées
Rennes
France

Raounak LOUDAD
Université Le Havre Normandie
France

Abdelkader MOJTABI
Université Paul Sabatier Toulouse III
France

Innocent MUTABAZI
Université Le Havre Normandie
France

Jaafar Khalid NACIRI
Laboratoire de Mécanique
Université Hassan II
Casablanca
Morocco

Abdelaziz OMARI
École nationale supérieure des Arts et Métiers
Bordeaux
France

Michel POTIER-FERRY
Arts et Métiers ParisTech
Université de Lorraine
Metz
France

Antonio RODRIGUEZ DE CASTRO
École nationale supérieure
des Arts et Métiers
Bordeaux
France

Abdelghani SAOUAB
Université Le Havre Normandie
France

Bernard TRIBOLLET
Laboratoire Interfaces et Systèmes
Electrochimiques (LISE)
UMR 8235, CNRS
Sorbonne Université
Paris
France

Index

T

V, W, X

Other titles from

iSTE

in

Mechanical Engineering and Solid Mechanics

2024

LE VAN Anh
Contact in Structural Mechanics: A Weighted Residual Approach

LEROY Maurice
Rheology, Physical and Mechanical Behavior of Materials 2: Working at High Strain Rates, Forming, Sintering, Joining and Welding

2023

EL HAMI Abdelkhalak
Uncertainty and Artificial Intelligence: Additive Manufacturing, Vibratory Control, Agro-composite, Mechatronics
(Artificial Intelligence in Mechanics Set – Volume 1)

EL HAMI Abdelkhalak, EID Mohamed
Reliability-based Modeling of System Performance
(Reliability of Multiphysical Systems Set – Volume 19)

EL HAMI Abdelkhalak, DELAUX David, GRZESKOWIAK Henri
Applied Reliability for Industry 1: Predictive Reliability for the Automobile, Aeronautics, Defense, Medical, Marine and Space Industries
(Reliability of Multiphysical Systems Set – Volume 16)
Applied Reliability for Industry 2: Experimental Reliability for the Automobile, Aeronautics, Defense, Medical, Marine and Space Industries
(Reliability of Multiphysical Systems Set – Volume 17)
Applied Reliability for Industry 3: Operational Reliability for the Automobile, Aeronautics, Defense, Medical, Marine and Space Industries
(Reliability of Multiphysical Systems Set – Volume 18)

LEROY Maurice
Rheology, Physical and Mechanical Behavior of Materials 1: Physical Mechanisms of Deformation and Dynamic Behavior

2022

BAYLE Franck
Product Maturity 1: Theoretical Principles and Industrial Applications
(Reliability of Multiphysical Systems Set – Volume 12)
Product Maturity 2: Principles and Illustrations
(Reliability of Multiphysical Systems Set – Volume 13)

EL HAMI Abdelkhalak, DELAUX David, GRZESKOWIAK Henri
Reliability and Physics-of-Healthy in Mechatronics
(Reliability of Multiphysical Systems Set – Volume 15)

LANNOY André
Reliability of Nuclear Power Plants: Methods, Data and Applications
(Reliability of Multiphysical Systems Set – Volume 14)

LEDOUX Michel, EL HAMI Abdelkhalak
Heat Transfer 3: Convection, Fundamentals and Monophasic Flows
(Mathematical and Mechanical Engineering Set – Volume 11)
Heat Transfer 4: Convection, Two-Phase Flows and Special Problems
(Mathematical and Mechanical Engineering Set – Volume 12)

PLANCHETTE Guy
Cindynics, The Science of Danger: A Wake-up Call
(Reliability of Multiphysical Systems Set – Volume 11)

2021

CHALLAMEL Noël, KAPLUNOV Julius, TAKEWAKI Izuru
Modern Trends in Structural and Solid Mechanics 1: Static and Stability
Modern Trends in Structural and Solid Mechanics 2: Vibrations
Modern Trends in Structural and Solid Mechanics 3: Non-deterministic Mechanics

DAHOO Pierre Richard, POUGNET Philippe, EL HAMI Abdelkhalak
Applications and Metrology at Nanometer Scale 1: Smart Materials, Electromagnetic Waves and Uncertainties
(Reliability of Multiphysical Systems Set – Volume 9)
Applications and Metrology at Nanometer Scale 2: Measurement Systems, Quantum Engineering and RBDO Method
(Reliability of Multiphysical Systems Set – Volume 10)

LEDOUX Michel, EL HAMI Abdelkhalak
Heat Transfer 1: Conduction
(Mathematical and Mechanical Engineering Set – Volume 9)
Heat Transfer 2: Radiative Transfer
(Mathematical and Mechanical Engineering Set – Volume 10)

2020

SALENÇON Jean
Elastoplastic Modeling

2019

BAYLE Franck
Reliability of Maintained Systems Subjected to Wear Failure Mechanisms: Theory and Applications
(Reliability of Multiphysical Systems Set – Volume 8)

BEN KAHLA Rabeb, BARKAOUI Abdelwahed, MERZOUKI Tarek
Finite Element Method and Medical Imaging Techniques in Bone Biomechanics
(Mathematical and Mechanical Engineering Set – Volume 8)

IONESCU Ioan R., QUEYREAU Sylvain, PICU Catalin R., SALMAN Oguz Umut
Mechanics and Physics of Solids at Micro- and Nano-Scales

LE VAN Anh, BOUZIDI Rabah
Lagrangian Mechanics: An Advanced Analytical Approach

MICHELITSCH Thomas, PÉREZ RIASCOS Alejandro, COLLET Bernard, NOWAKOWSKI Andrzej, NICOLLEAU Franck
Fractional Dynamics on Networks and Lattices

SALENÇON Jean
Viscoelastic Modeling for Structural Analysis

VÉNIZÉLOS Georges, EL HAMI Abdelkhalak
Movement Equations 5: Dynamics of a Set of Solids
(Non-deformable Solid Mechanics Set – Volume 5)

2018

BOREL Michel, VÉNIZÉLOS Georges
Movement Equations 4: Equilibriums and Small Movements
(Non-deformable Solid Mechanics Set – Volume 4)

FROSSARD Etienne
Granular Geomaterials Dissipative Mechanics: Theory and Applications in Civil Engineering

RADI Bouchaib, EL HAMI Abdelkhalak
Advanced Numerical Methods with Matlab® 1: Function Approximation and System Resolution
(Mathematical and Mechanical Engineering Set – Volume 6)

Advanced Numerical Methods with Matlab® 2: Resolution of Nonlinear, Differential and Partial Differential Equations
(Mathematical and Mechanical Engineering Set – Volume 7)

SALENÇON Jean
Virtual Work Approach to Mechanical Modeling

2017

BOREL Michel, VÉNIZÉLOS Georges
Movement Equations 2: Mathematical and Methodological Supplements (Non-deformable Solid Mechanics Set – Volume 2)
Movement Equations 3: Dynamics and Fundamental Principle (Non-deformable Solid Mechanics Set – Volume 3)

BOUVET Christophe
Mechanics of Aeronautical Solids, Materials and Structures
Mechanics of Aeronautical Composite Materials

BRANCHERIE Delphine, FEISSEL Pierre, BOUVIER Salima, IBRAHIMBEGOVIC Adnan
From Microstructure Investigations to Multiscale Modeling: Bridging the Gap

CHEBEL-MORELLO Brigitte, NICOD Jean-Marc, VARNIER Christophe
From Prognostics and Health Systems Management to Predictive Maintenance 2: Knowledge, Traceability and Decision (Reliability of Multiphysical Systems Set – Volume 7)

EL HAMI Abdelkhalak, RADI Bouchaib
Dynamics of Large Structures and Inverse Problems (Mathematical and Mechanical Engineering Set – Volume 5)
Fluid-Structure Interactions and Uncertainties: Ansys and Fluent Tools (Reliability of Multiphysical Systems Set – Volume 6)

KHARMANDA Ghias, EL HAMI Abdelkhalak
Biomechanics: Optimization, Uncertainties and Reliability (Reliability of Multiphysical Systems Set – Volume 5)

LEDOUX Michel, EL HAMI Abdelkhalak
Compressible Flow Propulsion and Digital Approaches in Fluid Mechanics
(Mathematical and Mechanical Engineering Set – Volume 4)
Fluid Mechanics: Analytical Methods
(Mathematical and Mechanical Engineering Set – Volume 3)

MORI Yvon
Mechanical Vibrations: Applications to Equipment

2016

BOREL Michel, VÉNIZÉLOS Georges
Movement Equations 1: Location, Kinematics and Kinetics
(Non-deformable Solid Mechanics Set – Volume 1)

BOYARD Nicolas
Heat Transfer in Polymer Composite Materials

CARDON Alain, ITMI Mhamed
New Autonomous Systems
(Reliability of Multiphysical Systems Set – Volume 1)

DAHOO Pierre Richard, POUGNET Philippe, EL HAMI Abdelkhalak
Nanometer-scale Defect Detection Using Polarized Light
(Reliability of Multiphysical Systems Set – Volume 2)

DE SAXCÉ Géry, VALLÉE Claude
Galilean Mechanics and Thermodynamics of Continua

DORMIEUX Luc, KONDO Djimédo
Micromechanics of Fracture and Damage
(Micromechanics Set – Volume 1)

EL HAMI Abdelkhalak, RADI Bouchaib
Stochastic Dynamics of Structures
(Mathematical and Mechanical Engineering Set – Volume 2)

GOURIVEAU Rafael, MEDJAHER Kamal, ZERHOUNI Noureddine
From Prognostics and Health Systems Management to Predictive Maintenance 1: Monitoring and Prognostics
(Reliability of Multiphysical Systems Set – Volume 4)

KHARMANDA Ghias, EL HAMI Abdelkhalak
Reliability in Biomechanics
(Reliability of Multiphysical Systems Set –Volume 3)

MOLIMARD Jérôme
Experimental Mechanics of Solids and Structures

RADI Bouchaib, EL HAMI Abdelkhalak
Material Forming Processes: Simulation, Drawing, Hydroforming and Additive Manufacturing
(Mathematical and Mechanical Engineering Set – Volume 1)

2015

KARLIČIĆ Danilo, MURMU Tony, ADHIKARI Sondipon, MCCARTHY Michael
Non-local Structural Mechanics

SAB Karam, LEBÉE Arthur
Homogenization of Heterogeneous Thin and Thick Plates

2014

ATANACKOVIC M. Teodor, PILIPOVIC Stevan, STANKOVIC Bogoljub, ZORICA Dusan
Fractional Calculus with Applications in Mechanics: Vibrations and Diffusion Processes
Fractional Calculus with Applications in Mechanics: Wave Propagation, Impact and Variational Principles

CIBLAC Thierry, MOREL Jean-Claude
Sustainable Masonry: Stability and Behavior of Structures

ILANKO Sinniah, MONTERRUBIO Luis E., MOCHIDA Yusuke
The Rayleigh–Ritz Method for Structural Analysis

LALANNE Christian
Mechanical Vibration and Shock Analysis – 5-volume series – 3rd edition
Sinusoidal Vibration – Volume 1
Mechanical Shock – Volume 2
Random Vibration – Volume 3
Fatigue Damage – Volume 4
Specification Development – Volume 5

LEMAIRE Maurice
Uncertainty and Mechanics

2013

ADHIKARI Sondipon
Structural Dynamic Analysis with Generalized Damping Models: Analysis

ADHIKARI Sondipon
Structural Dynamic Analysis with Generalized Damping Models: Identification

BAILLY Patrice
Materials and Structures under Shock and Impact

BASTIEN Jérôme, BERNARDIN Frédéric, LAMARQUE Claude-Henri
Non-smooth Deterministic or Stochastic Discrete Dynamical Systems: Applications to Models with Friction or Impact

EL HAMI Abdelkhalak, RADI Bouchaib
Uncertainty and Optimization in Structural Mechanics

KIRILLOV Oleg N., PELINOVSKY Dmitry E.
Nonlinear Physical Systems: Spectral Analysis, Stability and Bifurcations

LUONGO Angelo, ZULLI Daniele
Mathematical Models of Beams and Cables

SALENÇON Jean
Yield Design

2012

DAVIM J. Paulo
Mechanical Engineering Education

DUPEUX Michel, BRACCINI Muriel
Mechanics of Solid Interfaces

ELISHAKOFF Isaac *et al.*
Carbon Nanotubes and Nanosensors: Vibration, Buckling and Ballistic Impact

GRÉDIAC Michel, HILD François
Full-Field Measurements and Identification in Solid Mechanics

GROUS Ammar
Fracture Mechanics – 3-volume series
Analysis of Reliability and Quality Control – Volume 1
Applied Reliability – Volume 2
Applied Quality Control – Volume 3

RECHO Naman
Fracture Mechanics and Crack Growth

2011

KRYSINSKI Tomasz, MALBURET François
Mechanical Instability

SOUSTELLE Michel
An Introduction to Chemical Kinetics

2010

BREITKOPF Piotr, FILOMENO COELHO Rajan
Multidisciplinary Design Optimization in Computational Mechanics

DAVIM J. Paulo
Biotribology

PAULTRE Patrick
Dynamics of Structures

SOUSTELLE Michel
Handbook of Heterogenous Kinetics

2009

BERLIOZ Alain, TROMPETTE Philippe
Solid Mechanics using the Finite Element Method

LEMAIRE Maurice
Structural Reliability

2007

GIRARD Alain, ROY Nicolas
Structural Dynamics in Industry

GUINEBRETIÈRE René
X-ray Diffraction by Polycrystalline Materials

KRYSINSKI Tomasz, MALBURET François
Mechanical Vibrations

KUNDU Tribikram
Advanced Ultrasonic Methods for Material and Structure Inspection

SIH George C., NAÏT-ABDELAZIZ Moussa, VU-KHANH Toan
Particle and Continuum Aspects of Mesomechanics

Printed and bound by CPI Group (UK) Ltd, Croydon, CR0 4YY
28/07/2026
14931342-0004